D0321802

IONIZATION, CONDUCTIVITY AND BREAKDOWN IN DIELECTRIC LIQUIDS

IONIZATION, CONDUCTIVITY AND BREAKDOWN IN DIELECTRIC LIQUIDS

Ignacy Adamczewski

Technical University of Gdańsk, Poland

TAYLOR & FRANCIS LTD

Red Lion Court, Fleet Street, London, E.C.4

1969

Originally published by Państwowe Wydawnictwo Naukowe, Warsaw
under the title JONIZACJA I PRZEWODNICTWO CIEKŁYCH DIELEKTRYKÓW

Copyright by Państwowe Wydawnictwo Naukowe, Warsaw, 1965

English edition first published 1969 by Taylor & Francis Ltd., Red Lion Court, Fleet Street, London, E.C.4.

Printed and bound in Great Britain by Taylor & Francis Ltd., Red Lion Court, Fleet Street, London, E.C.4.

SBN 85066 027 0

© Taylor & Francis Ltd. 1969
All Rights Reserved. No part of this publication
may be reproduced, stored in a retrieval system,
or transmitted, in any form or by any means,
electronic, mechanical, photocopying, recording
or otherwise, without the prior permission of
Taylor & Francis Ltd.

The Author, Professor IGNACY ADAMCZEWSKI
Head of the Physics Department of the Technical
University and Medical Academy of Gdańsk, Poland

Preface to the Polish Edition

DIELECTRICS in the three states of matter, solid, liquid, and gaseous, are a topic of great interest in modern scientific research, and particularly in certain spheres of physics, chemistry and engineering. This interest is expressed in the publication of monographs on the subject from time to time, with varying depth of treatment. There are, however, several gaps in the bibliography making it difficult to become completely acquainted with the subject; this in its turn slows up the progress of research and frequently results in work being unnecessarily repeated.

This book is intended to fill some of the important gaps, namely the mechanism of ionization, electrical conductivity and breakdown in dielectric liquids. These subjects are closely connected both with one another and with many other very important aspects of contemporary physics and engineering with the result that ever-increasing interest is being displayed in them. This is evidenced by the growing number of papers published all over the world, very often in journals which are difficult to obtain.

It is also hoped that this book will help publicize current work being carried out in the field, and discuss the results obtained with a view to the clarification of problems which as yet remain unsolved. It includes the results of research work published in several hundred experimental and theoretical papers up to the middle of 1963, and several dozen of the author's own articles written over the past 30 years; there is also a large number of contributions from the author's collaborators, including work taken from doctoral theses.

The author wishes to express his grateful thanks to the following persons who have assisted in the preparation of this book, or who have kindly furnished experimental results before publication elsewhere: Dr. S. Bernasik, Dr. E. Gazda, Dr. O. Gzowski, Dr. J. Terlecki, Dr. J. Dera, Dr. B. Jachym, Dr. T. Umiński, Dr. M. Chybicki, Dr. A. Januszajtis, Dr. W. Nowak and Mr. J. Kalinowski.

Special thanks are also due to Professors M. Jeżewski, J. L. Maksiejewski and C. Pawłowski for reading the manuscript, and for much helpful criticism and advice.

IGNACY ADAMCZEWSKI

Gdańsk, May, 1963.

Preface to the English Edition

The English Edition of this book has been considerably extended and brought up to date; a large amount of new material published between 1963 and 1968 has been added. At the end of the book are listed the titles of the papers held on the International Colloquium upon the "Phenomena of Conduction in Dielectric Liquids", Grenoble, France, September, 1968.

Very warm thanks are due to Messrs. Taylor & Francis Ltd, London, and particularly one of the directors, Mr. G. F. Lancaster, for the opportunity to bring out this book in English. A similar debt of gratitude is owed to Professor Rotblat for his initiative in recommending my work for publication in the English language and for reading the manuscript.

The author finally wishes to express his thanks to those who translated the English edition and corrected and edited the text with regard to both the language and the contents: Dr. Angerer, Dr. Butler, Dr. Januszajtis, Mr. Lachman, Mr. Niećko and Mrs Słuckin.

IGNACY ADAMCZEWSKI

Gdańsk, September, 1968.

Introduction to the English Edition

Professor Adamczewski has made a special study of ionization and conductivity in liquids for more than thirty years and this book brings together many of the results of his own and of his students' work as well as giving a wide survey of the field and the essential theoretical background. Dielectric liquids have many important practical uses and a better understanding of the processes of ionization and conduction in such liquids is much needed. In the literature of this subject there has been too wide a gap between fundamental theory and practical applications and the present volume should help to bridge this gap. It will also put readers into touch with the Polish, Russian and other literature on the subject.

J. W. BOAG

Contents

Preface to the Polish Edition vii
Preface to the English Edition viii
Introduction to the English Edition ix

PART I
GENERAL PHYSICOCHEMICAL PROPERTIES OF DIELECTRIC LIQUIDS

Chapter 1. CHARACTERISTICS OF DIELECTRIC LIQUIDS . . 1
   1. Introduction 1
   2. The Best Liquid Insulators 2
   3. Saturated Hydrocarbons 3
   4. Unsaturated Hydrocarbons 5
   5. Electron Configurations in Molecules. Chemical Bonds . 6
   6. Hydrocarbon Oils. Silicones 17

Chapter 2. MOLECULAR STRUCTURE AND THE MACROSCOPIC PROPER-
         TIES OF LIQUIDS 21
   1. Fundamental Quantities 21
   2. Liquid Viscosity 24
   3. Surface Tension. The Parachor . . . 28
   4. Molar Refraction. Dielectric Polarization . . 31

Chapter 3. THERMODYNAMIC THEORIES OF THE LIQUID STATE . 32
   1. Introduction 32
   2. Numerical Methods for Hydrocarbons . . . 32
   3. The Cell and Tunnel Theories of Liquids . . . 35
   4. Theories of Liquid Viscosity 42

Chapter 4. THE PURIFICATION OF LIQUIDS . . . 50
   1. Defining the Purity of a Liquid . . . 50
   2. Chemical Purification. Distillation . . . 50
   3. Zonal Melting and Crystallization Methods . . 53

References to Part I 57

PART II
THE ACTION OF IONIZING RADIATION ON MATTER

Chapter 5. THE PRINCIPAL EFFECTS OF IONIZING RADIATION . 61
   1. Introduction 61
   2. The Energy of Ionizing Radiation 62
   3. Photoelectric and Compton Effects. Pair Formation . . 62

Chapter 6. STOPPING OF IONIZING PARTICLES BY MATTER   .   .   67
   1. Stopping Power of Different Substances   .   .   .   67
   2. Mean Ionization Density. LET   68
   3. Electron Stopping   70

Chapter 7. THE ABSORPTION OF X-RAYS AND γ-RAYS   .   .   76
   1. The General Absorption Law   76
   2. Absorption Cross Section   76
   3. Components of the Absorption Coefficient.   .   .   78
   4. The Absorption Coefficient for Liquids   .   .   .   82

Chapter 8. PRINCIPLES OF IONIZING RADIATION DOSIMETRY   .   85
   1. Dosimetry Units   85
   2. Calculation of the Dose   86
   3. Dose Rate   87
   4. Other Dosimetric Methods   88

Chapter 9. MECHANISM OF THE INTERACTION OF IONIZING RADIA-
      TION IN LIQUIDS   90
   1. Types of Interaction Effect   90
   2. Energy Levels in a Molecule. Excitation and Ionization   .   91
   3. The Radiolysis of Water   96
   4. Free and Solvated Electrons in Liquids   .   .   .   101
   5. Ionization Effects in Time and Space   .   .   .   105
   6. The Radiation Chemistry of Hydrocarbons   .   .   108
   7. Physical and Chemical Changes in Hydrocarbons due to
      Radiation   112
   8. Industrial Applications of Hydrocarbon Radiochemistry   .   120

References to Part II   122

PART III
ELECTRICAL CONDUCTION IN DIELECTRIC LIQUIDS
AT LOW FIELDS

Chapter 10. MEASURING APPARATUS   125
   1. Introduction   125
   2. Different Kinds of Liquid-filled Ionization Chambers   .   126
   3. Methods of Measuring very low Electric Currents   .   130

Chapter 11. ELECTRICAL CONDUCTION IN DIELECTRIC LIQUIDS AT
      LOW FIELDS   139
   1. Natural Conduction (self-conduction)   .   .   .   139
   2. The Influence of Cosmic Rays   143
   3. Ionization or Induced Conduction in Gases   .   .   145
   4. Ionization or Induced Conduction in Dielectric Liquids   .   149
   5. Research on Emission and the Injection of Electrons into
      Dielectric Liquids   161
   6. The Influence of Temperature on the Conductivity of
      Dielectric Liquids   175

Chapter 12. THE MOBILITY OF IONS 185
  1. Introduction 185
  2. Methods of Measurement 185
  3. The Dependence of the Mobility of Ions on the Viscosity of
       the Liquid 186
  4. Other Work on Ion Mobility 192
  5. The Dependence of the Mobility of Negative Charge Carriers
       (Electrons or Ions) on the Electric Field Strength . . 199
  6. The Dependence of the Mobility of Ions on Temperature.
       Activation Energy 205
  7. Measurements of the Mobilities of Ions in Highly Viscous
       Liquids 219
  8. Research on Ionization Conductivity and the Mobility of
       Ions in Liquefied Gases 228

Chapter 13. RECOMBINATION OF IONS 233
  1. General Law of Recombination 233
  2. Different Types of Recombination 234
  3. The Dependence of the Coefficient of Ion Recombination on
       the Viscosity of Liquids 235
  4. New Research on the Phenomenon of Ion Recombination . 238

Chapter 14. THE DIFFUSION OF IONS 243
  1. The General Law of Diffusion 243
  2. Different Types of Diffusion 243
  3. Methods of Measuring the Coefficient of Self-diffusion . 244
  4. The Relationship between the Coefficient of Self-diffusion on
       both the Viscosity and Temperature of the Liquid . . 248
  5. The Measurement of the Coefficient of Ion Diffusion . . 250

Chapter 15. A SUMMARY OF THE RESULTS OF EXPERIMENTS ON THE
       MOBILITY, RECOMBINATION AND DIFFUSION OF IONS.
       THE ACTIVATION ENERGY OF THESE PROCESSES . 254
  1. Results of Experiments 254
  2. A theoretical Analysis of the Energy of Activation . . 261
  3. The Activation Energy of Complex (Competing) Processes . 263

Chapter 16. THE THEORETICAL PRINCIPLES OF LOW-FIELD
       ELECTRICAL CONDUCTION IN DIELECTRIC LIQUIDS . 266
  1. Introduction 266
  2. A General System of Equations for Ionizing Conduction . 266
  3. The Application of the Similarity Principle . . 269
  4. Theories of Ion Mobility 270
  5. Jaffe's Theory 278
  6. Lea's Theory 281
  7. Plumley's Theory 284
  8. Frenkel's Theory 286
  9. Adamczewski's Theoretical Conception . . . 289

10. Summary of the Formulae 293
11. The Possibilities of Applying the Theory of Semiconductors to Explain the Mechanism of Conduction in Dielectric Liquids 294

Chapter 17. THE APPLICATION OF LIQUID-FILLED IONIZATION CHAMBERS TO PROBLEMS OF DOSIMETRY . . . 303
1. The Necessity of Introducing Liquid-filled Ionization Chambers 303
2. Chemical Liquid Dosimeters 306
3. Liquid-filled Ionization Chambers as Dosimeters . . 308
4. Different Kinds of Dosimetric Ionization Chambers filled with Liquids 315
5. Other Possibilities of Applying Liquid-filled Ionization Chambers 321

References to Part III 325

PART IV
THE ELECTRICAL CONDUCTIVITY OF DIELECTRIC LIQUIDS AT HIGH FIELDS AND ELECTRICAL BREAKDOWN

Chapter 18. ELECTRICAL CONDUCTIVITY AT HIGH FIELDS . . 335
1. Introduction 335
2. Current–Voltage Characteristics 336
3. Results of Experimental Work 338
4. Conclusions from Experimental Work . . . 349

Chapter 19. ELECTRICAL BREAKDOWN 353
1. Introduction 353
2. Breakdown in a Vacuum in Gases and in Vapours . 353
3. Research Methods 354
4. Pre-breakdown Phenomena. Current Fluctuation . . 356
5. Experimental Results 359
6. The Effect of Electrode Material and Condition of its Surface on Electrical Breakdown 361
7. Dependence of Electric Strength on the Number of Breakdowns 363
8. Effect of Temperature and Hydrostatic Pressure on Electrical Strength 364
9. Dependence of Breakdown Stress on Other Factors . . 366
10. The Effect of Additives and Impurities . . . 367
11. Effect of Oxygen on Breakdown Stress . . . 368
12. Other Methods of Studying Liquids Subjected to High Electric Fields 369
13. Electrical Breakdown in High-frequency Fields . . . 372

Chapter 20. THEORY OF CONDUCTION IN DIELECTRIC LIQUIDS AT HIGH FIELDS AND ELECTRICAL BREAKDOWN . . 377
1. Basic Theoretical Problems 377

2. Energy Levels of Electrons at the Metal–Liquid Boundary . 377
3. Distribution of Potential Inside a Dielectric Material . . 380
4. Collision Ionization 382
5. Breakdown Mechanisms. Hypotheses . . . 385
6. Ward-Lewis Statistical Theory of Breakdown . . . 394
7. Effect of Field Distribution on Breakdown . . . 397
8. Swan's Theory 400
9. Watson and Sharbaugh's Theory 400
10. Theory of Breakdown due to Solid Particles and other
      Impurities 401
11. Kao's Theory 401
12. Krasucki's Theory 402
13. General Remarks 403

References to Part IV 409

Author Index 419

Subject Index 427

PART I

General Physicochemical Properties of Dielectric Liquids

Chapter 1

Characteristics of Dielectric Liquids

1.1. *Introduction*. The last few years have seen a rapid advance in studies on the mechanism of ionization, electrical conductivity and breakdown in liquid dielectrics. This is due to their fundamental importance in many branches of contemporary physics, chemistry, engineering and radiobiology. These studies are closely linked with other very important fields of physics such as the physics of plasma in gases, semiconductors, the dosimetry of ionizing radiation, the resistance of materials to electrical breakdown, etc.

These studies are also very important in physical chemistry and some branches of radiation chemistry. Direct measurements of certain basic quantities connected with electrical conductivity such as coefficients of mobility, recombination and diffusion of ions, or activation energy in various chemical compounds enable the microscopic values to be linked with the microstructure of liquids (in particular with the structure of molecules) and the electron distribution in a molecule, etc.

In many instances, the investigation of ionization effects in liquids and the electrical conductivity of free electrons produced in them can be conducted in simple dielectrics only. These liquids must also be very good insulators, since in other liquids in which molecular dissociation phenomena occur, ionization effects are damped by the much stronger electrolytic effects (stronger by several orders of magnitude).

This is why research into the mechanism of ionization and conductivity in liquid dielectrics is also very vital to radiobiology, health physics and medicine since it helps to fill a wide gap in what we know about the ionization of gases and liquids, and that of tissues and living organisms as a whole. Our knowledge in this field is now of considerable importance both in radiobiology and other topics of a more general nature connected with the effect of ionizing radiation on matter.

Many difficult problems, their roots reaching back to elementary physical phenomena, are still waiting for an explanation. It suffices to say that we still lack a clear view of what fraction of the free electrons and ions produced during the ionization process influences other molecules in the liquid and its electrical conductivity, and what fraction is destroyed immediately as a result of a special type of ionic recombination. At present, there are a number of different theories all inconsistent with each other, and none of them giving a full explanation of all the known experimental facts. Many fundamental problems are involved here, such as the determination of the dosimetric values and units which are transferred directly from gases into tissue, the use of dielectric liquids as detectors of ionizing radiation, etc.

This book reviews the existing situation in research on ionization, conductivity and breakdown in liquid dielectrics on the basis of results from a large number of papers by various workers and several dozen of the author's own papers written by him since 1932 and also his colleagues in the years 1945–1966.

There are still many gaps in our present knowledge of the mechanism of ionization and conductivity in dielectric liquids. The author considers that it is difficult to provide an adequate explanation of unsolved problems without being familiar with the latest advances in certain other related fields. This is why a brief review of current knowledge on the physicochemical properties of dielectric liquids, their molecular structure, electron distribution in molecules, etc., is inserted in Parts I and II. This review also includes fundamental advances in the study of the effect of ionizing radiation on matter, and radiation chemistry as a whole. Special attention is given to the radiation chemistry of the substances discussed later in Parts III and IV.

Since many experimental facts in this field are still waiting for adequate theoretical explanation and certain misunderstandings exist regarding the interpretation of the mechanism of a number of fundamental processes, the results of experimental and theoretical studies have been grouped separately.

It is already possible to arrive at satisfactory explanations for published experimental results (see § 16), but the author expects that the material collected in the course of extensive experimental work may be suitable for further theoretical treatment and contribute to a more thorough understanding of the phenomena.

1.2. *The Best Liquid Insulators.* In general the term 'dielectric' is given to substances characterized by very low electrical conduction, that is those in which the number of free carriers of electrical charge (free electrons, positive and negative ions) capable of motion in an electric field is very small. We are chiefly concerned here with the physicochemical properties of the best liquid insulators, i.e. those in which no molecular dissociation occurs and, consequently, in which the electrochemical phenomena usually accompanying dissociation are not observed. Such insulators behave more like gases at high pressures. Their most important physical property is specific conductivity σ_s determined by measuring the electrical current i in a flat capacitor, the distance between electrodes being d, and the area over which the electrical charge is measured, S. If U is the voltage applied to the plates of the capacitor, σ_s is given by the formula:

$$\sigma_s = \frac{id}{US} \quad (\Omega^{-1}\,\text{cm}^{-1}).* \tag{1.1}$$

The reciprocal of σ_s is called the resistivity of the liquid. Certain dielectric liquids, after undergoing thorough physical and chemical purification, can be obtained in a purity for which the specific electrical conductivity is 10^{-19}–$10^{-20}\,\Omega^{-1}\,\text{cm}^{-1}$; this may be compared with commercially available liquids of high-purity analysis grade, where the conductivity is 10^{-10}–$10^{-12}\,\Omega^{-1}\,\text{cm}^{-1}$.

A value of $10^{-10}\,\Omega^{-1}\,\text{cm}^{-1}$ is usually taken as the maximum limit for defining dielectrics; sometimes water is counted as a dielectric since, if very pure, its intrinsic conductivity can be reduced to a minimum value of about $10^{-8}\,\Omega^{-1}\,\text{cm}^{-1}$, when the molecules, even in this state of purity, are only subject to insignificant electrolytic dissociation.

The best liquid insulators include liquefied gases, saturated and aromatic hydrocarbons, certain cyclic and unsaturated hydrocarbons, paraffin, silicon and vaseline oils, CCl_4 and others. The best and most exhaustively studied insulators are discussed in greater detail below.

1.3. *Saturated Hydrocarbons.* Saturated hydrocarbons of the type C_nH_{2n+2} (paraffin hydrocarbons) are among the most extensively examined dielectric liquids. Their name alone indicates that they are saturated with hydrogen atoms, and this is why they have the lowest chemical affinity of all hydrocarbons; hence the name 'paraffin' (*parum affinis*). The first few compounds of the series have the following symbols:

CH_4	methane	$CH_3CH_2CH_2CH_2CH_3$	
CH_3CH_3	ethane	or	pentane
$CH_3CH_2CH_3$	propane	$CH_3(CH_2)_3CH_3$	
$CH_3CH_2CH_2CH_3$	butane	$CH_3(CH_2)_4CH_3$	hexane

Compounds having the same number of carbon and hydrogen atoms can still differ from each other by the configuration of the atoms; they are then called isomers. The number of isomers of a compound depends on the number of carbon atoms in the molecule: heptane, for example, has 9 isomers, octane 18, nonane 35, decane 75 and $C_{20}H_{42}$ 366 319.

Hydrocarbons having the form of long chains (without branches) are known as *normal*, and designated n-pentane, n-hexane, etc. Those with branches have names which indicate the length of chain and number of branchings, as well as the number of the carbon atom at which the

---

* In this book the physical units have been used after the authors of original works. At the end of the book in the table on p. 418 they are compared with the SI units (Systeme Internationale) lately introduced in many countries.

branching begins. Thus, for example, the following symbols and names apply to the isomeric group of hexanes:

$$CH_3CH_2CH_2CH_2CH_2CH_3 \qquad \textit{n}\text{-hexane}$$

$$
\begin{array}{l}
\quad\ \ CH_3 \\
\quad\ \ | \\
CH_3.CH.CH_2CH_2CH_3 \qquad \text{2-methylpentane}
\end{array}
$$

$$
\begin{array}{l}
\qquad\quad CH_3 \\
\qquad\quad | \\
CH_3.CH_2.CH.CH_2CH_3 \qquad \text{3-methylpentane}
\end{array}
$$

$$
\begin{array}{l}
\quad\ \ CH_3 \\
\quad\ \ | \\
CH_3.C.CH_2CH_3 \qquad \text{2,2-dimethylbutane} \\
\quad\ \ | \\
\quad\ \ CH_3
\end{array}
$$

$$
\begin{array}{l}
\qquad CH_3CH_3 \\
\qquad |\quad\ | \\
CH_3.CH.CH.CH_3 \qquad \text{2,3-dimethylbutane}
\end{array}
$$

Saturated hydrocarbon molecules of type C_nH_{2n+2} display certain general features in all the three states of matter. They have the shape of long zig-zag chains (figs. 1.1 and 1.2) with an almost identical cross section for all the molecules in the homologous series. The length h increases in proportion to the number of CH_2 groups, i.e. the number of carbon atoms in the molecule, n:

$$h = h_0 + a(n-1), \tag{1.2}$$

where $h_0 = 4$ Å, and $a = 1\cdot 22$ Å. There is one CH_3 group at each end of the chain.

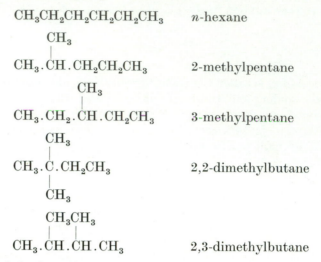

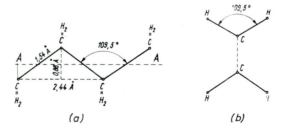

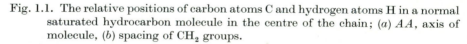

(a)   (b)

Fig. 1.1. The relative positions of carbon atoms C and hydrogen atoms H in a normal saturated hydrocarbon molecule in the centre of the chain; (a) AA, axis of molecule, (b) spacing of CH_2 groups.

In the solid state each molecular chain lies in one fixed plane; in the liquid state, and even more so in the gaseous state, the molecule takes the form of a cylinder (straight or bent), because individual CH_2 groups can turn around on their bond axes, and C—C groups rotate about the axis of the molecule. These rotations do not, however, change the basic dimensional constants of the molecule, that is the C—C bond length of

1·54 Å, and the angle between bonds in the chain (109·5°). One can imagine these molecules as cylinders with a constant base area but a height varying according to the number of carbon atoms in the molecule (fig. 1.2). The mean distance between the molecules is 4·9–5·2 Å. Every new CH_2 group added to the chain increases the height of the cylinder by 1·22 Å.

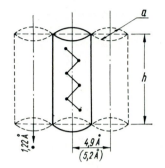

Fig. 1.2. Molecular spacing in normal saturated hydrocarbons (a hexane molecule, C_6H_{14}, is shown); *a* is the diameter of the base of the cylinder occupied by one molecule; *h* is the height of the cylinder.

The main source of detailed numerical data about the dimensions and structure of these molecules has been x-ray examinations of long saturated hydrocarbons ($C_{17}H_{36}$ to $C_{35}H_{72}$), and work on hydrocarbons in the liquid and gaseous states carried out by Stewart [1] and Wierl [2].

It is of interest that as early as 1899 Lord Rayleigh had established the basic structure of these molecules using the thin-layer technique. According to this method, a small amount of liquid when poured on the surface of water forms a molecular layer arranged such that the molecules take up positions with their bases on the surface of the water and the axes of the molecular cylinders perpendicular to the surface. The principle was originally used for determining the basic dimensions of molecules in fatty acids and alcohols, and results obtained by this simple method were fully confirmed later by x-ray examinations.

More data were supplied from work carried out by Kohlrausch and Koppl [3] on the Raman effect, and other studies on the absorption of infra-red radiation in liquids [4, 5]. Further comprehensive data about saturated hydrocarbons and other dielectric liquids may be found in works by Karrer [6], Egloff [7], Eyring [8], Frenkel [9], Hirschfelder *et al.* [10], Reid and Sherwood [11], Nikuradse [12], Tatevskii [13], Adamczewski [25, 73, 83], Rossini [84], Tilikher [85], Skanavi [86], Charlesby [87], Bogoroditskii [88], Lewis [89], Watson [90], Morant [91] and Sharbaugh and Watson [98].

1.4. *Unsaturated Hydrocarbons.* Unsaturated hydrocarbons are those having less hydrogen atoms per molecule than saturated hydrocarbons with the same number of carbon atoms. These include compounds with general formulae C_nH_{2n}, C_nH_{2n-2}, C_nH_{2n-4}, C_nH_{2n-6} and C_nH_{2n-8}. They are divided into a number of specific groups:

(1) ethylenic, C_nH_{2n},
(2) type C_nH_{2n-2} with double bonds

$$C{=}C{=}C, \quad C{=}C{-}C{=}C, \quad C{=}C{-}(CH_2)_x{-}C{=}C,$$

(3) acetylenic, C_nH_n, with acetylene C_2H_2 being the most important in this group,
(4) cyclic and ring-type; these are chiefly aromatic hydrocarbons including benzene and its derivatives, toluene, xylene, etc.

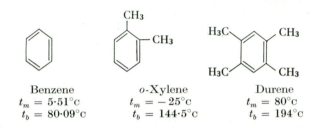

Benzene   *o*-Xylene   Durene
$t_m = 5{\cdot}51°c$ $t_m = -25°c$ $t_m = 80°c$
$t_b = 80{\cdot}09°c$ $t_b = 144{\cdot}5°c$ $t_b = 194°c$

(5) multi-ringed cyclic, such as diphenyl,
(6) naphthalene, anthracene, phenanthrene and others with condensed rings:

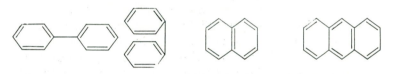

Diphenyl, $C_{12}H_{10}$ Naphthalene, $C_{10}H_8$ Anthracene, $C_{14}H_{10}$

1.5. *Electron Configurations in Molecules. Chemical Bonds.* In order to obtain a clear understanding of certain physicochemical properties of dielectric liquids, and particularly liquid hydrocarbons, it is necessary to relate them to contemporary views on quantum theory, the spatial distribution of electronic charge in various atomic and molecular states, and the nature of the chemical bond. The fundamentals of the electronic theory of organic compounds are presented below, but we are here restricted to a discussion of the simplest models of liquid molecules. A more extensive treatment as well as the fundamentals of quantum mechanics closely connected with the problem can be found in reference sources such as Karagounis [14], Gumiński [15], Coulson [16], Kittel [92–94], Kondratiev [95], Haissinski [96] and Pauling [97].

According to contemporary concepts of quantum physics and chemistry, the configuration of electrons in hydrogen and carbon atoms can be represented as in fig. 1.3. The hydrogen atom has one electron in the 1*s* state, and the carbon atom has six electrons distributed as follows: two electrons in the 1s state, two electrons in the 2s state and two electrons in the 2p state. This configuration is written $1s^2, 2s^2, 2p^2$.

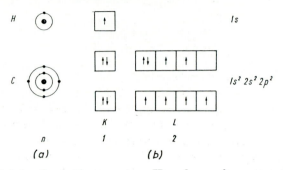

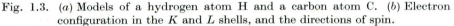

Fig. 1.3. (a) Models of a hydrogen atom H and a carbon atom C. (b) Electron configuration in the K and L shells, and the directions of spin.

According to wave mechanics, the state of an electron in an atom can be defined by a wave function ψ, which is usually a complex function of the space variables x, y, z and time t. Since ψ can take both positive and negative values, ψ^2 (or, more exactly, the product $\psi\psi^*$, where ψ^* is the conjugate value) is used in physical interpretations. Born defines ψ^2 (or $\psi\psi^*$) as the probability of finding an electron at a given point in space, while $\psi^2\,dv$ is the probability of finding an electron in the volume element $dv = dx\,dy\,dz$, at a time t. The function ψ must be normalized, that is:

$$\int_{-\infty}^{+\infty} |\psi(x,y,z,t)|^2\,d\tau = 1\,;$$

this occurs in Schrödinger's differential wave equation, which, for a carbon atom in the bound state, can be written:

$$\nabla^2\psi + \frac{8\pi^2 m}{h^2}\,(E - V)\,\psi = 0, \tag{1.3}$$

where

$$\nabla^2\psi = \frac{\partial^2\psi}{\partial x^2} + \frac{\partial^2\psi}{\partial y^2} + \frac{\partial^2\psi}{\partial z^2}, \quad V = -\frac{e^2}{r}. \tag{1.4}$$

In the above equations, m is the mass of an electron and h is Planck's constant.

For a single hydrogen atom, where the potential energy distribution has spherical symmetry, it is convenient to introduce a system of polar coordinates r, ϑ, φ; r is the distance from the centre of the atomic nucleus, φ is the azimuthal angle, and angle ϑ is the third coordinate. ψ is then the product of the three quantities $\psi = R(r)\,\theta(\vartheta)\,\Phi(\varphi)$, each of them being a function of one variable only. Schrödinger's equation then takes the form:

$$\frac{\partial^2\psi}{\partial r^2} + \frac{2}{r}\frac{\partial\psi}{\partial r} + \frac{8\pi^2 m}{h^2}\left(E + \frac{e^2}{r}\right)\psi = 0. \tag{1.5}$$

The simplest solution to eqn. (1.5) is:

$$\psi(r) = \exp(-ar), \tag{1.6}$$

where $a = 4\pi me^2/h^2$. The solution of this equation yields a number of eigenvalues for the energy:

$$E_n = -\frac{2\pi^2 me^4}{h^2}\frac{1}{n^2}, \tag{1.7}$$

where n is the main quantum number.

The maximum number of electrons $2n^2$ at any given energy level is defined by Pauli's principle, according to which there is only one electron which possesses the particular quantum numbers n, l, m and σ present in an atom; n is the main quantum number, l the first secondary quantum number, m the magnetic number and σ the spin quantum number.

The main quantum number may adopt the values of natural numbers $(1, 2, 3, \ldots)$. The first secondary number l determines the orbital momentum of the electron as it moves around the nucleus; it has values from 0 to $n-1$. The number m takes values from $l-1$ to $l+1$, and the value 0 (altogether $2l+1$ different values). The number σ characterizing the electron's motion around its own axis, its spin, can have only two values, $+\frac{1}{2}$ or $-\frac{1}{2}$ (we say that the spins are parallel or anti-parallel). The states for which the quantum number l has values 0, 1, 2 and 3 are called, respectively, the s, p, d and f states.

For s states, for which the quantum number $l = 0$, the function ψ depends only on r (θ and Φ are constant) and the electron clouds therefore have a spherical shape. The probability $P(r)$ of finding an electron at a distance r from the nucleus in an element of volume dv is $P(r) = 4\pi r^2 \psi^2 \, dr$.

Figure 1.4 (a), (b), (c) shows the shape of an electron cloud in an atom and the shape of the function ψ for the $1s$, $2s$ and $2p$ states. Figure 1.5 is a diagram of the probability:

$$P(r) = \int_r^\infty \psi^2 \, 4\pi r^2 \, dr = \exp(-2r/a_0)\left[1 + (2r/a_0) + (2r^2/a_0^2)\right],$$

as a function of the distance r from the nucleus of the atom; $a_0 = h^2/4\pi^2 me^2$. As is seen from the figure, the greatest probability of finding an electron in a hydrogen atom in the $1s$ state is when $r = 0\cdot5292$ Å; this corresponds to the first orbit radius in Bohr's model. At higher energy states, e.g. the $2s$ state, the maximum probability of finding an electron (maximum electron cloud density) is further from the nucleus, but there is also a certain probability of finding an electron at smaller distances. For these states, there are nodal planes (i.e. planes in which the probability of finding the electron is zero) surrounding the centre of an atom, the number of planes being $n-1$, where n is the main quantum number. In this case the secondary quantum number is 0 ($l = 0$); this is characteristic of s states. If l is not zero, the electron configuration in an atom is no longer spherically symmetrical. The axial symmetry is still maintained for $l = 1$ when, depending on the value of the magnetic quantum number m, the axis of symmetry can be the z axis (for $m = 0$), the x axis (for $m = +1$) or the y axis (for $m = -1$). The charge distribution is in the shape of dumb-bells, with one node at the intersection of the coordinate axes.

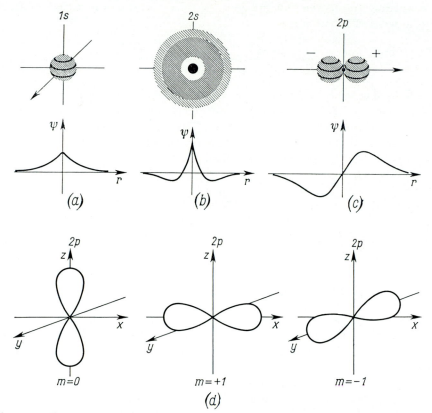

Fig. 1.4. Electron cloud distribution in an atom (above) and the shape of the wave function ψ (below); (a) in the $1s$ state, (b) in the $2s$ state, (c) in the $2p$ state, (d) cloud distribution for the $2p$ state and for various values of m. Ref.: Karrer [6].

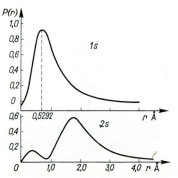

Fig. 1.5. The probability $P(r) = 4\pi r^2 \psi^2 \, dr$ of finding an electron in the $1s$ and $2s$ states at a distance r from the nucleus of the atom.

States for which $l = 1$ are called p levels. As the quantum numbers n and l increase, the shape of the cloud of charge becomes more complex. For $l = 2$, even more complicated d states arise, with which we shall not be concerned here.

According to the above definitions, the electron states in a carbon atom can be written in the form:

$$\underbrace{1s\downarrow\uparrow}_{K} \quad \underbrace{2s\downarrow\uparrow \ (2p_x)\uparrow \ (2p_y)\uparrow \ (2p_z)}_{L} .$$

The K shell is occupied by two s electrons with spins in opposite directions. The second shell L is occupied by two s electrons with opposite spins and two $2p$ electrons with parallel spins. The $2p_z$ state is unoccupied. A carbon atom ought thus to be bivalent because the valence of an atom is determined by the number of electrons in the outer shell. It is known,

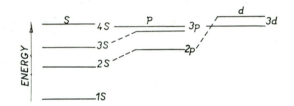

Fig. 1.6. The energy level distribution in a carbon atom.

however, that carbon occurs in chemical compounds in a tetravalent form. It must therefore be assumed that one of the electrons in the $2s$ state moves to the $2p$ state, this move being facilitated by the fact that the energy difference between these two states is very small (see fig. 1.6). The electron configuration can then be written $1s^2 2s^1 2p_x^1 2p_y^1 2_z^1$, which in effect gives four identical electron clouds of a mixed nature with an angle of $109°$ between them. It is known as a hybrid bond orbital and is designated in quantum chemistry sp^3. The four equivalent sp^3 hybrids are directed towards the corners of a regular tetrahedron (fig. 1.7) and result in identical chemical bonds.

Fig. 1.7. sp^3 hybrid electron density distribution in a carbon atom; each cloud is occupied by one electron. Ref.: Karrer [6].

In the formation of hybrids, i.e. when one electron moves from the $2s$ to the $2p$ state, the energy given out is 96 kcal/mole. The sp^3 hybrid has a vectorial nature; when superimposed on another sp^3 hybrid or a $1s$ electron of a hydrogen atom, the result is a C—C or C—H covalent bond. These bonds are cylindrically symmetrical about the C—C axis, which means that the atoms can rotate freely about the axis. This type of bond is known as a σ bond. In their simplest form they occur in the CH molecule, whose dipole moment is 0·3 debyes. The bond direction is from the carbon atom to the hydrogen atom. A significant rôle is played by sp^3 orbitals in the behaviour of carbon bonds in aliphatic compounds.

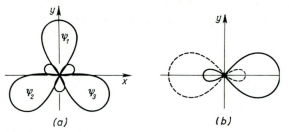

(a)   (b)

Fig. 1.8. Hybrid electron density distributions; (a) sp^2 hybrid, (b) sp hybrid. Ref.: Karagounis [14].

Besides sp^3 orbitals, there are also sp^2 and sp orbitals. The sp^2 orbitals are formed by superimposing the wave function of an s electron on two p electron functions. This combination results in three identical sp^2 valency hybrids lying in the same plane, in directions such that they form angles of 120°. The configuration is thus trigonal (fig. 1.8). The fourth $2p_z$ electron lies outside this plane and perpendicularly to it. Its distribution function has the shape of dumb-bells.

For sp^2 hybrid covalent bonds, the electron clouds arranging themselves in cylindrical symmetry along the bond axis give a σ bond, while the single $2p_z$ electron clouds of two different carbon atoms interpenetrate and exhibit an anti-symmetrical arrangement in relation to the nodal plane (fig. 1.9). These electron bonds are called π bonds.

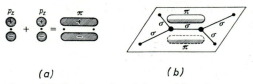

(a)   (b)

Fig. 1.9. (a) Molecular orbital of a π bond formed by the mutual superimposition of two atomic π orbitals, (b) a model of σ and π bonds in ethylene. Refs.: (a) Karagounis [14], (b) Karrer [6].

The σ and π bond arrangement is known as a double bond. It differs from other bonds in that it does not permit rotation of the atoms about the C—C axis and displays unsaturated properties, partly because here the electrons are only coupled to each other due to the spins being in opposite directions.

The double bond is only 1·7 times stronger than a single σ bond; this means that the π bond is weaker than the σ bond. Bonds of this type occur in the ethylene molecule C_2H_4, where three hybridized orbitals in the same plane form σ bonds (one C—C bond and two C—H bonds), while the single remaining $2p_z$ electron forms a π bond with a similar electron from the other carbon atom. Figure 1.9 (b) shows a model of this bond system in ethylene: two C atoms are bound by sp^2 hybrid states. Six atomic nuclei lie in one plane together with the s electrons surrounding them and some of the p electrons, these latter forming five σ bonds. The remaining p electrons extend symmetrically outside the plane in directions perpendicular to it, making it possible to form bonds with other molecules by π bonding. Figure 1.9 (a), (b) shows the arrangement of these σ and π bonds.

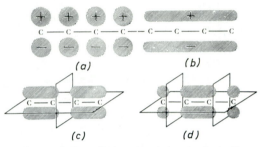

Fig. 1.10. Configurations of the electronic states in butadiene C_4H_6 shown diagrammatically; (a), (b) and (c) in the normal state, (d) in the excited state. Ref.: Karrer [6].

We shall also consider the distribution of the electronic states in a butadiene molecule C_4H_6. The main skeleton of the molecule is composed of σ bonds lying in one plane and is formed from four trigonal sp^2 hybrids and six s hydrogen states. Perpendicular to this plane lie the π bond clouds. Disregarding the zig-zag shape of the molecule, it can be represented as in fig. 1.10. The strongest bond is formed from that state in which two electrons have anti-parallel spins (fig. 1.10 (a), (b)). These electrons can move freely along the entire molecular chain. The two remaining electrons are in a higher state in the molecule and divide the electron cloud in the way shown in fig. 1.10 (c), or remain in the excited state as in fig. 1.10 (d). These states combine two carbon atoms, dividing the molecule into two parts along its centre. It can be said that σ bonds are usually formed from orbitals symmetrical with respect to the molecular axis, that is from s–s or p_x–p_x orbitals, while the linear combination of p_z–p_z or p_y–p_y orbitals results in a π bond.

Bond orbitals also differ from each other in the component of electron momentum along the axis of the molecule. In σ orbitals, this component is 0; in π orbitals it is 1, and there are also δ orbitals for which this component has the value 2.

The s states have two σ electrons, p states have two σ electrons and four π electrons, and d states have two σ electrons, four π electrons and four δ electrons. The differences between these states can be summarized as follows:

σ state	π state
Rotational symmetry of the electron cloud about the bond axis; no nodal planes	Lack of rotational symmetry; nodal planes divide electron clouds equally in two

There is yet another type of hybridized orbital besides sp^3 and sp^2 orbitals, namely the sp orbital. It is formed as a result of interaction between a $2s$ and a $2p_x$ orbital, while the two orbitals $2p_y$ and $2p_z$ remain unaffected. The spatial distribution of an sp orbital looks like two dumb-bells parallel to each other (fig. 1.8 (b)). One of them is the result of the mutual interaction of the electron clouds forming a σ bond, the other as a result of the interaction of $2p_x$ and $2p_y$ electrons.

Hybridized orbitals of the sp type are those chiefly instrumental in forming triple bonds, which occur in acetylene, for example. In a covalent bond formed from two such hybrids, four atoms (2C and 2H) are arranged linearly along the axis joining them; this structure is evident from the lack of *cis/trans* isomers among acetylene derivatives. The atoms are joined by σ bonds, and also have two π bonds. A triple bond is stronger than two single bonds; its unsaturated character is also more clearly evident than that of a double bond.

All the bonds discussed so far are localized. Each bond can also be defined by the number of electrons per atom which form it. This number determines the order of the bond. Two electrons are required for each order. Special symbols are often used for their designation:

	Bond order	Number of electrons
Single bond $(\sigma sp^3)^2$ (aliphatic)	1	2
Double bond $(\sigma sp^2)^2 (\pi_z 2p)^2$ (ethylenic)	2	4
Triple bond $(\sigma sp)^2 (\pi_z 2p)^2 (\pi_y 2p)^2$ (acetylenic)	3	6

The order of a bond is not always a whole number. In benzene, for example, the order of the C—C bond is 1·5, and in graphite it is 1·33. This means that a certain proportion of electrons making up the bond must every now and then shift in the direction of other bonds, giving a certain average value for all interactions.

The stronger the bond, the less is the distance between the atoms. The distances between atoms in organic molecules have been determined accurately by experimental methods such as x-ray analysis (mainly for solids), electron beam interference (for gases), the examination of complex infra-red absorption spectra and microwave techniques.

Table 1.1 gives the distances between atoms for various bonds and the corresponding bond energy.

Table 1.1. Distances between atoms and corresponding bond energies

Bond	Distance (Å)	Bond energy (kcal/mole)	Substance
C—C	1·54	83·6	ethane
C=C	1·35	145	ethylene
C≡C	1·20	196	acethylene
C—H	—	98·7	
C—O	—	86	
C=O	—	165·6–178·6	
C—N	1·47	62	
C=N	1·28	129·6	
C≡N	1·15	187·3	
C—S	—	59·6	
C—P	—	120	

The distribution of electrons in aromatic compounds and compounds conjugate with them is of considerable interest. In benzene or naphthalene, the σ bonds formed from sp^2 orbitals are localized in one plane (fig. 1.11). The π bonds formed from $2p_z$ orbitals are not localized. In benzene, six π electrons rotate above and beneath the molecular plane in a ring with dimensions appropriate to the configuration of the six carbon

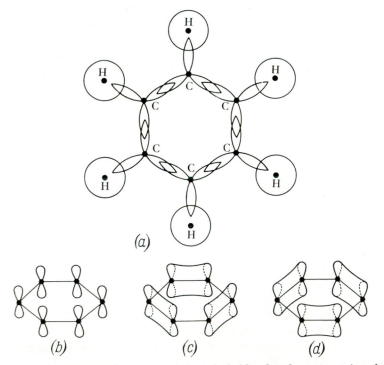

Fig. 1.11. Schematic representation of the π hybrids of carbon atoms in a benzene molecule; (a) in one plane, (b), (c) and (d) side-view of individual orbitals and double bonds in two different states of benzene. Ref.: Coulson [16].

atoms in the molecule (fig. 1.12). The order of the C—C bond in benzene is 1·53, and the distance between atoms is 1·39 Å. The dynamic equilibrium between the π electrons can result in external factors being able to change the chemical properties of the molecule. It explains, for example,

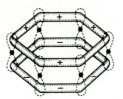

Fig. 1.12. Delocalized molecular π orbitals in benzene. Ref.: Coulson [16].
(● Carbon Atoms)

why the benzene molecule displays characteristics of double bonds (as in (*a*) or (*b*)) in a number of compounds. A similar phenomenon may also occur with other more complex aromatic compounds and substances conjugate with them.

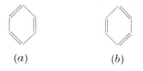

(*a*)   (*b*)

The non-localized π electrons are significant factors in the phenomena of electrical conductivity of metals, graphite, semiconductors, etc., as well as in biological and physiological phenomena (the transmission of redox waves).

The π electrons are also important in the mechanism of the absorption of light; a molecule which has absorbed a quantum of light with an energy $h\nu$ moves from a lower level to a higher one acquiring an amount of energy $\Delta E = h\nu = h^2(N+1)\,8mL^2$, where N is the number of π electrons in the molecule. A wavelength

$$\lambda_m = \frac{8mc}{h}\frac{N^{3/2}}{N+1} \tag{1.8}$$

corresponds to this amount of energy. The wavelengths of light absorbed depend on the number of π electrons in a molecule, and consequently upon the number of double bonds.

A more detailed analysis of the distribution of electron energy in hydrocarbon molecules was carried out by Kuhn *et al.* [17, 18] in 1959–60; they inserted various types of potential in the Schrödinger equation, carrying out the calculations on an analogue computer. They considered the three potential models shown in fig. 1.13. In (*a*), the electron can move in a one-dimensional box of length L equal to the length of the molecular chain. The energy level of a π electron is then equal to $E_n = h^2 n^2/8mL^2$, where h is Planck's constant, m is the mass of an electron and n is a natural number. The energy levels for $n = 1, 2, 3, 4$ and 5 are shown in fig. 1.14. The wave function adopts the form:

$$\psi_n = \left(\frac{2}{L}\right)^{1/2} \sin \frac{ns}{L} \pi, \qquad (1.9)$$

which is shown in fig. 1.15 for a number of values of n. The probability of finding a π electron in the ground state of the molecule is then:

$$P(s) = \sum_{n=1}^{j} 2\psi_n^2(s), \qquad (1.10)$$

the wave function $\psi_n(s)$ being normalized thus:

$$\int_{-\infty}^{+\infty} \psi_n^2 \, ds = 1.$$

In fig. 1.13 (b), the electron moves in a periodically varying potential, whose minimum value occurs at the points where the carbon atoms are situated. Model (c) is two-dimensional; the potential distribution is shown in fig. 1.13 (c) and the corresponding energy levels in fig. 1.14 (c).

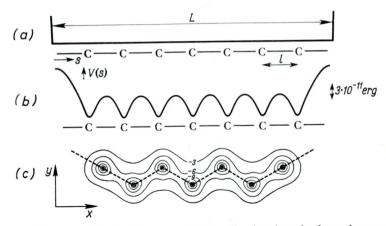

Fig. 1.13. Three models for the potential distribution in a hydrocarbon molecule. Ref.: Kuhn *et al.* [17].

Kuhn *et al.* adopted the following assumptions for their computations: firstly, that the charge on a carbon atom is shielded by an electron shell such that its interaction is limited to a sphere of radius 1·39 Å; secondly, inside this sphere, an electron is influenced by the effective charge, which for carbon is 3·25e. The resultant potential at each point is a result of the superposition of the potentials of each atom. The numerical values which they obtained theoretically for the absorption spectra of a number of hydrocarbons are in exact agreement with experimental values.

McCubbin and Gurney [19] have investigated electrical conductivity in paraffin polymers; Lorquet [20] has also determined the intra-molecular energy levels for δ and π orbitals, and the following properties for paraffins:

		ev
(a)	Self-energy of C—H bonds	− 13·8
(b)	Interaction energy between a C—H bond and a carbon atom	− 1·57
(c)	Self-energy of C—C bonds	− 17·64
(d)	Interaction energy between C—C and C—H bonds and a carbon atom (as above)	− 2·49
(e)	Interaction energy between a C—C bond and a carbon atom	− 3·79

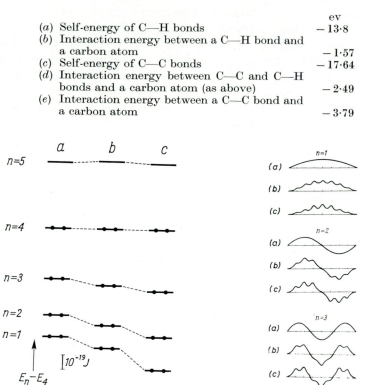

Fig. 1.14. Energy levels calculated for models (a), (b) and (c) in fig. 1.13. Ref.: Kuhn *et al.* [17].

Fig. 1.15. The wave function ψ along the zig-zag line of the carbon atoms for models (a), (b) and (c) in fig. 1.13. Ref.: Kuhn *et al.* [17].

It was assumed that the width of the valency band is about 10 ev and occurs at about − 15 ev, the upper limit of the band being at about − 10 ev (fig. 1.16).

Since they have a direct bearing on the interpretation of the electrical properties of dielectric liquids, these diagrams of the wave mechanics of hydrocarbon molecules can be of great assistance in explaining the division of hydrocarbons into groups having different values of electrical conductivity.

1.6. *Hydrocarbon Oils. Silicones.*   Part of the research into ionization, conduction and breakdown in dielectric liquids has been carried out using very high viscosity oils. A brief description of the most important properties of insulating hydrocarbon oils may therefore be of interest here.

Oils of this kind are obtained either from natural crude oil or from gases by synthetic methods. In both cases the liquid oil is a mixture of various substances containing hundreds of different compounds. Aromatics, paraffins, cycloparaffins, naphthenes and olefin hydrocarbons occur in the greatest proportions. Non-hydrocarbon compounds containing oxygen, sulphur, nitrogen and traces of metals, such as iron, copper, aluminium, sodium, etc., can also occur in naphtha derivatives and can considerably change the electrical properties of the oils. The relative occurrence of different compounds in the oil depends to a great extent on where the crude oil comes from. The paraffin content is often fairly large (sometimes up to 80%), while other crude oils contain mostly aromatics and similar substances.

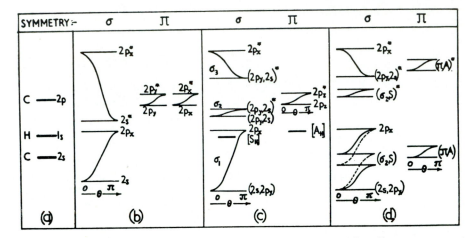

Fig. 1.16. Schematic development of intramolecular bands. All ordinates represent orbital energies; abscissae θ (in each column) represents the phase shift of a particular Bloch function between adjacent carbon atoms of the chain. The curves should not be considered unique or quantitative. (*a*) Free-atom orbital; (*b*) linear chain of carbon atoms (both $2p_y$ and $2p_z$ bands have π symmetry); (*c*) zig-zag chain of carbon atoms (energy levels of symmetric $[S_H]$ and antisymmetric $[A_H]$ hydrogen orbitals are also indicated); (*d*) complete molecule. Ref.: McCubbin and Gurney [19].

From the physical point of view, the oils are usually characterized by their density, molecular weight, refractive index, viscosity, specific resistance and loss angle. The density is within the range 0·87–0·96 g/cm³, the molecular weight from 230 to 290, refractive index 1·47–1·54, viscosity 0·01 P–several P, and specific resistance 10^{15}–10^{18} Ω cm. The dielectric constant is about 2·3. Another property of these liquids often measured is the breakdown strength, which is to a large extent dependent on the purity of the oil, particularly on the water content; for top-grade technical oils its value is about 400–500 kv/cm. The tangent of the loss angle is calculated from the ratio of the leakage (loss) current to the charging current. All the above values refer to measurements at room temperature. Some of them, particularly viscosity, specific resistance and breakdown strength, decrease rapidly with rise in temperature.

The absorption spectra of hydrocarbon oils show very characteristic absorption bands, especially in the infra-red region. The band for C—H bonds occurs at about $3 \cdot 5 \, \mu$; the aromatic bands for C—C bonds occur at $6 \cdot 4 \, \mu$; the bands corresponding to CH_2 and CH_3 group bonds at $6 \cdot 8$–$7 \cdot 2 \, \mu$; the bands for *iso*paraffins at $8 \cdot 7 \, \mu$; the naphthalene band between 10 and $11 \, \mu$; the band for aromatic compounds in the range $11 \cdot 7$–$13 \cdot 4 \, \mu$, and the band for paraffin chains somewhere about $13 \cdot 8 \, \mu$.

Prolonged action of an electric field results in an absorption band occurring in the oil near $5 \cdot 8 \, \mu$, corresponding to the C=O bond. As a result of strong irradiation with x-rays or neutrons, non-saturated compounds are formed. Hydrocarbon oils are of great practical importance in the electrical industry as they are used in transformers, capacitors, pressure devices, cables, etc.

Another distinct group of hydrocarbon oils are liquid polyethylenes with the following structure:

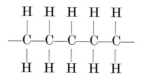

Depending on the length of the chain, polyethylene molecules are oily, waxy or have a solid consistency. They are obtained mainly by synthetic methods from ethylene at high pressures (1000–2000 kG/cm²) and fairly high temperatures. Their chief application as dielectrics is in high-frequency cables (e.g. in radar equipment).

In the past few years, the conductivity and breakdown of silicon compounds have become the subject of many investigations. Like saturated hydrocarbons, the silicones feature a chain-like molecular structure. The similarities and differences in the structure are shown below:

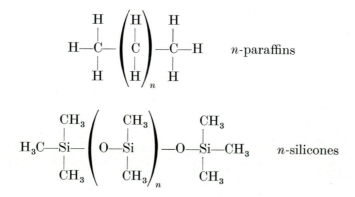

The first compounds in the silicone homologous series are:

	n	Density (g/cm³)
(1) Dimer	0	0·761
(2) Trimer	1	0·818
(3) Tetramer	2	0·852
(4) Pentamer	3	0·871

The specific gravities of the higher compounds extend up to 0·97. Their boiling points range from 99·5 to over 250°C, freezing points from -68 to -40°C, viscosities from 0·65 to 10^5 CP, and refractive indices from 1·38 to 1·403.

Silicones have all the features of good dielectrics; they have a very high specific resistance and are not hygroscopic (steam can accumulate only on their surfaces). They are chemically inert and do not damage materials with which they come into contact; the physicochemical properties vary only slightly with temperature over the range -90 to $+300$°C. They also undergo relatively minor chemical changes (compared with organic substances) when subjected to large radiation doses.

Comprehensive data on the properties of the various oils can be found in works by Birks [21] and Kok [22].

Chapter 2

Molecular Structure and the Macroscopic Properties of Liquids

2.1. *Fundamental Quantities.* In the past two decades a great deal of work has been published concerning the relationship between macroscopic quantities characterizing various types of liquids and their molecular structure. The commonest method of investigation has been a simple tabulation of experimental data according to molecular structure. Correlating equations have been derived either from theoretical considerations based on the laws of thermodynamics and statistical mechanics or, in more complex treatments, by semi-empirical methods. Several examples will be quoted in Chapter 3.

The studies are of great importance in that they enable a number of relationships to be determined between molecular structure and quantities closely connected with electrical conductivity and breakdown in liquids, such as ionic mobility coefficients, and recombination and diffusion coefficients. These relationships can be useful towards explaining the very complex mechanism of electron and ion transfer in liquids.

Since saturated hydrocarbons are the most frequently studied liquid dielectrics, they are treated here in the most detail. Egloff [7] and Landolt and Börnstein [23] are extremely comprehensive sources of data on hydrocarbons. A paper by Kurata and Isida [24] should also be mentioned here; it contains one of the most successful attempts to explain theoretically the relationships between fundamental physicochemical properties and molecular structure for liquid saturated hydrocarbons (of type C_nH_{2n+2}). They obtained the following expressions for the critical temperature T_c and the boiling point T_b:

$$T_c = \frac{1}{0 \cdot 000702 + (0 \cdot 00419/n^{2/3})}, \qquad (2.1)$$

e.g. for $n = 6$, $T_c = 507 \cdot 4°K$.

$$T_b = \frac{1}{0 \cdot 000570 + (0 \cdot 007753/n^{2/3})}, \qquad (2.2)$$

e.g. for $n = 6$, $T_b = 342 \cdot 0°K$.

They also derived a relationship between T_c and T_b, viz.:

$$\frac{1}{T_c} = \frac{0 \cdot 5404}{T_b} + 0 \cdot 000394. \qquad (2.3)$$

These formulae are in agreement with experimental values within 1–3% over the range $n = 1$ to $n = 17$. They are derived in § 3.2.

Using a semi-empirical method, the present author has derived the following relationships for the melting point T_m and boiling point T_b:

$$T_m = 360[1 - 0 \cdot 56 \exp(-0 \cdot 104 n)], \tag{2.4}$$

$$T_b = 146\sqrt{(n)} - 17, \tag{2.5}$$

e.g. for $n = 6$, $T_b = 341°\text{K}$.

A simpler form of equation (2.4) is:

$$T_m = 67\sqrt{(n)} + 18, \tag{2.6}$$

e.g. for $n = 6$, $T_m = 178°\text{K}$.

Equations (2.5) and (2.6) are in agreement with experiment over the range $5 \leqslant n \leqslant 17$. For $n = 1, 2, 3$ and 4, that is for gaseous hydrocarbons, the deviations under normal conditions are fairly large, whereas similar deviations are not observed when equations (2.1) and (2.2) are used.

Figure (2.1) shows curves of the melting and boiling points, T_m (t_2) and T_b (t_1) for paraffinic hydrocarbons. The liquid state corresponds to the area contained between the two curves.

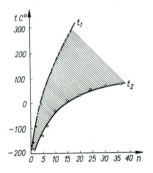

Fig. 2.1. Curves of boiling point t_1 and melting point t_2 plotted against the number of carbon atoms n for paraffin hydrocarbons.

A detailed study of the relationship between the melting points and molecular structure of saturated hydrocarbons has shown that the melting points of compounds with n even are slightly lower than those taken from a smoothed curve as in fig 2.1; similarly T_m for compounds with n odd is slightly above the smoothed values. The differences involved can be as high as several degrees and can be explained by the fact that in the two cases for molecules in the solid state, the types of crystal lattice are slightly different, with chemical bonds between two adjacent molecules of different strengths. This is partly due to the shifting of the axis of the principal moment of inertia in the two types of molecules.

Other values from the literature of interest are bond energies: the C—H energy is $98 \cdot 7$ kcal/mole and the C—C energy is $83 \cdot 6$ kcal/mole. Further, the relationships between the heat of fusion H_f, the heat of evaporation at the boiling point H_e, the heat of combustion H_c and the number of carbon atoms n in a hydrocarbon molecule can be written:

$$H_f = (n-3) \text{ kcal/mole},$$

$$H_e = (0\cdot75n + 2\cdot5) \text{ kcal/mole},$$

$$H_c = (157n + 60) \text{ kcal/mole}.$$

The increase in heat of combustion per CH_2 group is 157 kcal/mole. Numerical values for, say, hexane according to these formulae are: $H_f = 3$ kcal/mole; $H_e = 7\cdot0$ kcal/mole, and $H_c = 1002$ kcal/mole. These values are in agreement with experiment to within 1%.

Adamczewski [25], using a semi-empirical method, has derived formulae relating the density and viscosity of these liquids to the number of carbon atoms n in the molecule and the temperature. These formulae are important starting points for further investigation of the mechanism of conductance and electrical breakdown. Taking the molecular weight to be given by $M = 14\cdot026n + 2\cdot01626$, the density at room temperature can be calculated from:

$$\rho = \frac{14\cdot026n + 2\cdot01626}{16(n + 2\cdot27)} \text{ g/cm}^3, \tag{2.7}$$

and the density at any other temperature from:

$$\rho = \frac{14\cdot026n + 2\cdot01626}{15\cdot6(n + 2\cdot27)\left(1 + \dfrac{0\cdot0194}{n + 8}t\right)} \text{ g/cm}^3. \tag{2.8}$$

Examples of the use of these equations are given in tables (2.1) and (2.2).

Table 2.1. Densities of paraffinic hydrocarbons (from eqn. (2.7))

n	Formula	Density (g/cm³)	
		Experimental	Theoretical
5	C_5H_{12}	0·626	0·623
6	C_6H_{14}	0·6595	0·652
7	C_7H_{16}	0·6837	0·678
8	C_8H_{18}	0·7028	0·695
9	C_9H_{20}	0·7179	0·713
10	$C_{10}H_{22}$	0·7298	0·725
11	$C_{11}H_{24}$	0·7404	0·736
12	$C_{12}H_{26}$	0·7493	0·748
13	$C_{13}H_{28}$	0·7568	0·756
14	$C_{14}H_{30}$	0·7636	0·762
15	$C_{15}H_{32}$	0·7688	0·768
16	$C_{16}H_{34}$	0·7749	0·778

Table 2.2. Densities of nonane and heptane at various temperatures (from eqn. (2.8))

t °c	Nonane C_9H_{20} Densities (g/cm^3)		t °c	Heptane C_7H_{16} Densities (g/cm^3)	
	Experimental	Theoretical		Experimental	Theoretical
-50	0·7726	0·7726	-90	0·7738	0·780
-10	0·7417	0·736	-50	0·742	0·742
10	0·7260	0·725	0	0·7004	0·700
30	0·7105	0·704	25	0·6796	0·678
70	0·6775	0·675	50	0·6577	0·656
110	0·6445	0·647	70	0·6395	0·640
150	0·6096	0·615	90	0·621	0·625

2.2. *Liquid Viscosity.*   For the viscosity η of a liquid, Adamczewski [25] has derived two expressions for saturated hydrocarbons, one for compounds where $5 \leqslant n \leqslant 17$ having the form:

$$\eta = 4{\cdot}7 \times 10^{-3}\,kT \exp\left[\frac{-0{\cdot}116\,nkT + 0{\cdot}0093(n+5{\cdot}05)}{kT}\right], \qquad (2.9)$$

and another for those where $28 \leqslant n \leqslant 64$, which are solids under normal conditions:

$$\eta = 1{\cdot}095 \times 10^{-4} \exp\left[\frac{0{\cdot}0046\,nkT + 0{\cdot}00154(n+86{\cdot}5)}{kT}\right], \qquad (2.10)$$

k is Boltzmann's constant ($= 8{\cdot}6167 \times 10^{-5}$ ev/°κ) and T is the absolute temperature.

Table 2.3 shows the viscosities of a number of saturated hydrocarbons at various temperatures calculated using eqns. (2.9) and (2.10) compared with experimental values taken from Landolt and Börnstein's tables [23].

Table 2.3. Viscosities of saturated hydrocarbons

Compound	Temperature	Viscosity (poises)	
		Experimental	Theoretical
C_5H_{12}	273·74	0·002869	0·00256
	293	0·002425	0·00222
	303	0·002151	0·00202
C_6H_{14}	298	0·00329	0·00328
	308	0·00283	0·00306
	323	0·00248	0·00263
	338	0·00218	0·00232
C_7H_{16}	283	0·00546	0·00466
	293	0·00481	0·00429

Table 2.3. (*continued*) Viscosities of saturated hydrocarbons

Compound	Temperature	Viscosity (poises)	
		Experimental	Theoretical
C_8H_{18}	273·25	0·00703	0·00752
	293	0·00524	0·0057
	328	0·003724	0·00387
	371·5	0·002544	0·00266
	395	0·002126	0·00222
C_9H_{20}	273	0·00969	0·00967
	283	0·00825	0·00841
	293	0·00711	0·00720
	333	0·00438	0·00440
	353	0·00360	0·00361
	373	0·00299	0·00302
$C_{10}H_{22}$	293	0·0092	0·00936
	313	0·0069	0·00704
	333	0·0054	0·00560
	373	0·0036	0·00366
$C_{11}H_{24}$	273	0·0172	0·01745
	283	0·0142	0·01425
	293	0·0117	0·01210
	313	0·00864	0·00882
	333	0·00662	0·00678
	353	0·00529	0·00536
	373	0·00435	0·00431
$C_{14}H_{30}$	293	0·0218	0·0264
	313	0·0147	0·0180
$C_{17}H_{36}$	293·2	0·04209	0·0513
	273·2	0·01018	0·0119
	423·2	0·005984	0·00627
	473·2	0·003938	0·003978
	573·2	0·002030	0·00195
$C_{28}H_{58}$	373·2	0·0287	0·0284
	423·2	0·0146	0·0150
	473·2	0·00885	0·00908
	523·2	0·00592	0·00585
	573·2	0·00417	0·00428
$C_{36}H_{74}$	373·2	0·0488	0·0431
	423·2	0·02315	0·0218
	473·2	0·01344	0·0127
	523·2	0·00874	00·0815
	573·2	0·00608	0·00568
$C_{64}H_{130}$	373·2	0·1785	0·1845
	423·2	0·0714	0·0801
	473·2	0·0375	0·0412
	523·2	0·0229	0·0240
	573·2	0·0152	0·0155

Differences between the theoretical and experimental results are of the order of 5–10%.

The formulae quoted above have the same form as a number of others to be found in the literature, viz.:

$$\eta = A \exp(W_1/kT) \tag{2.11}$$

and

$$\eta = BT \exp(W_2/kT). \tag{2.12}$$

A and B are constants for a given liquid, and W_1 and W_2 are the activation energies for the viscous process in the liquids. The physical meaning of the terms is as follows: the factors A^{-1} and $(BT)^{-1}$ are proportional to the number of molecular collisions at a given temperature, whilst the exponential factor $\exp(W/kT)$ characterizes the collision efficiency, in other words the probability that the molecular viscous process will occur. The numerical value of the exponential term at room temperature is about 10^{-29} for $W = 40$ kcal/mole, 10^{-15} for $W = 20$ kcal/mole, and 10^{-7} for $W = 10$ kcal/mole. The viscosity relationship expressed by eqn. (2.11), i.e. neglecting the effect of temperature, is the one most frequently encountered in the literature.

Unfortunately, none of the known theories of viscosity is completely satisfactory for correlation purposes, and theoretical values obtained from these theories sometimes differ by several orders of magnitude from experimental values. Some correlating equations are given below:

$$\eta = A \exp \frac{W}{kT} \quad \text{(Andrade–Guzman [26]),} \tag{2.13}$$

$$\eta V^{1/3} = B \exp \frac{C}{VT} \quad \text{(Andrade [27]),} \tag{2.14}$$

$$\eta = \frac{kT\tau_0}{\pi a \delta^2} \exp \frac{W}{kT} \quad \text{(Frenkel [9]),} \tag{2.15}$$

$$\eta = \frac{N\delta}{V} (2\pi mkT)^{1/2} \exp\left(\frac{E}{kT}\right) \quad \text{(Eyring [8]),} \tag{2.16}$$

$$\eta = A \exp \frac{B}{f} \quad \text{(Doolitle [28]),} \tag{2.17}$$

where τ_0 is the period of free vibration of the molecule; a, the molecular radius; δ, the distance between two adjacent molecules; and f, the relative free volume between molecules:

$$f = \frac{V - V_0}{V_0} + \frac{1}{V_0} \frac{dV}{dT} (T - T_0). \tag{2.18}$$

In eqn. (2.18), V is the volume of liquid; V_0, the volume of the molecules; and T is temperature.

Glasstone *et al.* [29] have derived a more complex viscosity correlating equation:

$$\eta = \frac{\lambda_1}{\lambda\lambda_2\lambda_3}\frac{v_h}{v-v_s}(2\pi mkT)^{1/2}\exp(\Delta e_f/kT), \qquad (2.19)$$

where λ_1 is the distance between two layers of liquid moving relative to each other under the influence of external forces; λ, the distance between two adjacent equilibrium positions in the direction of motion; λ_2, the distance between two neighbouring molecules in the direction of motion;

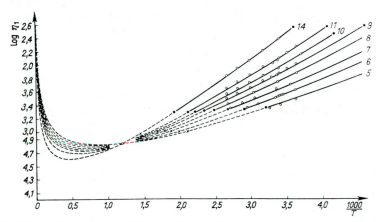

Fig. 2.2. Plot of $\log\eta$ against T (eqn. (2.9)). Experimental data are shown as open circles. Full circles are melting points (above) and boiling points (below). Critical temperatures are marked with crosses. The number by each curve denotes the number of carbon atoms in the molecule.

λ_3, the mean distance between two adjacent molecules in the moving layer in a direction perpendicular to the direction of motion; v, volume per molecule; v_h, volume of one hole; v_s, specific volume of the molecule; and Δe_f, the activation energy required for a single molecule to flow into a hole already available.

Figure 2.2 is a plot of eqn. (2.9). The solid lines are the portions of the curves valid for real liquids. The common point at which all the curves cross lies, from eqn. (2.9), in the purely mathematical region above the critical temperature.

According to Eyring [8], the activation energy for viscous flow is composed of two parts: the energy required to form a hole, and the energy required to shift a molecule into the hole. The viscous energy is found to be approximately half the molar energy of evaporation:

$$\Delta W_e/\Delta W_{\text{visc}} = 2\cdot45.$$

Telang [30] has published several papers on the mechanism of liquid viscosity and the relationship between the activation energy W, the frequency factor Z and the surface tension and structure of the liquid.

He obtained the following equation for viscosity:

$$\eta = Z \exp(W/RT), \qquad (2.20)$$

where

$$Z = \left(\frac{hN}{V^{3/2}}\right)\left[\frac{b}{(V-b)^{4/3}}\right]$$

and

$$W = 1{\cdot}091 N^{1/3}(M/\rho)^{2/3}\,\sigma.$$

In the above equations, h is Planck's constant, N is Avogadro's number, V is the molar volume, M, the molecular weight, ρ, the density, and $b = Nv_s$ (see eqn. (2.19)). The dependence of the frequency factor on activation energy has also been studied by Ruetschi [31].

Macedo and Litowitz [32] have derived a formula for liquid viscosity:

$$\eta = A_0 \exp(E_v/RT + \gamma v_0/v_f), \qquad (2.21)$$

where A_0 is a temperature dependent variable; E_v, the height of the potential barrier between equilibrium positions; γ, a constant; v, free volume; and v_f, average free volume per molecule. The equation can be used for a very wide range of liquids from silicon to liquid argon while accounting for hydrogen bonding and van der Waals' bonds.

Much recent research work has been devoted to the effect of molecular rotation on liquid viscosity. Davies and Matheson [33] state that this effect can be used to explain the three different types of viscosity–temperature relationship observed with certain liquids. For example, the Arrhenius relationship, $\log \eta = A + W/RT$, is valid over the entire temperature range for liquefied inert gases such as neon, argon, krypton, *neo*pentane, etc. These molecules possess a spherical structure and are therefore completely free to rotate in all directions in space. For other liquids, the Arrhenius equation is true only above a certain temperature. These liquids include propane, toluene, etc.

A further group of liquids such as *iso*propylbenzene behave according to the Arrhenius equation at high temperatures, while at lower temperatures feature two different temperature dependences over different ranges, both inconsistent with the Arrhenius equation.

In general, molecules with a limited rotational ability in space (e.g. long chains) have a higher viscosity. Sometimes this is the reason for the discontinuity of the dependence of physicochemical properties such as density, viscosity and specific heat on temperature. This may be observed, for example, in paraffin hydrocarbons from $203°\text{к}$ to $288°\text{к}$.

2.3. *Surface Tension. The Parachor.* An important property connected with the structure of liquids is the value of the parachor P; this features additive properties with regard to individual atoms and bonds in the molecules and is closely related to the surface tension σ.

Surface tension can be determined in a simple manner by calculating the amount of work necessary to divide a system of cross-sectional area of

1 cm² into two parts, i.e. the work necessary to increase the cross-sectional area to 2 cm².

According to Lennard-Jones [34, 35], the force between two molecules is made up of two components, an attractive force F_1:

$$F_1 = -\frac{a}{r^7}, \tag{2.22}$$

and a repulsive force F_2:

$$F_2 = \frac{b}{r^{13}}, $$

where r is the separation of the molecules. The work needed to bring the two molecules together will then be:

$$A_m = \int_r^{\infty} \left(-\frac{a}{r^7}\right) dr = \frac{a}{6r^6}, \tag{2.23}$$

and the amount of work A_s required to separate two bodies with a cross-sectional area of 1 cm² is:

$$A_s = \frac{a}{6r^6}\frac{1}{r^2} = \frac{a}{6r^8}. \tag{2.24}$$

From the above definition of surface tension:

$$A_s = 2\sigma,$$

therefore:

$$\frac{a}{6r^8} = 2\sigma \quad \text{or} \quad a = 12\sigma r^8. \tag{2.25}$$

Surface tension can be measured either by the capillary tube method or by using a stalagmometer and counting the number of drops obtained from 1 cm³ of the liquid under test. The first method is based on the simple relationship that the weight of a column of water of height h in a capillary tube of radius r is balanced by the force due to surface tension:

$$\sigma 2\pi r = \pi r^2 h\rho g, \tag{2.26}$$

where ρ is the density of the liquid. Hence:

$$\sigma = \frac{rh\rho g}{2}. \tag{2.27}$$

In the second (drop-weight) method, surface tension measurement is based on the principle that a drop breaks off from the end of the thin tube of the stalagmometer when its weight $q = v\rho g/n$ just exceeds the force of surface tension $k\sigma$ acting round the circumference of the tube. Thus:

$$\sigma = \frac{vg\rho}{nk}, \tag{2.28}$$

where n is the number of drops in a volume v and k is a constant for the apparatus.

Surface tension can also be calculated from the velocity v of ultrasonic waves in the liquid. It is possible to show that

$$v = \frac{\lambda \varphi}{2} \sqrt{\left(\frac{72\sigma}{m}\right)}, \qquad (2.29)$$

where λ is the wavelength, φ is the phase shift between two points and m is the molecular weight.

Surface tension is also connected with the activation energy of transport processes and molecular migration in liquids, i.e. with viscosity and ionic mobility, recombination and diffusion coefficients.

There is a relationship between the parachor P and the surface tension:

$$P = \frac{M}{\rho - \rho_0} \sigma^{1/4}, \qquad (2.30)$$

where M is the molecular weight, ρ is the liquid density and ρ_0 is the density of the vapour. From this formula it is evident that at temperatures well removed from the critical temperature (i.e. when $\rho \gg \rho_0$), the parachor takes the form:

$$P = V_m \sigma^{1/4}, \qquad (2.31)$$

where $V_m = M/\rho$, the molar volume. For $\sigma = 1$, therefore, the parachor is equal to the molar volume. Values of the parachor for various atoms and bonds are listed in table 2.4.

Table 2.4. Values of the parachor P

Element	P	Bond	P
C	4·8	Triple	46·6
H	17·1	Double	23·2
N	12·5	Three-member ring	16·7
O	20·0	Four-member ring	11·6
S	48·2	Five-member ring	8·5
Cl	54·3	Six-member ring	6·1

The additive nature of the parachor can be seen in the example of benzene for which, according to Kekulé,

$$P = 6(\text{C}) + 6(\text{H}) + 3(\text{double bonds}) + 1(\text{ring}) = 207 \cdot 1;$$

experiment gives a value of 206·2.

The parachors of saturated hydrocarbons can be calculated from the equation:

$$P_{\text{th}} = 39n + 34 \cdot 2, \qquad (2.32)$$

which gives values agreeing with those obtained experimentally (P_{ex}) to an accuracy of a few per cent. For example, for $n = 2$, $P_{\text{ex}} = 110 \cdot 5$; $P_{\text{th}} = 112 \cdot 2$; for $n = 6$, $P_{\text{ex}} = 270$; $P_{\text{th}} = 268 \cdot 2$; for $n = 32$, $P_{\text{ex}} = 1322$; $P_{\text{th}} = 1282$, etc.

A general equation for any paraffin is obtained by adding the parachors of the appropriate number of carbon and hydrogen atoms. Thus for the CH_3 group, $P = 56 \cdot 1$, and for the CH_2 group, $P = 39$; for a compound $CH_3(CH_2)_xCH_3$:

$$P = 2(CH_3) + x(CH_2).$$

2.4. *Molar Refraction. Dielectric Polarization.* Another property of atoms with an additive character is molar refraction R, defined by the equation:

$$R = \frac{(v^2 - 1) M}{(v^2 + 1) \rho}, \qquad (2.33)$$

where v is the refractive index, M is the molecular weight, and ρ is the density. Numerical values of molar refraction for saturated hydrocarbons can be calculated from the data available for carbon and hydrogen atoms. For example, for the yellow sodium line $R_C = 2 \cdot 418$ and $R_H = 1 \cdot 11$; hence for a CH_2 group, $R = 4 \cdot 638$, and for the CH_3 group, $R = 5 \cdot 748$. For a compound $CH_3(CH_2)_xCH_3$:

$$R = 2(CH_3) + x(CH_2) = 4 \cdot 64x + 2 \cdot 2.$$

Molar refraction is closely connected with the polarizability of a molecule; certain dielectrics possess a dipole moment. The electric charge in the molecule is not distributed uniformly, and the degree of dissymmetry is defined by the product of the dipole charge q and a distance l, known as the dipole moment. Its unit is the debye, where

$$1 \text{ debye} = 10^{-18} \text{ e.s.u.}$$

A dipole moment is induced in every volume element of a dielectric placed inside a charged capacitor and is related to the electrical polarization. Clausius and Mosotti showed that the formula for molar polarization can be written:

$$P_M = \frac{(\varepsilon - 1) M}{(\varepsilon + 2) \rho} = \frac{4\pi N \alpha}{3}, \qquad (2.34)$$

where ε is the dielectric constant of the molecule; α, the mean polarizability; M/ρ, the molar volume; and N, the Avogadro number. Debye [36] later extended eqn. (2.34) to include two terms:

$$P_M = \frac{(\varepsilon - 1) M}{(\varepsilon + 2) \rho} = \frac{4\pi N}{3} \left(\alpha + \frac{\mu^2}{3kT} \right). \qquad (2.35)$$

The first term $(4\pi N\alpha/3)$ gives the polarization resulting from induction or deformation, and the second, $[4\pi N(\mu^2/3kT)/3]$, due to the permanent dipole moment $\mu \; (= ql)$ possessed by the molecule, is the orientational polarization.

Substances without a permanent dipole are called non-polar, whereas those possessing permanent dipoles are known as polar compounds. For non-polar liquids, the molar refraction is equal to the molar polarization of the dielectric, $R = P_M$.

Chapter 3

Thermodynamic Theories of the Liquid State

3.1. *Introduction.* The majority of the formulae quoted in §§ 2.1, 2.2 and 2.3 have been derived by semi-empirical methods, due to the fact that a satisfactory and exhaustive theory of liquids comparable with the kinetic theory of gases or the theory of solids is still lacking. The problem is not easy because of the complex structure of liquids, the diverse nature of the forces between atoms in the molecules, and the rotation, vibration and oscillation of atoms and electrons. There are a number of theoretical treatments, including the extremely comprehensive theory of liquids developed by Frenkel [9], but none has satisfactorily explained the properties of liquids with reference to their elementary molecular structure. The theories put forward by Andrade and Guzman [26], Barker [37–39], Eyring [8], Kurata and Isida [24] and Rice and Allnatt [40] are perhaps the most important, but good results have been obtained only for liquids with a very simple molecular structure such as the liquefied inert gases He, Ar, Xe, Kr and monatomic metals in the liquid state. In other cases, a semi-empirical treatment had to be resorted to at a certain stage in the reasoning.

3.2. *Numerical Methods for Hydrocarbons.* Tatevskii *et al.* [13] have described a number of methods for calculating physicochemical properties of saturated normal and branched-chain hydrocarbons. The procedure was based on certain experimentally confirmed facts whereby physical quantities can be calculated in a simple manner from the number of C—C and C—H bonds in the molecule, irrespective of its structure. Denoting, therefore, some macroscopic quantity P, its value can be expressed

$$P = \sum_{i=0}^{3} n_i p_i + \sum_{i \leqslant j=1}^{4} n_{ij} p_{ij} \qquad (3.1)$$

where n_i is the number of C—H bonds of the type C_i—H; n_{ij}, the number of C—C bonds of the type C_i—C_j; and p_i and p_{ij}, the values of P corresponding to, respectively, C_i—H and C_i—C_j bonds. The chemical structure of the molecule readily gives the values of n_i and n_j; for example, the 2,2-dimethyl-3-ethylheptane $(C_{11}H_{24})$ molecule has the following structure:

$$
\begin{array}{c}
C_1 \\
| \\
C_1\text{—}C_4\text{—}C_3\text{—}C_2\text{—}C_2\text{—}C_2\text{—}C_1 \\
| \quad\quad | \\
C_1 \quad C_2 \\
\quad\quad | \\
\quad\quad C_1
\end{array}
$$

The figures 1, 2, 3 and 4 denote the number of C—C bonds for each carbon atom, i.e. atoms marked C_1 have one C—C bond and three C—H bonds, etc. One may write therefore:

$$n_1 = 15, \quad n_2 = 8, \quad n_3 = 1, \quad n_{12} = 2, \quad n_{13} = 0, \quad n_{14} = 3, \atop n_{22} = 2, \quad n_{23} = 2, \quad n_{24} = 0, \quad n_{33} = 0, \quad n_{34} = 1, \quad n_{44} = 0,} \right\} \tag{3.2}$$

and eqn. (3.1) can be expressed in the form:

$$P = 15p_1 + 8p_2 + p_3 + 2p_{12} + 3p_{14} + 2p_{22} + 2p_{23} + p_{34}. \tag{3.3}$$

The following relationships between n_i and n_{ij} and between P_{ij} and p_{ij} emerge:

$$\left. \begin{array}{l} n_1 = 3n_{12} + 3n_{13} + 3n_{14}, \\[4pt] n_2 = 2n_{22} + n_{12} + n_{23} + n_{24}, \\[4pt] n_3 = \tfrac{2}{3}n_{33} + \tfrac{1}{3}n_{13} + \tfrac{1}{3}n_{23} + \tfrac{1}{3}n_{34}, \end{array} \right\} \tag{3.4}$$

$$P_{ij} = p_{ij} + \frac{4-i}{i}\, p_i + \frac{4-j}{j}\, p_j, \tag{3.5}$$

where P_{ij} is the quantity referring to the particular combination of C—C and C—H bonds. The practical use of these equations is that, having determined isolated values for several known hydrocarbons, they can be extended to other normal and isomeric compounds. Additive quantities which can be calculated from eqn. (3.1) include molar volume, molar refraction, heat of formation, heat of combustion, heat of evaporation, magnetic permeability, the log of the vapour pressure and free energy (less accurately). The molar volume, for example, can be expressed by the equation:

$$V_m = \sum_{i \leqslant j=1}^{4} n_{ij} V_{ij}, \tag{3.6}$$

and the molar refraction R_m as

$$R_m = \sum_{i \leqslant j=1}^{4} n_{ij} R_{ij}, \tag{3.7}$$

which corresponds to the well-known experimental equation:

$$R_m = (n-1) R_{\text{C-C}} + (2n+2) R_{\text{C-H}}. \tag{3.7a}$$

The heat of evaporation is:

$$L^0 = \sum_{i \leqslant j=1}^{4} n_{ij} L_{ij} \tag{3.8}$$

or

$$\lambda^0 = \sum_{i \leqslant j=1}^{4} n_{ij} \lambda_{ij}, \tag{3.9}$$

where

$$\lambda^0 = L^0/4\cdot575, \quad \lambda_{ij} = L_{ij}/4\cdot575.$$

3

The vapour pressure equation can be written:

$$\ln p = -\frac{L^0}{RT} + B \quad \text{or} \quad \ln p = -\frac{\lambda^0}{T} + b, \tag{3.10}$$

where

$$\lambda^0 = L^0/4 \cdot 575.$$

$B \ (= 2 \cdot 3b)$ can be calculated from

$$B = \sum_{i \leqslant j = 1}^{4} n_{ij} b_{ij}. \tag{3.11}$$

The heat of formation, heat of combustion and free energy of formation can similarly be calculated from:

$$\Delta H^0 = \sum_{i \leqslant j = 1}^{4} n_{ij} H_{ij}. \tag{3.12}$$

The density is calculated from the equation:

$$\rho = \frac{M}{V_m} = \frac{M}{\sum_{i \leqslant j = 1}^{4} n_{ij} V_{ij}}, \tag{3.13}$$

and the boiling point from:

$$t_b = \frac{\lambda^0}{b - \log P} - 273 \cdot 16 = \frac{\sum_{i \leqslant j = 1}^{4} n_{ij} \lambda_{ij}}{\sum_{i \leqslant j = 1}^{4} n_{ij} b_{ij} - \log P} - 273 \cdot 16. \tag{3.14}$$

Table 3.1. quotes examples of quantities calculated in this manner by Tatevskii *et al.* [13], or measured experimentally. They have also decribed two other calculation methods suitable for saturated hydrocarbons. The first is based on the assumption that some homologous series

Table 3.1. Properties of normal saturated hydrocarbons

Liquid	n	V_m (cm³)	t_b (°c)	L^0 (cal)	ρ (gm/cm³)	ν_0†	R_m
Pentane	5	115·20	36·07		0·626	1·357	25·26
Hexane	6	130·95	68·74	~ 7200	0·659	1·374	29·90
Heptane	7	146·56	98·43	~ 8000	0·684	1·387	34·55
Octane	8	162·59	125·66	~ 8700	0·702	1·397	39·19
Nonane	9	178·71	150·80	9500	0·717	1·405	43·84
Decane	10	194·93	174·12	~ 10210			
Undecane	11	210·96	195·89				
Dodecane	12	226·96	216·28				

† ν_0 is the refractive index.

have certain properties either which change linearly with the number of carbon atoms n, or which remain constant, or which are additive. The second is an extension of this; both methods give excellent agreement with experiment. A comparison of their equations with those quoted in §2.2 shows that for normal hydrocarbons, certain of them are exactly equivalent. The methods described here are very useful for calculating properties of unusual isomers. Two of Tatevskii's examples for complex hydrocarbons are shown in table 3.2, with a comparison of experimental and theoretical values.

Table 3.2.   Properties calculated theoretically by Tatevskii *et al.* [13] compared with experimental values for two hydrocarbons

		Hydrocarbon	
		2,2,3-Trimethyl-pentane ($n = 8$)	2,4-Dimethyl-3-ethylpentane ($n = 9$)
V_m (cm³)	Experimental	159·53	173·80
	Theoretical	159·21	173·16
ρ (gm/cm³)	Experimental	0·719	0·738
	Theoretical	0·720	0·740
R_m	Experimental	38·92	43·40
	Theoretical	38·98	43·35
t_b (°C)	Experimental	114·8	136·7
	Theoretical	113·2	138·8
H_e (cal)	Experimental	—	—
	Theoretical	10075	8838

3.3. *The Cell and Tunnel Theories of Liquids.* From among the modern theories of liquids, the cell and tunnel theories are particularly successful as they allow a number of macroscopic relationships to be derived using simple models of molecular structure. The fundamentals of the cell theory were first put forward by Lennard-Jones and Devonshire [34, 35], using the assumption that liquid molecules move independently of each other inside a cell formed by neighbouring molecules. The potential energy $u(R)$ at a distance R from the centre of the cell is given by the equation:

$$u(R) = 4\varepsilon \left[\left(\frac{\sigma}{R} \right)^{12} - \left(\frac{\sigma}{R} \right)^{6} \right], \tag{3.15}$$

while the probability $p(R)\,dR$ that a molecule is to be found in the element at a distance from R to $R+dR$ from the centre of the cell is defined as:

$$p(R) = R^2 \frac{\exp\left[-V(R)/kT \right]}{\displaystyle\int R^2 \exp\left[-V(R)/kT \right] dR}. \tag{3.16}$$

$n(R)$, which defines the density of molecules at a distance R from the centre of the cell, is expanded as follows:

$$n(R) = (16\pi)^{-1} \sum_i \frac{m_i}{d_i R} \int_0^\infty dy \int_{y-a_i}^{1/2a_i} \frac{p(r)\,dr}{r} \int_{y-R}^{1/2a_i} \frac{p(r')\,dr'}{r'}, \qquad (3.17)$$

where m_i and a_i denote, respectively, the number of molecules at the ith distance from the cell centre. A graph of this equation yields a curve with several maxima (similar to that in fig. 3.1). The maxima correspond to the greatest probability of finding a nearest-neighbour molecule at that distance, and the size of these maxima decreases rapidly with increasing distance.

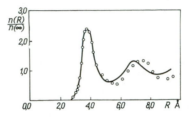

Fig. 3.1. Average molecular density as a function of the distance R from a central molecule. Refs.: Barker [39] and Henshaw [41].

Barker's [37–39] model of the liquid state is known as the 'tunnel' model. It is more adaptable to mathematical calculations and gives better agreement with experiment for the cases considered so far. The theory has been developed further by Rice and Allnatt [40]. It assumes that the molecules in a liquid are arranged in a line and move chiefly inside a two-dimensional tunnel. The average distance l between molecules in the line is equal to the distance between adjacent lines, while the spacing of molecules along different lines and their longitudinal and transverse spacing in a given line are independent of each other. Assuming the Lennard-Jones form of the potential $u(R)$ for the mutual interaction of two molecules (eqn. (3.15)), Barker derived the following expression for the molecular density $n(R)$ at a distance R from a given molecule:

$$n(R) = \left(\frac{\alpha}{2\pi R}\right) \int_0^R \exp\left[\frac{-\alpha(R^2 - z_1^2)}{2}\right] g(z_1)\,dz_1$$

$$+ \sum_{n=1}^\infty \left(\frac{\alpha N_n}{2\pi R l}\right) \int_0^\infty \exp\left[\frac{-\alpha(a_n^2 + R^2 - z_1^2)}{2}\right] I_0[\alpha a_n (R^2 - z_1^2)^{1/2}]\,dz_1, \quad (3.18)$$

where the first term refers to the molecules in the same tunnel and the second accounts for molecules from neighbouring tunnels; n denotes the number of the tunnel, N is the number of molecules in this tunnel, and a_n is its distance from the axis of the tunnel under consideration. I_0 is a modified Bessel function. Barker performed calculations for liquid argon at 84°K where, in eqn. (3.15), $\varepsilon/k = 119\cdot8°$K and $\sigma = 3\cdot405$ Å; the reduced volume at this temperature, $V/N\sigma^3$, is of the order of 1·184, $l = 1\cdot11\sigma$, $\alpha = 73\cdot5/\sigma^2$ and z is the coordinate along the tunnel axis. Figure 3.1 is a graph of $n(R)/n(\infty)$ as calculated by Barker [39], where $n(\infty)$ is the average number density at large distances from the central molecule. There are significant contributions from neighbours in the same tunnel as the central molecule, and from molecules in the first, second and third-neighbour tunnels. The solid curve shows the experimental results of Henshaw [41] from studies on neutron diffraction in liquids; the open circles denote results obtained by the tunnel theory. Barker claims that the cell model gives less satisfactory agreement with experiment than the tunnel model.

Much theoretical and experimental work on equilibrium molecular distributions in various hydrocarbons, silicone oils and other compounds has been carried out by Levy and van Wazer [42], van Wazer and Groenweghe [43], Carmichael *et al.* [44–46] and other workers. They have determined equilibrium states in chain–chain and chain–ring systems, and calculated equilibrium constants and the thermodynamic functions S, H and F for these systems.

Many theoretical papers [47–55] have recently been published on the statistical mechanics of molecules in a liquid with reference to structure and bond energies of individual atoms, and on the influence of molecular configurations upon liquid properties such as surface tension, viscosity, density, refractive index, etc. These topics are to only a limited extent connected with dielectric liquids; extensions of the cell theory are of more relevance, but only those which have assisted in solving problems relating to saturated hydrocarbons are described here.

In Kurata and Isida's [24] treatment of the theory of normal paraffins, it is assumed that the entire space V filled with N molecules of a given liquid can be divided into cells of identical volume τ; each molecule of the liquid can contain x such cells so that the volume $xN\tau$ is filled by the molecules of the liquid, and the remaining part, $V - xN\tau$, forms the 'free volume'. This free volume is also divided into cells called 'holes', and plays an important part in relating temperature and pressure changes to viscosity, diffusion, etc. The number of these cells is clearly defined by $(V/\tau - xN)$.

This type of theory was proposed as early as 1944 by Flory [56], and later by Grunberg and Nissan [57, 58], Prigogine *et al.* [59], and Lennard-Jones and Devonshire [35]. The Helmholz free energy F may be expressed by:

$$F = Nf + kT\left[\left(\frac{V}{\tau} - xN\right)\ln\left(1 - \frac{\tau x N}{V}\right) + N\ln\frac{\tau x N}{V}\right] + \frac{z\psi}{2}(V - \tau x N)\frac{\tau x N}{V}, \quad (3.19)$$

where z is the number of nearest-neighbour cells, f is the free energy connected with intra-molecular freedom, ψ is the increase in free energy for the following rearrangement: a pair of occupied cells plus a pair of holes changes into a cell occupied twice plus two holes.

Defining the volume of one cell as $v = V/N$, and denoting pressure by P and the chemical potential of the molecule by μ, the following relationships are obtained:

$$P = -\left(\frac{\partial F}{\partial V}\right)_{T,N} = \frac{-kT}{\tau}\left[\ln\left(1 - \frac{x\tau}{v}\right) + (x-1)\frac{\tau}{v}\right] - \frac{z\psi}{2\tau}\left(\frac{x\tau}{v}\right)^2, \qquad (3.20)$$

$$\mu = \left(\frac{\partial F}{\partial N}\right)_{T,V} = f + kT\left[\ln\left(\frac{x\tau}{v}\right) - x\ln\left(1 - \frac{x\tau}{v}\right) - (x-1)\right] + \frac{zx\psi}{2}\left(1 - \frac{2x\tau}{v}\right). \qquad (3.21)$$

For large v, eqn. (3.20) becomes the perfect gas equation. From the above relationships, Kurata and Isida derived a number of equations for the critical temperature, boiling point and melting point. As is well known, at the critical point the conditions:

$$\left(\frac{\partial P}{\partial v}\right)_T = 0 \quad \text{and} \quad \left(\frac{\partial^2 P}{\partial v^2}\right)_T = 0 \qquad (3.22)$$

apply. Equation (3.21) may be therefore rearranged to give:

$$T_c = \left(\frac{z\psi_c}{k}\right)\left[\frac{x}{(x^{1/2}+1)^2}\right], \qquad (3.23)$$

$$\frac{\tau x}{v_c} = \frac{1}{(x^{1/2}+1)} \qquad (3.24)$$

and

$$\frac{\tau P_c}{kT_c} = \ln\left(1 + \frac{1}{x^{1/2}}\right) - \frac{2x^{1/2}-1}{2x}, \qquad (3.25)$$

where T_c, P_c and v_c are, respectively, the critical temperature, pressure and volume.

Assuming that the energy ε and entropy S are independent of temperature, one may write:

$$\psi = \varepsilon - TS \quad \text{and} \quad \psi_c = \varepsilon - T_c S. \qquad (3.26)$$

Substitution into eqn. (3.24) yields:

$$\frac{1}{T_c} = \left[\left(\frac{k}{z\varepsilon} + \frac{S}{\varepsilon}\right) + \frac{2k}{z\varepsilon}\left(\frac{1}{2x} + \frac{1}{x^{1/2}}\right)\right]. \qquad (3.27)$$

Experimental values are seen to give very good agreement with

$$\frac{1}{T_c} = 0\cdot000543 + 0\cdot002909\left(\frac{1}{2n} + \frac{1}{n^{1/2}}\right), \qquad (3.28)$$

where n is the number of carbon atoms in the molecule. From this last expression, numerical values of the individual terms in eqn. (3.27) can be determined. For x, Kurata and Isida did not choose the value of n which would result from a comparison of eqns. (3.27) and (3.28), but instead took $x^{3/2} = n$; the reason was that with $x = n$, the density $\rho_c = n/v_c$ would be reduced to zero, which is in disagreement with experiment where it takes the same constant value for all paraffins. The expression:

$$\rho_c = \frac{n\tau\rho_0}{v_c} = \frac{\rho_0\, n^{1/3}}{n^{1/3}+1} \qquad (3.29)$$

allows this discrepancy to be avoided.

The relationship $1/T_c = f(n)$ also gives excellent agreement with experiment, the individual terms having the following numerical values:

$$\left(\frac{k}{z\varepsilon} + \frac{S}{\varepsilon}\right) = 0.000047 \qquad (3.30)$$

and

$$\frac{2k}{z\varepsilon} = 0.002755. \qquad (3.31)$$

Equation (3.27) can therefore be reduced to the form:

$$\frac{1}{T_c} = 0.000702 + \frac{0.00419}{n^{2/3}}. \qquad (3.32)$$

The values of T_c, T_b and T_m are given in table 3.3. As can be seen, the agreement between the theoretical and experimental values is very good.

The expansion of eqn. (3.25) into a series in $1/x$ (with $x = n^{2/3}$) makes it possible to determine the relationships between T_c, P_c and v_c from other formulae. For example:

$$\frac{\tau P_c}{kT_c} = \frac{1}{3x^{3/2}} = \frac{1}{3n} \qquad (3.33)$$

or

$$\frac{nP_c}{T_c} = \frac{k}{3\tau} \qquad (3.34)$$

results in $k/3\tau = 0.337$ atm. cm.$^{-2}$ deg.$^{-1}$. Hence it can be deduced that

$$\frac{MP_c}{T_c} = \text{constant (independent of } n), \qquad (3.35)$$

which agrees with the results of Grunberg [58]. A combination of eqns. (3.34) and (3.35) gives the Kamerlingh–Onnes constant:

$$\frac{kT_c}{P_c v_c} = \frac{1}{(n^{1/3}+1)\,[n^{2/3}\ln(1+n^{-1/3}) - n^{1/3} + \frac{1}{2}]}. \qquad (3.36)$$

The ratio of observed to calculated value for this constant is about 1.40, and is constant over a wide range of n (from 4 to 19).

From the values of the terms in the above equations, parameters d and ε can be evaluated. They are shown in table 3.4 for various crystalline lattices. Kurata and Isida derived a formula for boiling point by assuming that at this point both pressures and chemical potentials are equal in the liquid and vapour phases:

$$P_g = P_l \quad \text{and} \quad \mu_g = \mu_l.$$

Table 3.3. Critical temperatures, boiling points and melting points of
normal saturated hydrocarbons

n	T_c (exp.)	T_c (theor.)†	T_b (exp.)	T_b (theor.)‡	T_b (theor.)§	T_m (exp.)	T_m (theor.)‖	T_m (theor.)¶
1	190·7	204·4	120·1	111·5	129·0	—	85·0	35·6
2	305·3	299·2	183·3	184·7	189·0	—	112·8	67·7
3	368·8	368·1	232·7	230·9	236·0	85·3	134·0	96·8
4	426·2	422·8	274·3	272·6	275·2	134·6	152·0	122·4
5	470·4	468·4	310·4	309·1	310·0	144·5	167·5	146·2
6	508·0	507·4	342·7	342·0	341·0	178·7	182·0	167·0
7	540·0	541·4	371·9	371·5	369·0	182·8	195·0	186·1
8	569·4	571·4	398·7	397·6	396·0	216·5	208·0	203·4
9	595·4	598·4	423·4	423·6	421·0	219·3	219·0	218·9
10	619·3	623·1	446·2	447·0	444·0	242·0	230·0	232·9
11	642·6	645·6	467·7	470·0	467·0	247·0	240·2	245·5
12	663·8	666·2	488·0	489·0	489·0	263·5	250·2	256·7
13	683·2	684·9	506·8	507·0	509·0	267·0	259·3	266·8
14	701·0	702·7	525·2	525·5	529·0	278·8	270·0	276·2
15	717·6	718·9	542·0	543·5	549·0	281·2	278·0	284·4
16	734·3	734·2	558·3	560·5	569·0	293·0	286·0	291·9
17	749·3	748·5	573·7	576·0	586·0	295·0	294·0	298·8
18	763·2	762·2	588·6	590·0	603·0	301·0	302·2	304·6
19	776·0	774·6	602·4	603·0	619·0	304·8	310·0	310·0
20	—	—	—	—	—	309·4	318·8	315·0
25	—	—	—	—	—	327·0	353·0	333·4
30	—	—	—	—	—	338·7	386·0	344·2
36	—	—	—	—	—	350·0	420·0	351·4
60	—	—	—	—	—	374·0	538·0	359·2
70	—	—	—	—	—	378·0	580·0	359·6
94	—	—	—	—	—	404·8	668·0	359·6

† From eqn. (2.1) or (3.32).   ‡ From eqn. (2.2) or (3.39).   § From eqn. (2.5).   ‖ From eqn. (2.6).   ¶ From eqn. (2.4).

Table 3.4. Values of d and ε for three lattices

Lattice type	z	d	d (Å)	ε/k (°K)
Diamond lattice	4	$0\cdot86\tau^{1/3}$	4·4	182
Simple cubic lattice	6	$1\cdot00\tau^{1/3}$	5·2	121
Face-centred cubic lattice	12	$1\cdot12\tau^{1/3}$	5·8	60

For purposes of simplification, they also assumed that v_g is much larger than v_l, that the density of the vapour can be neglected compared with the density of the liquid, and that at low temperatures the number of holes in the liquid is relatively small. They obtained the formula:

$$\ln P = \ln\left[\frac{kT}{x\tau}\right] + x\left(1 + \frac{zS}{2k}\right) - 1 - x\frac{z\varepsilon}{2kT}. \tag{3.37}$$

By substituting $x = n^{2/3}$, the boiling point T_b is given by:

$$\frac{1}{T_b} = \frac{2k}{z\varepsilon} + \frac{S}{\varepsilon} + \frac{A}{n^{2/3}}, \tag{3.38}$$

where A is a constant expressed as:

$$A = \frac{2k}{z\varepsilon}\left[\ln\left(\frac{kT_b}{n^{2/3}\tau}\right) - 1\right].$$

Comparison with experimentally determined values yields:

$$\frac{1}{T_b} = 0\cdot00057 + \frac{0\cdot007753}{n^{2/3}}. \tag{3.39}$$

The excellent agreement between theory and experiment can be seen in table 3.3.

For the heat of evaporation, the following equation was derived:

$$H_e = n^{2/3}\frac{z\varepsilon}{2}, \tag{3.40}$$

from which $z\varepsilon/2 = 2115$ cal/mole. For $z = 12$, $\varepsilon = 352\cdot5$ cal/mole, equal to the activation energy of a methane molecule.

Aspects of the cell theory have also been considered by Taylor [61] and Simha and Hadden [60]. For a long molecular chain, it is usual to divide the molecule into a number of segments (less than the number of carbon atoms in the molecule) in order to calculate the distribution of inter-molecular force potentials. Several other parameters must also be taken into consideration, viz. free rotation of each segment, the gyroscope effect, the vibration of the molecular chain as a whole, the effect of neighbouring molecules, and internal degrees of freedom associated with vibration along bonds and rotation about bonds without changing the angle between them. Simha and Hadden [60] have derived an equation connecting p, V and T with quantities such as the number of segments in a molecule, their size and the energies connected with individual bonds as functions of temperature and pressure:

$$\frac{pM}{RT\rho} = \frac{1+\dfrac{s}{3}}{1-2^{-1/6}\left(\dfrac{N_A v\rho s}{M}\right)^{1/3}} + \frac{6\left(1+\dfrac{s}{3}\right)\left[1{\cdot}011\left(\dfrac{N_A v\rho s}{M}\right)^2 - 1{\cdot}2045\right]}{1{\cdot}011\left(\dfrac{N_A v\rho s}{M}\right)^2 - 2{\cdot}409}$$

$$+ 2[s(z-2)+2]\frac{\varepsilon}{kT}\left(\frac{N_A v\rho s}{M}\right)^2\left[1{\cdot}011\left(\frac{N_A v\rho s}{M}\right)^2 - 1{\cdot}2045\right], \quad (3.41)$$

where ρ is the density of the liquid; v, the size of the segment; ε, the maximum depth of the potential energy well for a segment pair; $r = 2^{1/6}v^{1/3}$ is their separation; M is the molecular weight; s is the number of segments in the molecule; and p, R, T and N_A have the usual meanings. The value of s can be obtained from:

$$s = \frac{n+1}{2} = \frac{M+12}{28}. \quad (3.42)$$

The first two terms in eqn. (3.41) express the relationship between volume and lattice vibration, whilst the factor $(s+3)$ is the sum of $(s-3)$ internal rotations and six degrees of freedom in a single chain. The third term represents field variation in space. Simha and Hadden used this equation to determine the dependence of certain parameters on chain length, temperature and pressure for paraffins (normal and branched-chain). Their numerical results are quoted in tables 3.5 and 3.6.

Table 3.5. Values of energy and volume for paraffins

Series	$N_A v$ (cm³/mole)	$\varepsilon \times 10^{16}$ (erg/mole)	$N_A \varepsilon$ (kcal/mole)
n-C	32·4	294	0·42
R₁CR₂	31·9	261	0·38
R₃			
R₁C...CR₄	31·7	243	0·35
R₂ R₃			

Table 3.6. Enthalpies of evaporation

Substance	ΔH_e (kcal/mole) (theoretical)	ΔH_e (average)
n-C₁₃	17·3 at 37·8°c	14·2 from 55–107°c
n-C₂₀	26·3 at 37·8°c	19·5 from 137–197°c
C.(C₄)₃	15·4 at 37·8°c	12·9 from 39–92°c
(C₄)₂.C.C₈C.(C₄)₂	27·4 at 37·8°c	21·2 from 172–235°c

3.4. *Theories of Liquid Viscosity.* The viscosity of liquids is an important topic connected with electrical conductivity, ion transfer, ionic activation energies, etc. The work of Andrade and Guzman [26], Doolitle [28, 62], Eyring [8], Frenkel [9], Hall [63] and Kurata and Isida [24] should be mentioned in particular. The Eyring theory [8] is probably the best known.

From among recent theoretical studies of liquid viscosity, the treatment by Hirai and Eyring [64], which considers two different mechanisms of viscosity, shear viscosity and bulk viscosity, is discussed here in some detail. Shear viscosity arises when liquid layers having different velocities move relative to each other and is observed most frequently in normal viscosity measurements using, for example, an Ostwald viscometer. Bulk viscosity arises during local pressure changes and plays an important part in the absorption of ultrasonic waves in a liquid. Bulk viscous resistance is due to part of the translational energy being converted into oscillatory, rotational and vibrational motion, or due to changes in the structure of associated liquids.

Shear viscosity is most important in gases and non-associated liquids and has been investigated by Herzfeld and Rice [65] and Kneser [66]. The mechanism of bulk viscosity is chiefly applicable to associated liquids, and the theory for water has been developed by Hall [63]. The Hirai–Eyring theory is closely connected with Hall's treatment.

Both theories are based on the assumption that the liquid consists of a number of molecules N_0, each having a volume v, and a number of holes N_h each of volume v_h. The thermodynamic equilibrium between the numbers of molecules and holes at a given pressure and temperature changes with pressure, and if the pressure rises, some holes diffuse to the surface of the liquid. The changes in energy and volume that accompany this process are transferred through the liquid by elastic vibrations of the lattice and elastic waves. If the internal pressure decreases, holes diffuse inwards from the surface of the liquid to its bulk.

Figure 3.2 illustrates the energy changes in a liquid in two states, A and B, produced by a rise in pressure from p_1 to p_2. In state A, a group of molecules contains one hole; in state B, under the influence of pressure, this hole disappears. Both the volume and energy of the system in state A are greater than in state B by amounts equal to the volume of the hole v_h and the energy ε_h. There is a potential barrier of the order of $\varepsilon_j(B)$ between states A and B, and a certain activation energy is required to pass from state A to state B.

The decrease in the number of holes with time is given by:

$$-\frac{dN_h}{dt} = K_{h0} N_h - K_{0h} N_0, \tag{3.43}$$

where K_{h0} and K_{0h} are, respectively, the probabilities of holes disappearing and appearing, defined by:

$$K_{h0} = \frac{kT}{h} \frac{F^+(B)}{F_h} \exp - \left[\frac{\varepsilon_j(B) - \mu p v_h}{kT}\right] \tag{3.44}$$

and

$$K_{0h} = \frac{kT}{h}\frac{F^+(B)}{F_0}\exp\left[-\frac{\varepsilon_h + \varepsilon_j(B) + (1-\mu)\,pv_h}{kT}\right],\qquad (3.45)$$

where F_h, F_0 and F^+ denote the distribution functions in state A, state B and in the excited state respectively; μ is the symmetry parameter defining the energy increase in state A (μpv_h) and the energy change in state B $[-(1-\mu)\,pv_h]$.

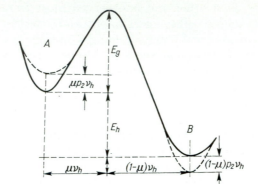

Fig. 3.2. Energy states of a group of molecules (A) containing a hole, and (B) without a hole; E_g is the activation energy. Ref.: Hirai and Eyring [64].

At pressure equilibrium $dN_h/dt = 0$ and, instead of (3.44), we may write:

$$\frac{N_{h2}}{N_0} = \frac{K_{0h}}{K_{h0}} = \frac{F_h}{F_0}\exp\left(-\frac{\varepsilon_h + pv_h}{kT}\right).\qquad (3.46)$$

The expression for the number of holes at equilibrium is therefore:

$$N_h = N_{h2}\left\{1 + \exp\left[\frac{(p_2 - p_1)\,v_h}{kT} - 1\right]\exp\left(-K_{h0}t\right)\right\}.\qquad (3.47)$$

Since $(p_2 - p_1)\,v_h/kT$ is usually small, eqn. (3.47) can be simplified to:

$$N_h - N_{h2} = N_{h2}(p_2 - p_1)\frac{v_h}{kT}\exp\left(-K_{h0}t\right).\qquad (3.48)$$

The relaxation time τ_B is expressed as:

$$\tau_B = \frac{1}{K_{h0}} = \frac{h}{kT}\exp\frac{\varepsilon_j(B) - \mu pv_h - TS_j(B)}{kT}\qquad (3.49)$$

and the bulk viscosity entropy of activation $S_j(B)$ as:

$$S_j(B) = k\ln\frac{F^+(B)}{F_h};\qquad (3.50)$$

the following expression for the volume change is therefore obtained:

$$V - V_2 = (V_1 - V_2)\exp\left(-t/\tau_B\right).\qquad (3.51)$$

When deriving these formulae, Hirai and Eyring used the well-known relationship for the increase in free energy when a hole is introduced into a liquid:

$$\Delta F = kT \left[N_h \ln \frac{N_k}{n N_0 + N_h\, N_0} + N_0 \ln \frac{n N_0}{n N_h} + N_h(\varepsilon_h + p v_h - T S_h) \right], \quad (3.52)$$

where $n = v_0/v_h$ (it can generally be assumed that the volume of the molecule is five or six times greater than that of the hole).

It is assumed that at equilibrium the change in free energy has a minimum in eqn. (3.52):

$$\ln \frac{N_h}{n N_0 + N_h} + 1 - \frac{1}{n} + \frac{\varepsilon_h + p v_h - T S_h}{kT} = 0 \qquad (3.53)$$

or

$$N_h = \frac{N_0\, n}{\sigma} \exp\left(-\frac{\varepsilon_h + p v_h}{kT} \right), \qquad (3.54)$$

where $\ln \sigma = 1 - (1/n) - (S_h/k)$ (since $n N_0 \gg N_h$ at normal temperatures) and $(p_2 - p_1) v_h \ll kT$. Remembering that the total volume of the liquid is the sum of the volumes of molecules and holes, we can write:

$$\alpha = \frac{1}{V}\left(\frac{\partial V}{\partial t}\right)_p, \quad \beta = -\frac{1}{V}\left(\frac{\partial V}{\partial p}\right)_T, \qquad (3.55)$$

$$\alpha_s = \frac{1}{V_0}\left(\frac{\partial V_0}{\partial t}\right)_p, \quad \beta_s = -\frac{1}{V_0}\left(\frac{\partial V_0}{\partial p}\right)_T, \qquad (3.55a)$$

where α is the coefficient of expansion, β is the coefficient of compressibility, and α_s and β_s are these coefficients for compression of the lattice structure (of the same order of magnitude as for solids). Similar coefficients α_h and β_h are those due to a changing number of holes in the liquid:

$$\alpha_h = \frac{1}{\sigma} \frac{\varepsilon_h}{kT^2} \exp\left(-\frac{\varepsilon_h + p v_h}{kT} \right) \qquad (3.56)$$

and

$$\beta_h = \frac{1}{\sigma} \frac{v_h}{kT} \exp\left(-\frac{\varepsilon_h + p v_h}{kT} \right). \qquad (3.56a)$$

For liquids it can be assumed that

$$\frac{\alpha_s}{\alpha} \approx \frac{\beta_s}{\beta} \approx 0{\cdot}2. \qquad (3.57)$$

By numerical calculation, the following was obtained:

$$n = \frac{v_0}{v_h} \approx \frac{\varepsilon_0}{\epsilon_h} = 5 \sim 6 \quad \text{and} \quad \sigma \approx 1,$$

ε_0 being the heat of evaporation. For methyl alcohol at 0°c, for example, $v_h = 10{\cdot}5\ \mathrm{cm}^3$, $\varepsilon_h = 1{\cdot}2$ kcal and $\sigma = 0{\cdot}7$.

When $(p_2 - p_1) v_h / kT$ is not small, the treatment has to be modified; this was discussed by Hirai and Eyring [64] in their extension for the bulk viscosity in non-Newtonian flow. From a comparison of eqn. (3.51) and the equation:

$$-\eta_B \frac{1}{V}\frac{dV}{dt} + \frac{1}{\beta_h}\frac{V_1 - V}{V} = p_2 - p_1, \qquad (3.58)$$

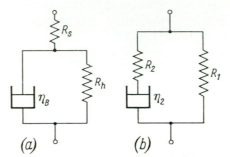

(a)   (b)

Fig. 3.3. Models of the mechanism of bulk density. Ref.: Hirai and Eyring [64].

they obtained the following final expression for the bulk viscosity η_B:

$$\eta_B = \frac{h}{v_h}\exp\frac{\varepsilon_h + \varepsilon_j(B) + (1 - \mu)pv_h - TS_j(B)}{kT};$$

$$\tau_B = \eta_B \beta_h \quad \text{and} \quad K_h = \frac{1}{\beta_h} = \frac{kT}{v_h}\exp\frac{\varepsilon_h + pv_h}{kT}, \qquad (3.59)$$

where $K = 1/\beta$ denotes the modulus of compressibility:

$$K_1 = \frac{K_s K_h}{K_s + K_h}, \quad K_2 = \frac{K_s^2}{K_s + K_h} \qquad (3.60)$$

and

$$\eta_2 = \left(\frac{K_s}{K_s + K_h}\right)^2 \eta_B; \quad \tau_2 = \frac{\eta_2}{K_2} = \frac{K_h}{K_s + K_h}\tau_B. \qquad (3.61)$$

Figure 3.3 shows diagrams of models for the mechanism of the two types of viscosity process. The resultant modulus K is the sum of two components: modulus K_h resulting from the presence of holes accounting for delayed resistance to elasticity, and modulus K_s, resulting from lattice deformations, which acts immediately in the elasticity phenomenon. Hirai and Eyring concluded that the most frequently calculated ultrasonic wave absorption coefficients in liquids should be written in the form:

$$\alpha_{\exp} = \frac{2\pi^2 f^2}{\rho c^3}\left(\tfrac{4}{3}\eta_s + \eta_2\right) \quad \text{instead of} \quad \alpha_{\mathrm{St}} = \frac{8\pi^2 f^2}{3\rho c^3}\eta_s, \qquad (3.62)$$

where f is the frequency, c is the velocity of sound in the liquid, ρ is the density of the liquid, η_s is the shear viscosity, and α_{St} and $\alpha_{\exp}$ are the Stokes and experimental absorption coefficients. The ratio of the two values of viscosity can be obtained by rearranging eqn. (3.62):

$$\frac{\eta_2}{\eta_s} = \frac{4}{3}\left(\frac{\alpha_{\exp}}{\alpha_{St}} - 1\right). \tag{3.63}$$

Experimental measurements of η_s, α_{St} and $\alpha_{\exp}$ substituted in eqn. (3.63) can be used to calculate η_2 for liquids.

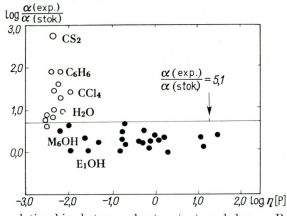

Fig. 3.4. The relationship between $\log(\alpha_{\exp}/\alpha_{St})$ and $\log\eta$. Ref.: Hirai and Eyring [64].

When the activation energy is identical for both types of viscosity process, the equation can be written:

$$\ln\frac{\eta_B}{\eta_s} = \ln n - \frac{S_j(B) - S_j(S)}{k} - \frac{\mu p v_h}{kT} \tag{3.64}$$

and if the entropies are equal, we obtain:

$$\alpha_{\exp}/\alpha_{St} \approx (\tfrac{3}{4}n + 1). \tag{3.65}$$

When $n = 5 \cdot 5$, $\alpha_{\exp}/\alpha_{St} \approx 5 \cdot 12$. It was considered that the value $\alpha_{\exp}/\alpha_{St} \approx 5$ divides the liquids into two groups which differ as regards the mechanism of their absorption; for values less than 5, the structural change mechanism is the most important.

Figure 3.4 shows the relationship between $\log(\alpha_{\exp}/\alpha_{St})$ and $\log\eta$ for a number of compounds; the horizontal straight line divides the liquids into two types: associated and non-associated (where $\alpha_{\exp}/\alpha_{St} > 5$) with different viscosities.

It was also shown that if the two viscosity processes have a similar temperature dependence, the theory would make us expect that pressure changes should result in different variations of viscosity for the two processes. This was investigated by Carnevale and Litovitz [67], who found the relaxation time to be:

$$\tau_2 = \frac{K_h}{K_0}\tau_B = \frac{1}{K_0}\frac{h}{v_h}\exp\frac{\varepsilon_h + \varepsilon_j(B) + (1-\mu)\,pv_h - TS_j(B)}{kT}, \tag{3.66}$$

and determined experimentally the dependence of $\log \eta$ on pressure. It can be deduced that τ_2 increases with pressure while τ_B decreases. An investigation of this relationship between viscosity and pressure is an interesting subject for further experimental work.

Doolitle's approach [28, 62] to the theory of free space in a liquid and Williams' [68] treatment of super-cooled liquids (polystyrenes and glass) were slightly different. Doolitle derived the following formula for viscosity:

$$\eta = A \exp (B/f), \tag{3.67}$$

where A and B are constants. Williams defined f, the relative free volume for a given temperature, by the equation $f = (v_g - v_0)/v_0$; its temperature dependence is given by the formula:

$$f = \frac{v_g - v_0}{v_0} + \frac{1}{v_0} \frac{dv}{dT} (T - T_g), \tag{3.68}$$

where v_0 denotes the volume that a non-associated liquid would attain if it could be reduced continuously, without phase change, to the absolute zero of temperature. T_g is the temperature of transition into the glassy state, and v_g the relative free volume at the temperature T_g. The expression for the viscosity will then become:

$$\log \eta = \log A + \frac{B}{2 \cdot 303 \left[\dfrac{v_g - v_0}{v_0} + \dfrac{1}{v_0} \dfrac{dv}{dT} (T - T_g) \right]}. \tag{3.69}$$

Fox and Flory [69] obtained numerous experimental data which were used for the verification of eqn. (3.69). They calculated values of T_g, v_g and dv/dT from the equations:

$$\left. \begin{aligned} T_g &= 100 - \frac{1 \cdot 0 \times 10^5}{M}, \\ v_g &= 0 \cdot 943 + 2 \cdot 4 \times 10^{-4} t_g, \\ \frac{dv}{dT} &= \left(5 \cdot 5 + \frac{643}{M} \right) \times 10^{-4}. \end{aligned} \right\} \tag{3.70}$$

According to Williams, for polystyrenes, where the molecular weight is between 1675 and 134 000, v_g varies from 0·0225 to 0·0287 for melting points from 40°c to 99°c. The coefficient of expansion:

$$\alpha = \frac{1}{v_0} \frac{dv}{dt}$$

for the same temperature range varies from $6 \cdot 3 \times 10^{-4}$ to $5 \cdot 85 \times 10^4$, with v_0 changing only very slightly from 0·934 to 0·940. This change can be expressed by the formula $v_0 = 0 \cdot 940 \exp (-11 \cdot 5/M)$. As can be seen, v_0 in this case increases with increase in M contrary to the behaviour for normal hydrocarbons for which, over the range of M from 72 to 900, v_0 decreases with an increase in M.

The constant A in formula (3.69) changes very sharply with a rise in M; two different correlations apply here:

$$\log A = -7{\cdot}08 + 1{\cdot}12 \log M \quad \text{for} \quad M \leqslant 35\,000$$

and

$$\log A = -12{\cdot}32 + 3{\cdot}38 \log M \quad \text{for} \quad M \geqslant 35\,000, \tag{3.71}$$

with the constant B in eqn. (3.69) equal to 0·91.

Equation (3.69) has a similar form to the Fox–Flory equation for $4000 \leqslant M \leqslant 38\,000$:

$$\log \eta_{217} = 1{\cdot}65 \log M - 5{\cdot}38. \tag{3.72}$$

Sometimes, instead of the relative free volume f being determined from eqn. (3.68), a constant a_T is introduced; this is the ratio of the relaxation time at an arbitrary temperature to that at a particular temperature characteristic of the given liquid (e.g. the melting point). The formula:

$$\log a_T = \frac{-17{\cdot}44(T - T_g)}{51{\cdot}6 + T - T_g} \tag{3.73}$$

can then be used instead of eqn. (3.68).

Bueche [70] suggested a more general equation containing an expression with the specific volume only:

$$\log a_T = \frac{-Bv_0}{2{\cdot}303(v_1 - v_0)} \frac{v_2 - v_1}{v_2 - v_0}, \tag{3.74}$$

where v_1 and v_2 are specific volumes at a reference temperature T_1 and an arbitrary temperature T_2.

The viscosity mechanism in high-viscosity mineral oils has recently been studied by Piotrowski [71]; he investigated the relationships between activation energy, enthalpy, free energy and liquid structure both theoretically and experimentally.

Chapter 4

The Purification of Liquids

4.1. *Defining the Purity of a Liquid.* The purity of a liquid is a very important factor in the study of electrical conductivity and breakdown in dielectric liquids. It is purity of a special kind, and one that can only be determined by measuring the intrinsic conduction. Dielectric liquids, supplied as analysis grade, frequently have a natural conduction of the order of $10^{-12} \, \Omega^{-1} \, cm^{-1}$, that is a million or so times too high to be used for investigating electrical conduction. Using special purification methods, a conduction of the order of 10^{-19}–$10^{-20} \, \Omega^{-1} \, cm^{-1}$ can be obtained, which is the lowest achieved by Nikuradse [12], Jaffe [72], Adamczewski [73], Pao [74], Gzowski and Terlecki [75] and others.

Purity of the liquids is of exceptional importance since no reproducible results can be obtained with liquids of insufficient purity, and the characteristic properties of a liquid are often masked by phenomena arising due to the presence of impurities. Traces of water and other electrolytic impurities are naturally the most important. The removal of these and small physical impurities, such as dust particles, is very important in all purification methods, especially when the liquids are to be studied in strong electric fields where there is the possibility of causing breakdown.

Experiments carried out with non-purified liquids are usually of no value as regards pure research, because the chemical properties are more dependent on the impurities contained in them than on their own molecular structure. It should, however, be stressed that results obtained using them are of great importance to industry, because liquids with this degree of purity are commonly used in industrial applications.

4.2. *Chemical Purification. Distillation.* As has been mentioned, saturated hydrocarbons are among the best dielectrics. The most common method of purification is by shaking them thoroughly with concentrated sulphuric acid. This precipitates unsaturated hydrocarbons in the form of a black deposit, which, together with the sulphuric acid residue, can easily be separated using a separating funnel because of the large difference in the specific gravity of hydrocarbons (0·63–0·75) and concentrated sulphuric acid (1·84). The procedure is repeated until the sulphuric acid leaves no deposit after shaking.

After separating off the acid, the liquid is washed with distilled water and then with caustic soda until completely free of acid. The caustic soda is then separated off and the remaining hydrocarbon is washed several

times with distilled water. The aqueous portion is readily separated from
the hydrocarbon by means of the separating funnel.

The liquid must then be carefully dried, because even the slightest
trace of water markedly increases the intrinsic conduction. Calcium
chloride or metallic sodium is usually used, but sometimes phosphorus
pentoxide. The liquid is then filtered through a double Schott filter (very
fine and of glass), and subjected to repeated fractional distillation. This
distillation removes all the remaining impurities, and the fraction required
can be separated over the appropriate boiling range.

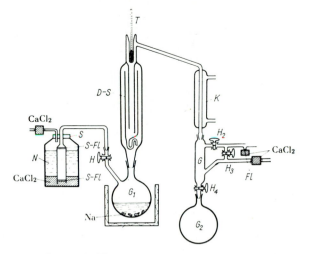

Fig. 4.1. Diagram of the distillation apparatus described in this chapter. Ref.:
Adamczewski [73].

Figure 4.1. shows a distillation apparatus used by Adamczewski [73].
It is connected to a filtration device consisting of a double Schott filter
S–Fl and a vessel *N* closed with a cork *S*. The distillation apparatus proper
consists of several sections with ground-glass joints and non-lubricated
stopcocks. The main part of the apparatus is the dephlegmator *D–S*
consisting of three concentric glass tubes filled with glass beads for
increasing the cooling surface; in the upper part there is a glass tube into
which a thermometer *T* is inserted. The dephlegmator is connected by
means of ground-glass joints to flask G_1 on one side and to *G* on the other
via the condenser *K*. *G* has three taps, H_2, H_3 and H_4, for collecting
fractions in the flask G_2; these taps enable the flask to be replaced even
when distillation is being carried out at reduced pressure. The apparatus
is also fitted with several drying tubes containing calcium chloride. After
the liquid has been dried, the distillation procedure is as follows: liquid
is poured into *N* containing a layer of calcium chloride or phosphorus
pentoxide. *N* is closed by a cork through which two tubes pass, one
connected to the Schott filter, and the other passing to the atmosphere
through a drying tube containing calcium chloride. Taps H_3 and H_4 are

closed, tap H_1 is opened and the pump is started. The reduced pressure inside the distillation apparatus causes the liquid to be drawn through the filter *S–Fl* to flask G_1. As the tube with the filter almost touches the bottom of the vessel N, the liquid has to pass through several centimetres of calcium chloride or P_2O_5. Metallic sodium is often placed inside flask G_1 and the liquid boiled until all the water has been removed. It may sometimes be necessary to use fractional distillation or special purification methods for certain kinds of substances.

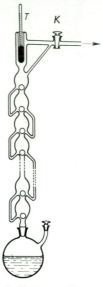

Fig. 4.2. Diagram of a ball-type column for fractional distillation.

A typical fractionating column for laboratory use is shown in fig. 4.2. The vapour from the boiling liquid passes to the upper portion of the column through a number of glass bulbs closed by glass spheres and interconnecting by means of side-arms. In very accurate work, the upper part of the column remains closed by tap K until equilibrium is attained and the vapour has reached the upper part. Tap K is then opened, and the lower boiling fraction is taken off. A Beckmann thermometer inserted in the upper part of the column enables vapour fractions to be separated with an accuracy of hundredths or even thousandths of a degree. If such a high accuracy for separating the individual fractions is not required, tap K can be kept open all the time but care should be taken that the distillation process is carried out slowly (so that the liquid falls in the form of drops). This is possible only when tall (at least ten-plate) columns are used.

There are many other designs of distillation columns, but those described give very good results and fully satisfy requirements as far as obtaining the highest purity liquids is concerned. The final stage in the process of the purification of a liquid is electrical purification, i.e. subjecting the sample of the liquid to prolonged action of an electric field.

Electrical purification is most frequently carried out in the ionization test chamber itself so that it does not involve any particular difficulties. It appears that in a liquid subjected to the action of an electric field, the electric conduction is reduced by as much as several orders of magnitude (sometimes in about 50 hours). The conduction does not return to its previous value unless impurities are introduced again; polarization in the liquid does not occur and there is usually symmetry of the positive and negative charges which gather at the collecting electrode of the ionization chamber.

The phenomenon of electrical purification is thought to be due to the removal of traces of water, electrolytic impurities, gas and vapour bubbles, dust and other suspensions by the electric field inside the chamber. It is supposed that the final stage of electrical purification is connected with the close ordering process of the molecules in the liquid. This is of particular importance in substances consisting of large molecules regularly arranged into thin layers as is the case for hydrocarbons where, over small areas, the structure appears pseudocrystalline. Further study of the process of electrical purification is required, particularly as regards the theory of semiconductors, discussed in § 16.9.

Nikuradse [12] and Pao [74] are further sources of information on the purification of dielectric liquids. The apparatus described, however, differs little from that in this section.

4.3. *Zonal Melting and Crystallization Methods* [76–78]. The zonal melting method was developed by Pfann [79] in 1952 in order to obtain metals of high purity; it was later used by a number of investigators for purifying semiconductors and other substances. Wilcox *et al.* [80] have published an extensive review of all the known methods, a comprehensive bibliography and a great deal of data on individual substances.

The zonal melting method depends on the fact that small amounts of impurities usually display different solubilities in the liquid and solid phases of the substance being purified because diffusion in a liquid is much more rapid than in a solid. If, then, a sample of a liquid which has been frozen in a tube is zonally melted, and if during melting this zone slowly shifts along the sample (see fig. 4.3), the concentration C of impurity at a certain distance x from the end of the sample can be expressed by the following equation:

$$C = C_0 \left[1 - (1-k) \exp\left(-\frac{kx}{l} \right) \right],\qquad (4.1)$$

where $k = k_s/k_l$; k_s is the concentration of impurity in the solid phase, and k_l the concentration in the liquid phase; l is the length of the molten zone.

Figure 4.4 illustrates the distribution of the concentration C of the impurity along the length x of the sample being treated after a single shift of the molten zone.

At the ends of the sample the concentration changes in a distinct manner, but inside the sample the concentration of the impurity remains constant. Figure 4.5 (a), (b) shows the dependence of the freezing and melting points on the concentration of the impurity in various cases.

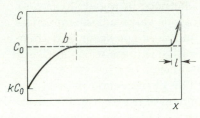

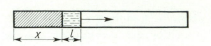

Fig. 4.3. Zonal melting; x is the solidified part of the sample, l is the length of the molten zone.

Fig. 4.4. Dependence of the concentration of the impurity C on the length of the solidified part x. Ref.: Pfann [79].

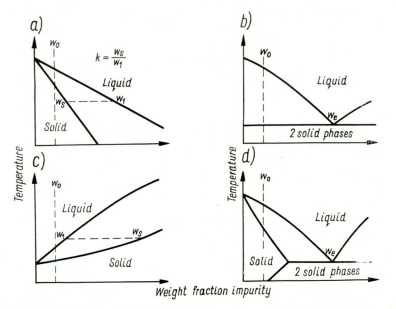

Fig. 4.5. Typical binary solid–liquid equilibrium phase behaviour; (a) solid solution with constant distribution coefficient k, (b) simple eutectic system, (c) solid solution with variable distribution coefficient, (d) partial solid solution with eutectic point. Ref.: Wilcox *et al.* [80]. (w_0—Original composition in weight fraction; w_e—eutectic composition in weight fraction; w_1—an average zone concentration; k—concentration of impurity in the solid phase; $w_s = kw_1$).

Figure 4.6 shows the curves for the distribution of impurity in the solid phase for various values of k, plotted from eqn. (4.1). The concentration is here defined by the function x/l. For $x = 0$ the concentration is:

$$C = C_0 k. \qquad (4.2)$$

Figure 4.7 shows the effect that purifying germanium has on its specific resistance R_s. As can be seen, the specific resistance is increased by a factor of about twenty. The zonal purification method has been particularly successful for the purification of metals such as germanium, tin, zinc, copper, etc.

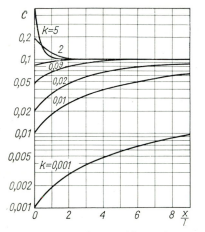

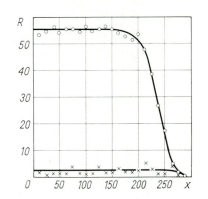

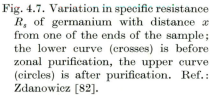

Fig. 4.6. Dependence of impurity concentration C on x/l for various values of k according to eqn. (4.1). Ref.: Pfann [79].

Fig. 4.7. Variation in specific resistance R_s of germanium with distance x from one of the ends of the sample; the lower curve (crosses) is before zonal purification, the upper curve (circles) is after purification. Ref.: Zdanowicz [82].

Herington *et al.* [76] have used the zonal melting method for purifying organic substances such as naphthalene, benzoic acid, pyrene, anthracene, chrysene, etc. The substances were introduced into a glass tube (fig. 4.8) about 120 cm long and 3·6 cm in diameter. A metal slider with a coil of iron wire for heating a small layer of the substance at any height was moved at a uniform velocity of about 4 cm per hour outside the tube. The drive for the slider was from a synchronous motor (fig. 4.9). In order to ensure that the purification proceeds correctly, a difference in temperature between that of the substance being purified and the ambient temperature is necessary; 20–30°C below the melting point is sufficient for this purpose, and excellent results were obtained for melting points of around 80°C. For melting points from 0–50°C, external cooling to the glass tube down to −25°C was applied.

The apparatus described has also been adapted for semi-micro work [81], using quantities of the order of 0·1–0·5 g. The glass tube in this case was 10–15 cm long, and from 2–6 mm in diameter depending on the amount of substance to be melted. Heat was provided by a beam of light from a 12 v, 100 w lamp focused using a 24 cm concave spherical mirror with a focal length of 17 cm. If the substance was light-sensitive, the entire glass tube had to be inserted into an opaque metal tube with a sight glass and a filter to transmit the infra-red rays. The metal tube was fixed,

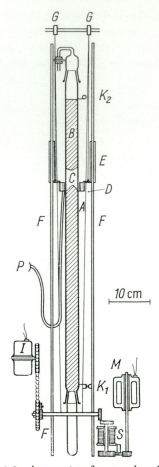

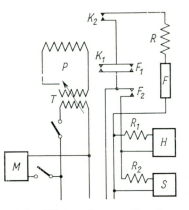

Fig. 4.8. Apparatus for zonal melting; *A*, 1·5 in. thick Pyrex tubing; *B*, re-solidified material; *C*, molten zone; *D*, bobbin-type wire-wound heater; *E*, brass tube, brazed to heater brackets, sliding along guide rod *F*; *G*, loose pulley carrying heater suspension cord; *I*, synchronous motor; K_1, contact switch; *M*, rewind motor; *S*, solenoid-operated clutch. Ref.: Herington *et al.* [76].

Fig. 4.9. Wiring diagram for the apparatus shown in fig. 4.8; *M*, synchronous motor; *T*, transformers; *S*, solenoid; *H*, rewind motor; F_1, F_2, relay contacts; *F*, relay; *R*, 22 kΩ resistance; R_1, 250 Ω resistance; R_2, 500 Ω resistance; *P*, heater element; K_1, microswitch normally open; K_2, microswitch normally closed. Ref.: Herington *et al.* [76].

while the glass tube containing the substance was moved uniformly upwards through a drive from a synchronous motor. Using this method, it was possible to reduce the melting range of liquids such as anthracene, chrysene, pyrene and benzoic acid from 2°c down to 0·5°c. The degree of purity of the liquid was also checked using fluorescent ultra-violet light. Pure anthracene, for example, changes the colour of fluorescent light from green to blue.

Terlecki *et al.* [78], using their own apparatus, have examined the variation of the intrinsic conduction of a liquid (cyclohexane) in various places in a molten sample. They found that the conduction decreases uniformly with increasing distance from the upper surface of the liquid (the sample was held vertically and the heating started from the top). They also found that at a certain distance, the rate of purification decreases quite sharply.

Methods of purification by fractional crystallization from solution are also used in research and industry. Its use is more widespread in industry and is particularly suitable for purifying certain organic substances such as phenol, naphthalene, xylene, benzene, etc. Molony and Roberts [77] have described a method whereby the crystallization efficiency can be increased from the usual 40–50% up to 95%.

References to Part i

[1] STEWART, G. W., 1928, *Phys. Rev.*, **31**, 174.
[2] WIERL, R., 1931, *Ann. Phys.*, 8, 521; 1932, *Ibid.*, **13**, 453.
[3] KOHLRAUSCH, L., and KOPPL, F., 1934, *Z. phys. Chem.*, **A26**, 209.
[4] THOMPSON, J. J., 1896, *Phil. Mag.*, **42**, 396; 1924, *Ibid.*, **47**, 337.
[5] KELLNER, L., 1951, *Proc. R. Soc.*, **64**, 521.
[6] KARRER, P., 1950, *Organic Chemistry* (New York: Elsevier).
[7] EGLOFF, G., 1946, *Physical Constants of Hydrocarbons* (New York: Reinhold Publishing Corp.).
[8] EYRING, H., 1936, *J. chem. Phys.*, **4**, 283.
[9] FRENKEL, J., 1947, *Kinetic Theory of Liquids* (Oxford: Clarendon Press).
[10] HIRSCHFELDER, J. O., *et al.*, 1954, *Molecular Theory of Gases and Liquids* (New York: McGraw-Hill).
[11] REID, R. C., and SHERWOOD, T. K., 1958, *The Properties of Gases and Liquids* (New York: McGraw-Hill).
[12] NIKURADSE, A., 1934, *Das Flüssige Dielektrikum* (Berlin: Springer).
[13] TATEVSKII, V. M., *et al.*, 1961, *Rules and Methods for Calculating the Physico-chemical Properties of Paraffinic Hydrocarbons* (Oxford: Pergamon Press).
[14] KARAGOUNIS, G., 1960, *Introduction to the Electronic Theory of Organic Compounds* (Warsaw: Polish Scientific Publishers) (in Polish).
[15] GUMIŃSKI, K., 1964, *Elements of Theoretical Chemistry* (Warsaw: Polish Scientific Publishers) (in Polish).
[16] COULSON, C. A., 1961, *Valence* (London: Oxford University Press).
[17] KUHN, H., *et al.*, 1960, *J. chem. Phys.*, **31**, 467.
[18] BÄR, F., *et al.*, 1960, *J. chem. Phys.*, **32**, 470.
[19] McCUBBIN, W. L., and GURNEY, I. D., 1965, *J. chem. Phys.*, **43**, 983.
[20] LORQUET, J. C., 1965, *Molec. Phys.*, **9**, 101.
[21] BIRKS, J. B., 1960, *Modern Dielectric Materials* (London: Heywood & Co. Ltd.).
[22] KOK, J. A., 1961, *Electrical Breakdown of Insulating Liquids* (Eindhoven: NV Philips' Gloeilampenfabrieken).
[23] LANDOLT, H., and BÖRNSTEIN, R., 1951, *Physikalisch-chemische Tabellen* (Berlin: Springer).
[24] KURATA, M., and ISIDA, S.-I., 1955, *J. chem. Phys.*, **23**, 1126.
[25] ADAMCZEWSKI, I., 1961, *Atompraxis*, **9**, 327.
[26] ANDRADE, E. N., and GUZMAN, J., 1914, *An. R. Soc. esp. Fís. Quím.*, **11**, 353.
[27] ANDRADE, E. N., 1934, *Phil. Mag.*, **17**, 497, 698.
[28] DOOLITLE, A. K., and DOOLITLE, D. B., 1957, *J. appl. Phys.*, **28**, 901.

[29] GLASSTONE, S., LAIDLER, K. J., and EYRING, H., 1941, *The Theory of Rate Processes* (New York: McGraw-Hill).
[30] TELANG, M. S., 1945, *J. phys. Chem.*, **49**, 579; 1946, *Ibid.*, **50**, 373; 1947, *J. chem. Phys.*, **15**, 525, 844, 885; 1949, *Ibid.*, **17**, 536.
[31] RUETSCHI, P., 1958, *Z. phys. Chem.*, **14**, 277.
[32] MACEDO, P. B., and LITOVITZ, T. A., 1965, *J. chem. Phys.*, **42**, 245.
[33] DAVIES, D. B., and MATHESON, A. J., 1966, *J. chem. Phys.*, **45**, 1000.
[34] LENNARD-JONES, J. E., 1924, *Proc. R. Soc.*, **A106**, 463.
[35] LENNARD-JONES, J. E., and DEVONSHIRE, A. F., 1939, *Proc. R. Soc.* **A169**, 317; 1939, *Ibid.*, **A170**, 464.
[36] DEBYE, P., 1912, *Phys. Z.*, **13**, 97.
[37] BARKER, J. A., 1960, *Aust. J. Chem.*, **13**, 187.
[38] BARKER, J. A., 1961, *Proc. R. Soc.*, **A259**, 442.
[39] BARKER, J. A., 1962, *J. chem. Phys.*, **37**, 1061.
[40] RICE, S. A., and ALLNATT, A. R., 1961, *J. chem. Phys.*, **34**, 2144.
[41] HENSHAW, D. G., 1960, *Phys. Rev.*, **119**, 9, 14; 1958; *Ibid.*, **111**, 1470.
[42] LEVY, M., and VAN WAZER, J. R., 1966, *J. chem. Phys.*, **45**, 1824.
[43] VAN WAZER, J. R., and GROENWEGHE, L. C. D., 1965, *Nuclear Magnetic Resonance in Chemistry* (New York: Academic Press).
[44] CARMICHAEL, J. B., and KINSINGER, J. B., 1964, *Can. J. Chem.*, **42**, 1996.
[45] CARMICHAEL, J. B., and WINGER, R., 1965, *J. Polym. Sci.*, **A3**, 971.
[46] CARMICHAEL, J. B., and HEFTEL, J., 1965, *J. phys. Chem.*, **69**, 2218.
[47] McCULLOGH, R. L., and HERMANS, J. J., 1966, *J. chem. Phys.*, **45**, 1941.
[48] SHONHORN, H., 1965, *J. chem. Phys.*, **43**, 2041.
[49] STRONG, S. L., and KAPLOW, R., 1966, *J. chem. Phys.*, **45**, 1840.
[50] JANNIUK, G., and SUMMERFIELD, G. C., 1966, *J. appl. Phys.*, **37**, 3953.
[51] KONO, R., *et al.*, 1966, *J. chem. Phys.*, **44**, 965.
[52] THOMAS, L. H., 1966, *Trans. Faraday Soc.*, **62**, 328.
[53] ISIHARA, A., 1966, *J. chem. Phys.*, **45**, 1855.
[54] BARLOW, A. J., *et al.*, 1966, *Proc. R. Soc.*, **A292**, 322.
[55] SCHOTT, H., and KAGHAN, W. S., 1965, *J. appl. Phys.*, **36**, 3399.
[56] FLORY, P. J., 1944, *J. chem. Phys.*, **12**, 425.
[57] GRUNBERG, L., and NISSAN, A. H., 1948, *Nature, Brit.*, **161**, 170.
[58] GRUNBERG, L., 1954, *J. chem. Phys.*, **22**, 157.
[59] PRIGOGINE, I., *et al.*, 1953, *J. chem. Phys.*, **21**, 559.
[60] SIMHA, R., and HADDEN, S. T., 1956, *J. chem. Phys.*, **25**, 702.
[61] TAYLOR, W. J., 1948, *J. chem. Phys.*, **16**, 257.
[62] DOOLITLE, A. K., 1952, *J. appl. Phys.*, **23**, 236.
[63] HALL, L., 1948, *Phys. Rev.*, **73**, 775.
[64] HIRAI, N., and EYRING, H., 1958, *J. appl. Phys.*, **29**, 810.
[65] HERZFELD, W. F., and RICE, F. O., 1928, *Phys. Rev.*, **31**, 691.
[66] KNESER, H. O., 1938, *Annls Phys.*, **32**, 277.
[67] CARNEVALE, E. H., and LITOVITZ, T. A., 1955, *J. acoust. Soc. Am.*, **27**, 547; 1955, *J. appl. Phys.*, **26**, 816.
[68] WILLIAMS, M. L., 1957, *Can. J. Phys.*, **35**, 134; 1958, *J. appl. Phys.*, **29**, 1395.
[69] FOX, T. G., and FLORY, P. J., 1948, *J. Am. chem. Soc.*, **70**, 2384; 1950, *J. appl. Phys.*, **21**, 581; 1951, *J. phys. Chem.*, **55**, 221; 1954, *J. Polym. Sci.*, **14**, 315.
[70] BUECHE, F., 1952, *J. chem. Phys.*, **20**, 1959; 1956, *Ibid.*, **25**, 599.
[71] PIOTROWSKI, K., 1960, *Bull. Sci. Assoc. Ingen.*, **73**, 759.
[72] JAFFE, G., 1913, *Annls Chim. Phys.*, **42**, 303.
[73] ADAMCZEWSKI, I., 1936, Mobility and Recombination of Ions in Dielectric Liquids (Akad. Nauk Techn., Doctoral Thesis) (in Polish).
[74] PAO, C. S., 1943, *Phys. Rev.*, **64**, 60; 1962, *Chem. Abstr.*, **56**, 1.
[75] GZOWSKI, O., and TERLECKI, J., 1959, *Acta phys. pol.*, **18**, 191.
[76] HERINGTON, E. F. G., *et al.*, 1956, *Chemy Ind.*, IV, 292.

[77] MOLONY, O. W., and ROBERTS, D., 1961, *J. appl. Chem.*, **11**, 283.

[78] TERLECKI, J., *et al.*, 1964, *Acta phys. pol.*, **26**, 1251.

[79] PFANN, W. G., 1952, *J. Metals, N.Y.*, **7**, 747.

[80] WILCOX, W. R., *et al.*, 1964, *Zone Melting of Organic Compounds* (California: Aerospace Corp.), p. 187.

[81] HANDLEY, R., and HERINGTON, E. F. G., 1956, *Chemy Ind.*, IV, 304.

[82] ZDANOWICZ, W., 1957, *Postępy Fiz.*, **8**, 147.

[83] ADAMCZEWSKI, I., 1937, *Annls Phys.*, **8**, 309.

[84] ROSSINI, F. D., *et al.*, 1953, *Selected Values of Properties of Hydrocarbons* (Washington: Nat. Bur. Stand.).

[85] TILIKHER, M. D., 1954, *Physicochemical Properties of Hydrocarbons* (in Russian) (Moscow: Gosgortekhisdat).

[86] SKANAVI, T. U., 1958, *The Physics of Dielectrics* (Moscow) (in Russian).

[87] CHARLESBY, A., 1960, *Atomic Radiation and Polymers* (Oxford: Pergamon Press).

[88] BOGORODITSKII, N. P., *et al.*, 1965, *Theory of Dielectrics* (Moscow) (in Russian).

[89] LEWIS, T. J., 1959, *Prog. Dielect.*, **1**, 97.

[90] WATSON, P. K., 1963, *Nature, Brit.*, **199**, 646.

[91] MORANT, M. J., 1963, *Br. J. appl. Phys.*, **14**, 469.

[92] KITTEL, C., 1956, *Introduction to Solid State Physics* (New York and London: John Wiley).

[93] KITTEL, C., 1963, *Quantum Theory of Solids* (New York and London: John Wiley).

[94] KITTEL, C., 1946, *J. chem. Phys.*, **14**, 614.

[95] KONDRATIEV, V. N., 1964, *La Structure des Atomes et des Molécules* (Paris: Masson Editeur) (in French).

[96] HAISSINSKI, M., 1967, *La Chimie Nucléaire et ses Applications* (Paris: Masson Editeur) (in French).

[97] PAULING, L., 1966, *Nature of the Chemical Bond* (Cornell University Press).

[98] SHARBAUGH, A. H., and WATSON, P. H., 1962, *Prog. Dielect.*, **4**, 199.

ADDITIONAL BIBLIOGRAPHY TO PART I

ADAMCZEWSKI, I., 1958, *Postępy Fiz.*, **9**, 49, 261.

ALTENBURG, K., 1950, *Kolloid Zeit.*, **197**, 153; 1952, *Ibid.*, **199**, 69.

COULSON, C. A., and MOFFITT, W. E., 1949, *Phil. Mag.*, **40**, 1.

DIENES, G. J ''1953, *J. appl. Phys.*, **24**, 779.

GRINDLAY, J., 1961, *Proc. phys. Soc.*, **77**, 1001.

ISIHARA, A., and HANKS, R. V., 1962, *J. chem. Phys.*, **36**, 2.

KRĘGLEWSKI, A., 1961, *J. phys. Chem.*, **65**, 1050.

MACY, R., *Organic Chemistry Simplified* (New York: Chemical Publishing Co. Inc.).

NAKANISHI, K., *et al.*, 1960, *J. chem. Engng*, **2**, 210.

NICHOLAS, J. F., 1959, *J. chem. Phys.*, **31**, 922.

PIGOŃ, K., 1956, *Wiad. chem.*, **9**, 454.

ROSIŃSKI, K., 1954, *Postępy Fiz.*, **5**, 305; 1955, *Ibid.*, **6**, 66.

ROWLINSON, J. S., and CURTISS, C. F., 1951, *J. chem. Phys.*, **19**, 1519.

STEGINSKY, B., 1961, *Am. J. Phys.*, **29**, 605.

TAMURA, M., *et al.*, 1961, *Bull. chem. Soc., Japan*, **34**, 684; 1962, *Chem. Abstr.*, **56**, 1.

PART II

The Action of Ionizing Radiation on Matter

Chapter 5

The Principal Effects of Ionizing Radiation

5.1. *Introduction.* Research into the mutual interaction of ionizing radiation and matter in all its states: isolated elementary particles, the gaseous, liquid and solid states, and human organisms, is of great importance in many fields of contemporary physics, especially nuclear physics, nuclear radiation dosimetry, plasma physics, health physics and certain fields of engineering. It also provides a basis for other sciences such as radiation chemistry, radiochemistry, radiobiology, radiology and related branches. The construction of very powerful sources of ionizing radiation, such as accelerators and atomic reactors, has opened up the possibility of making use of this radiation for industrial purposes, especially in the synthesis of plastic materials and for the acceleration of certain chemical reactions. This section deals with the basic phenomena of ionization in dielectric liquids, using the small doses which are employed in investigations on electrical conductivity and breakdown in liquids. A number of textbooks [1–5], including one by the present author [6], have treated certain basic aspects, that is the description of fundamental phenomena, the definition of magnitudes and units and the methods for measuring ionizing radiation effects (dosimetry).

The mechanism of the action of ionizing radiation on gases is much better known than that of the action on liquids and solids. The main difficulty with liquids is that the majority are electrolytes featuring, to a greater or lesser extent, molecular dissociation into positive and negative ions. The action of ionizing radiation is therefore usually blurred by the much stronger electrolytic action, and it is evident that the investigation of electrical conduction in ionized liquids should be carried out with those in which electrolytic dissociation does not occur, i.e. with dielectric liquids which are also the best liquid insulators.

It is also important that liquids used should not contain even the slightest traces of any electrolytic contamination. Some dielectric liquids, however, have a high electrical conductivity even when chemically as pure as possible. This is because their molecular structure and types of chemical bonds lend themselves very well to the detachment of electrons from atoms and molecules and their displacement in an electric field. These aspects are dealt with in greater detail in later sections of this book.

5.2. *The Energy of Ionizing Radiation.* Ionizing radiation can be classified into two types: electromagnetic radiation (x and γ), and corpuscular radiation (electrons, mesons, protons, neutrons, deuterons, α particles, fragments, etc.). The action on matter depends on the nature and energy of the radiation and on the properties of the material. Radiation energy can vary over a very wide range. A brief indication of the various energy values is given below:

0·025 ev	Energy of particles in thermal motion
12398/λ ev	Quantum of radiation energy of wavelength λ (in Å)
0·005–0·01 ev	Energy of molecule due to rotational motion
$\sim 0\cdot1$ ev	Quantum energy of atoms due to oscillatory motion
1·5–3·0 ev	Quantum energy of electron oscillation
1·5 ev	Upper limit of infra-red radiation
1·5–3 ev	Range for visible radiation
3–6 ev	Limits of ultra-violet radiation
0·01–0·3 мev	x-Rays produced normally in x-ray tubes
0·01–8·7 мev	β-Rays from radioactive substances
2–14 мev	α-Rays from radioactive substances
0·18–2·2 мev (average 1·76)	γ-Rays from radium source
1·17–1·33 мev (average 1·25)	γ-Rays from ^{60}Co
1·7–2 мev	γ-Rays from atomic reactors
5–12 мev	α-Particle radiation from radioactive substances; neutron radiation from the (α, n) reaction
(8–28) × 10^3 мev	Radiation from synchrophasotrons, cosmotrons and bevatrons
10^3–10^{14} мev	Cosmic radiation
10^3–10^5 мev	Main portion of the cosmic ray spectrum

5.3. *Photoelectric and Compton Effects. Pair Formation.* The action of electromagnetic radiation on matter is that its energy is partly absorbed by the matter. The main effects here are the photoelectric effect, Compton effect and the formation of pairs of negative and positive electrons. These phenomena are accompanied by secondary effects such as ionization, the stimulation of free vibration in atoms or molecules, new bonds between atoms, the formation of radicals and so on. These phenomena play a large part, particularly in liquids.

The simplest effect of radiation is the ionization of atoms or molecules. This ionization depends on the detachment of an electron from an electrically neutral atom; as a result, the atom becomes a positive ion, and the negative free electron, after a number of collisions, combines with other neutral atoms or gas molecules to form a negative ion. Thus pairs of negative and positive ions are formed in the irradiated substance. These pairs take part in the thermal motion of atoms and molecules. If, during this motion, two ions of the opposite sign collide, their charges become neutralized, the ions become neutral particles and the ionization effect disappears. This phenomenon is called ion recombination.

The ionization energy is different for various elements and substances (see table 8.1, p. 86). It is assumed that for complex substances, such as those in human organisms, the amount of energy required for the formation of an ion pair is the same as that for air, i.e. 34 ev. The ultimate physical effect of ionizing radiation on matter is often measured by the ionization effect alone, that is by the number of pairs formed in 1 cm³ or in 1 g of the substance. It should, however, be stressed that this ultimate effect is the result of the combination of various physical phenomena, sometimes very complex, whose nature is dependent on the type and energy of the radiation.

The photoelectric effect occurs when an electron is knocked out of an atom or a molecule by a photon; in this process the photon energy $h\nu$ is partly used up in the work P to separate an electron from the atom, the remaining energy being given to the electron in the form of kinetic energy $mv^2/2$. Thus the basic equation for the photoelectric effect is as follows:

$$h\nu = P + \frac{mv^2}{2}. \tag{5.1}$$

In this process the photon is completely destroyed. If an atom is combined in a crystal lattice (in a metal for example), the work P is increased by the amount required for the electron to escape from the crystal. As can be seen from eqn. (5.1), the photoelectric effect can occur only when $h\nu > P$, that is when the incident radiation frequency is greater than a certain frequency ν_0 characteristic of the atoms of the given substance, $\nu_0 = P/h$.

The Compton effect occurs when a photon is scattered over an electron of an atom or a molecule, where the frequency of the scattered photon ν is lower than that of an incident photon ν_0 and the scattered electron is ejected at a given velocity v. The change in frequency ν or wavelength λ of the incident photon depends on the scattering angle ϑ, but it does not depend on the wavelength:

$$\Delta\lambda = \lambda - \lambda_0 = \frac{c}{\nu} - \frac{c}{\nu_0} = \frac{h}{m_0 c}(1 - \cos\vartheta) = \frac{2h}{m_0 c}\sin^2\frac{\vartheta}{2}. \tag{5.2}$$

Compton showed that the effect occurs as if a quantum of radiation 'collided' with an electron in an elastic manner; some of its energy is transferred to the electron and the remainder is scattered at a certain angle to the original angle of incidence in the form of a quantum with lower energy.

Equation (5.2) is obtained from the laws of conservation of energy and momentum, assuming that the quantum has a mass $m = h\nu/c^2$ and a momentum $p = h\nu/c$, while taking into consideration the possibility of a relativistic change in the mass of an electron with a change in velocity:

$$m = \frac{m_0}{\sqrt{\left(1 - \dfrac{v^2}{c^2}\right)}}.$$

(5.3)

The photoelectric and Compton effects for various quanta of x and γ-rays are shown in fig. 5.1. As can be seen, the photoelectric effect F prevails for small quantum energies of 10–40 kev, whilst the Compton effect C is the most important for large quantum energies of the order of a million ev.

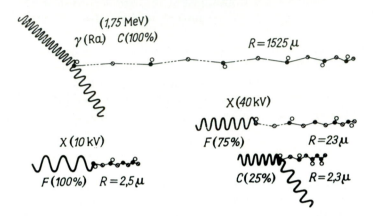

Fig. 5.1. Diagrams of the photoelectric effect (F) and the Compton effect (C) for various photon energies. R is the maximum range of the knocked-out electrons in water. (In lower diagrams the x-ray voltage is given.)

Figure 5.2 shows the paths of Compton electrons produced by γ-rays in a Wilson cloud chamber. A beam of γ-rays lies along the horizontal axis of the plate. In the Compton effect, the energy of the electrons ejected by the photon varies over the range from 0 to $2h\nu/(mc^2 + 2h\nu)$, where $h\nu$ is the photon energy and mc^2 the energy of an electron. The average energy of these ejected electrons is about half the energy of the photon.

Figure 5.3 shows the relationship between the energies of photo-electrons, Compton electrons and the energy of x or γ-rays. The greater the energy of the quanta of radiation, the higher is the energy of the photoelectrons and Compton electrons produced by these quanta.

Electron pair formation is a further phenomenon which plays an important rôle in the action of high-energy electromagnetic radiation on matter. This effect depends on the ability of quanta with an energy $E > 1\cdot02$ MeV to form, in the vicinity of the atomic nucleus, electron pairs (a negatron and a positron), which take up the total mass and energy of the quantum. The requirement that the energy of the quantum should be greater than $1\cdot02$ MeV comes from the fact that this energy is equivalent to the rest mass of two electrons (for one electron $E = m_0 c^2 = 0\cdot51$ MeV).

If a quantum has a higher energy than this critical energy, the surplus is transferred to the electrons as kinetic energy. A positron formed in this process usually disappears very quickly by combining with a negatron to give two quanta of electromagnetic radiation each with an energy of 0·51 MeV. The energy of these quanta is then used up in producing the photoelectric effect or Compton effect.

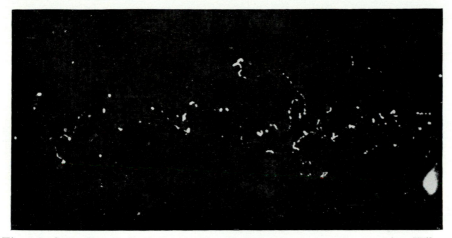

Fig. 5.2. Secondary ionization caused by the Compton electrons in argon in a Wilson chamber. Ref.: Charlesby [8].

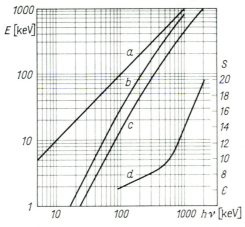

Fig. 5.3. Relationship between electron energy E and photon energy $h\nu$ from primary γ or x-rays: (a) for photoelectrons; (b) for Compton electrons with the maximum energy; (c) for Compton electrons with average energy; (d) the path in water (in cm) for a Compton collision. Ref. Charlesby [8].

In all these three effects a free electron is formed which, as it proceeds, ionizes the atoms or molecules of the substance to give a certain number of pairs, this number being dependent on its energy. The photoelectric effect prevails at low energies, the Compton effect at medium energies (fig. 5.1) and pair formation at very high energies.

If the energy of the primary or scattered quantum is too low to cause ionization (1–5 ev), it can still stimulate the atom to emit light; with lower values (0·1–1 ev) it can produce additional oscillation and rotational vibration of the atoms in the molecule or of whole molecules, and cause a rise in temperature of the surroundings.

Chapter 6

Stopping of Ionizing Particles by Matter

6.1. *Stopping Power of Different Substances.* Ionizing particles with very high energies generally lose their energy in passing through matter due to (1) ionization, (2) the formation of photons of electro-magnetic radiation accompanied by incomplete slowing-down and, (3) with very high energies, cascade processes involving large groups of particles and secondary photons. The last two processes are important for energies higher than 150 Mev, that is for cosmic radiation and particles produced in the largest accelerators. For lower energies, i.e. for the majority of cases we are interested in, the particles lose their energy almost exclusively in ionization processes. Since, on average, a certain constant amount of energy is required to form a single ion pair (about 34 ev in air), by knowing the energy of a particle, we can easily calculate the number of ions it would produce as it passes through matter. The number of ion pairs is, however, not constant and depends on the energy of the particle; the highest linear ion density (the number of ions per centimetre of the path) is in the final section of the path where the particle energy falls to zero.

Let us consider the special case of an α particle passing through air, and assume that the energy of this particle is 8·77 Mev (i.e. the energy of certain α particles in a C' thorium path). By experiment it has been found that initially these particles form two ion pairs over a path of 1 micron in air (under normal conditions), and that with a reduction in energy this number increases to an average 3·2 ion pairs per micron, i.e. 32 000 ion pairs per centimetre of an α particle path in air. Thus the mean energy loss along a 1 cm path is 34 ev × 32 000 = 1·08 Mev; the range of the particle in air should therefore be 8·77/1·08 = 8·1 cm. This result is in agreement with the results obtained experimentally.

If the retarding matter is not air but some other substance with a density n times that of air, the number of ion pairs per centimetre of the particle path will be approximately n times greater, while the particle range is n times shorter. We can therefore say that the slowing-down effect in this substance is n times greater than in air. Thus, for example, the slowing-down effect in water for α particles is about 773 times greater, and in photographic film emulsion about 1800 to 2000 times greater, than in air. The range of these particles in water is therefore about 0·1 mm and in film emulsion about 46·1 microns. The energy loss of a particle passing through matter is proportional to the product nZ where Z is the number of electrons in the atom and n the number of atoms in 1 cm³ of the substance.

Figure 6.1 shows the relationship between the energy loss and the energy of a particle. It is expressed as the number of ion pairs formed per mg/cm² of the substance for various ionizing particles including α particles, protons and electrons. The highest losses are incurred by α particles, then protons and electrons. It can also be seen that for low particle velocities the losses are much higher than for high velocities, and that they change with a variation in velocity.

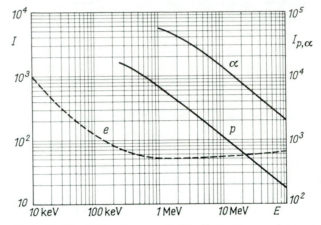

Fig. 6.1. Specific ionization I (I.p./mg cm⁻²) for electrons e, protons p and α particles in air as a function of their energy. The unit mg cm⁻² corresponds to a thickness of 10^{-3} cm multiplied by the density in gm cm⁻³. Ref.: Charlesby [8].

6.2. *Mean Ionization Density*. LET. Immediately after a substance has become ionized, the distribution of the resultant ion pairs is not uniform but depends on the type of radiation, its energy and the nature of the substance. For example, the distance between two adjacent pairs in water along the ionizing particle path is 2·5 Å for an α particle with an energy of 5·3 MeV (from ²¹⁰Po) and 1 Å for an energy of 1·47 MeV from the ¹⁰B(n, α)⁷Li reaction; it is 500 Å for β electrons with an energy of 5·65 MeV (from tritium), and 1 micron for γ rays with an energy of 1·25 MeV (from ⁶⁰Co).

The mean ionization density, i.e. the average number of ion pairs per unit length (e.g. 1 micron), is usually used as the linear measure of ionization and is evaluated from the equation:

$$I = \frac{\Delta E}{\Delta x} : W, \tag{6.1}$$

where ΔE is the energy loss over the length Δx, and W is the mean energy required to form one ion pair in the substance through which the particle passes.

The value 100 I.p./μ in water is taken as the mean ionization density for β electrons and for electrons produced by x and γ-rays. The international literature on radiation chemistry adopts a quantity called the linear energy transfer, abbreviated to LET. This is the ratio of the energy difference ΔE of a quantum or electron at the two ends of the path being transferred to the material, to the length Δx of this path.

$$\text{LET} = \frac{\Delta E}{\Delta x}.$$

The LET has units of $ev/\text{Å}$, kev/μ or J/m.

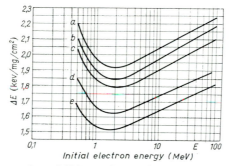

Fig. 6.2. Electron energy loss ΔE in various substances as a function of incident electron energy; (a) polyethylene; (b) water; (c) polymethyl methacrylate; (d) graphite; (e) aluminium. Ref.: Charlesby [8].

Figure 6.2 illustrates the dependence of electron energy loss in various substances on the initial energy of the electron. As can be seen, the energy loss is lowest for electron energies between 1·1 and 1·7 MeV.

The quantity W in eqn. (6.1) can vary over fairly wide limits for different substances and particle velocities. For air, W is usually equal to 34 ev; this value is found to be correct for electrons, α particles with energies greater than 4 MeV, and for protons with an energy of 1·5 MeV.

Figure 6.3 shows the change in the number of ion pairs formed by a beam of very fast deuterons (with an energy of 190 MeV) according to the distance travelled in the absorbing layer (curve a). We see that initially the number of ion pairs per unit length is almost constant but that it increases rapidly at the end of the deuteron path and then falls to zero. This curve is called the Bragg–Gray curve. The lower curve (b) shows the change in the number of particles in a beam. This beam disappears completely at a certain distance from the beginning of the absorbing layer.

As can be seen, the slowing down of ionizing particles in matter is a very complicated phenomenon. It is one of the most important aspects of contemporary physics, and much work has been done and written on the subject with the result that comprehensive material dealing with its mechanism has been collected. The most important experimental methods applied here are those employing ionization chambers, Geiger–Müller counters, proportional counters, Cherenkov counters, Wilson cloud chambers, diffusion chambers, bubble chambers and nuclear film emulsions. More detailed descriptions of these methods along with results obtained can be found in many special publications including *Handbuch der Physik* [7]. This volume is devoted solely to the problem, and includes an article by Glaser [7, 10] discussing experimental work and the theory of bubble chambers. These chambers are one of the most modern experimental methods in this particular field, permitting the ionizing particle paths in liquids, and the ionization effect produced, to be photographed.

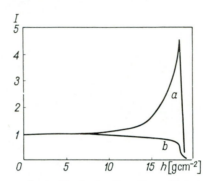

Fig. 6.3. (*a*) Dependence of specific ionization I on distance h from the source of radiation (Bragg–Gray curve) for deuterons; (*b*) dependence of the number of deuterons on distance h.

Many excellent photographs of elementary ionization processes in matter have been taken using the nuclear emulsion method, which adds new information to that obtained using Wilson chambers. Shapiro (in ref. [7]) has given a detailed description of this method. The Polish literature includes two articles by Adamczewski [11], in which the slowing-down of ionizing particles in nuclear film emulsions is discussed.

Figure 6.4 shows proton, meson and electron paths in a bubble chamber filled with propane. Path widths corresponding to different ion densities for the various kinds of particles can be observed on the photograph.

6.3. *Electron Stopping.* Electrons play an important rôle in ionizing radiation dosimetry since they occur in the secondary radiation of electromagnetic radiation as photoelectrons or Compton electrons, and as β-rays in the radiation of radioactive substances. They can also occur as δ-rays (secondary electrons knocked out of the atoms along the paths of α particles or other heavy atomic nuclei).

This book is mainly concerned with the action on matter of high-energy electrons with energies from fractions of мev to several hundred мev. Electrons with these high energies have velocities approximately equal to that of light; for example, electrons with an energy of 0·5 мev have a velocity of about 0·9c (c is the velocity of light equal to $\sim 3 \times 10^{10}$ cm/sec), and electrons with an energy of 2 мev a velocity of 0·98c. The energy of these electrons is evaluated from the relativity equation $E = mc^2$, where

$$E = m_0 c^2 + E_k \quad \text{and} \quad m = \frac{m_0}{\sqrt{\left[1 - \left(\dfrac{v}{c}\right)^2\right]}}, \tag{6.2}$$

and E_k is the kinetic energy of an electron. From these relationships we obtain the following equation for the velocity of an electron:

$$v = c\sqrt{\left[1 - \left(\frac{0\cdot511}{E_k + 0\cdot511}\right)^2\right]}, \tag{6.3}$$

E_k is here expressed in MEV.

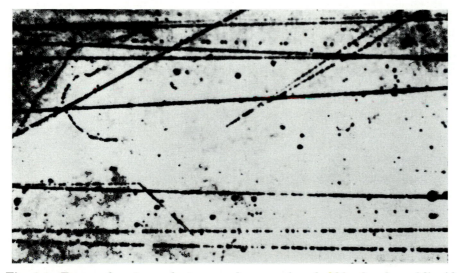

Fig. 6.4. Traces of protons, electrons and mesons in a bubble chamber of liquid propane. The direction of motion of the particles is from left to right. Ref.: Glasser [15].

High-energy electrons lose energy in passing through matter by the following three processes:

(1) Elastic scattering on atomic nuclei and on shell electrons.

(2) Atomic or molecular ionization of the medium.

(3) The production of electromagnetic radiation (Bremsstrahlung) in non-elastic collisions with atomic nuclei.

If the collision is elastic, the electrons in fact keep their kinetic energy but change direction and for low energies the deviation from the initial direction may be very large. The repeated scattering of an electron can, among other things, cause its path in the substance to be 1·5 to 4 times shorter than the actual path (a straightened zigzag path). This phenomenon calls for caution in estimating numbers of electrons, for example while measuring the activity of β preparations, as there is the possibility of electrons being knocked off from the preparation base.

The largest amount of energy is lost by electrons as a result of their ionizing action during collisions with electrons which belong to the atoms of a substance, and due to this interaction along the electron path, a number of positive and negative ions are formed. To form one ion pair in air, an electron loses energy of about 34 ev; thus, an electron with an energy of 0·34 Mev will form about 10 000 ion pairs along its path. The length of the path along which an electron loses its energy in the ionization process depends on its velocity v and on the nature of the substance. The evaluation of the stopping power of a given substance on electrically charged particles has been investigated theoretically many times.

One of the best known equations for the energy loss dE/dx of a particle with an electric charge ze, velocity v and mass m in a substance, having N atoms per cubic centimetre and an atomic number Z, was given by Bethe [12] and Möller [13] in the following form:

$$-\frac{dE}{dx} = \frac{4\pi e^2 (ze)^2}{mv^2} NZ \left[\log \frac{2mv^2}{I} - \log(1-\beta^2) - \beta^2 \right], \qquad (6.4)$$

where I is the mean excitation potential of the atoms in the inhibiting substance and $B = v/c$ (c being the velocity of light).

From eqn. (6.4) it follows that the energy loss per unit length is lowest for energies of about $2mc^2$ and that it increases slowly for higher energies and much more rapidly for lower energies, particularly those close to zero. For high energies (above 10 Mev) a correction for the screening effect should be introduced into eqn. (6.4); this is due to polarization of atoms of the substance along the path of the moving particle. This effect occurs mainly in solids and liquids, and results in increased energy losses.

For low electron energies (up to about 50 kev), a simplified equation due to Miller is used [14]:

$$-\frac{dE}{dx} = \frac{2\pi e^4}{E} NZ \ln \frac{E}{I} \sqrt{\frac{e}{2}}, \qquad (6.4\,a)$$

where e is the base of natural logarithms.

Table 6.1 shows the stopping powers (in Mev) for a 1 g/cm² thick layer of a number of substances.

Table 6.1. Electron stopping power of various substances [4]

Substance	Initial electron energy [ev]				
	10^4	10^5	10^6	10^7	10^8
Air	19·5	3·6	1·6	1·8	2·1
Water	21·7	4·1	1·8	2·0	2·3
Lead	10·0	2·2	1·1	1·2	2·7

To calculate the energy loss of an electron with any given energy over a 1 cm path, the figures in table 6.1 have to be multiplied by the density of the substance.

It follows from eqn. (6.4) that the energy loss may be expressed as:

$$-\frac{dE}{dx} \sim \frac{(ze^2)}{v^2} NZ. \tag{6.5}$$

Since

$$N = \rho \frac{6 \cdot 02 \times 10^{23}}{A}$$

(ρ being the density of the substance and A the mass number of the element), an electron with a velocity low compared with that of the speed of light loses, in a layer of thickness dx, an amount of energy expressed by:

$$-\frac{dE}{dx} = \frac{Z}{v^2 A} \rho\, dx. \tag{6.5 a}$$

The ratio Z/A varies slightly with a change in Z (from 0·4 to 0·5, except for hydrogen), and therefore it is convenient to define the thickness of the absorbing layer in units of ρx (g cm^{-2} or mg cm^{-2}) because the stopping powers of different substances for electrons with the same energy will then be almost identical.

The process of electron stopping during inelastic collisions with atomic nuclei causes high-energy electromagnetic radiation to be produced, as is the case with the formation of x-rays in x-ray tubes. The maximum energy of photons in this radiation corresponds to the maximum electron energy.

As a result of the three processes outlined above, an electron beam passing through a substance is subjected to slowing down and attenuation. The degree of slowing down can be described in two ways; either by plotting a curve of beam attenuation and determining the 50% attenuation layer as for γ-rays (eqn. (7.5)), or by evaluating the maximum range R of electrons in the given substance. R is defined by the equation:

$$R = \int_0^E \frac{dE}{dE/dx}, \tag{6.6}$$

where E is the electron energy and $\Delta E/\Delta x$ the loss in energy per unit path length.

Because of the zigzag shape of the electron path in a substance, it is impossible to evaluate theoretically the maximum range of an electron. Semi-empirical equations defining range as a function of energy, or vice versa, are usually used. One of the two equations most frequently encountered is:

$$R = 0 \cdot 542E - 0 \cdot 133 \quad \text{(Feather's equation)}, \tag{6.7}$$

where R is the range in g cm^{-2} in Al, and E is energy in мev. This equation is valid for energies from 0·7–3 мev. For low electron energies (between 50 kev and 0·8 мev), the equation:

$$R = 0 \cdot 407E^{1 \cdot 38}, \tag{6.7 a}$$

or Flammersfeld's equation:

$$E = 1 \cdot 92[(R\rho)^2 + 0 \cdot 22R\rho]^2, \tag{6.8}$$

where ρ is the density of the inhibiting substance, is in better agreement with experiment.

A very close approximation for the range (in $\mathrm{g\,cm^{-2}}$) in Al is obtained if we take half the electron energy in MEV. For substances other than Al, the range is almost the same as for Al. Electrons with an energy of 1 MEV have a range in Al of about 500 mg cm^{-2} (about 400 cm in air), and those with energy of 10 MEV a range of 4500 mg cm^{-2} (about 40 metres in air).

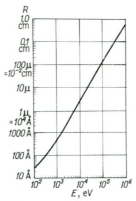

Fig. 6.5. Maximum range R of electrons in tissue ($\rho = 1$) as a function of energy E.
Ref.: Lea [16].

Table 6.2 shows data on the stopping of electrons with various energies. The approximate number of ion pairs created by an electron with velocity v in air under normal conditions over a 1 cm path is $45/(v/c)^2$, where c is the velocity of light.

Table 6.2. Ranges and ionizing powers for electrons with different energies and velocities

Energy (kev)	v/c	Range in air (cm)	Number of ion pairs per centimetre in air
0·15	0·024	0·0006	7700
2·55	0·1	0·04	2100
10·5	0·2	0·23	1000
46·6	0·4	3·4	250
127·8	0·6	17·9	130
341	0·8	83·5	70
1127	0·95	437	45
3114	0·99	1300	41

Figure 6.5 shows the dependence of the maximum range of electrons in tissue on their energy (on a log scale). As can be seen, it is almost a

straight line. Figure 6.6 shows the range R for α particles, protons, electrons and neutrons, and the half-thickness $D_{1/2}$ for γ-rays in various substances as a function of the radiation energy. Electrons emitted by radioactive substances (β-rays) are not monoenergetic. Their energy spectrum extends from zero to a certain maximum energy $E_{\max}$ (see fig. 6.7).

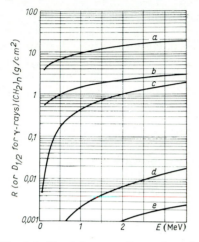

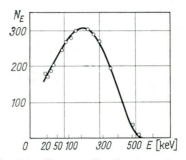

Fig. 6.6. Particle range R as a function of radiation energy E ((b) neutrons; (c) electrons; (d) protons; (e) α particles; (a) is the half-thickness $D_{1/2}$). The curves are for radiation in a number of substances similar to polyethylene; (a), (b) and (c) are in $(CH_2)_n$, and (d) and (e) in $C_{57}H_{110}O_6$. Ref.: Charlesby [8].

Fig. 6.7. Energy distribution of a β-radiation spectrum. N_E is the number of electrons for any energy E.

The half-thickness $D_{1/2}$ of the layer of absorbing substance required to reduce the intensity to one-half its value is defined approximately by the experimental equation:

$$D_{1/2} = 0 \cdot 095 \frac{Z}{A} E_{\max}^{3/2} \quad (\text{g cm}^{-2}), \tag{6.9}$$

which is valid for a large number of substances (from hydrogen to copper). The pseudo-exponential dependence observed for the attenuation of the electron beam intensity as it passes through matter is purely accidental, as the number of electrons passing through a certain thickness of the substance depends on many different factors such as the energy distribution, variation of range with energy, multiple scattering and the atomic and mass numbers of the absorber.

Chapter 7

The Absorption of X-rays and γ-rays

7.1. *The General Absorption Law.* The phenomena described in the preceding chapter result in a reduction in the intensity of electromagnetic radiation as it passes through a layer of a substance. The macroscopic absorption law, which has been formulated as a result of experimental investigations, is expressed as:

$$I = I_0 \exp(-\mu d), \tag{7.1}$$

where I_0 is the initial intensity of the x or γ-ray beam (see fig. 7.1), d is the thickness of the layer and I is the beam intensity when it emerges from the layer. Equation (7.1) can be written in a logarithmic form:

$$\ln I = \ln I_0 - \mu d, \tag{7.2}$$

where μ is the radiation intensity linear attenuation (extinction) coefficient, commonly termed the absorption coefficient. The magnitude of this coefficient can be derived from the elementary processes occurring during the interaction of ionizing radiation with matter, and can be linked with atomic quantities.

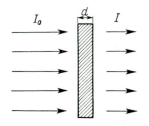

Fig. 7.1. Change in the intensity I of electromagnetic radiation (x or γ-rays) when passing through a layer of substance of thickness d (from left to right $I = I_0 \exp(-\mu d)$).

7.2. *Absorption Cross Section.* The degree to which x-ray or γ-ray energy is absorbed in a substance is determined by its active cross section. This is proportional to the probability that a photon, as it passes through the substance, is absorbed or interacts with an atom in the substance. The active cross section σ can be defined as follows: let us assume that a stream of photons strikes a plate of thickness d containing N atoms/cm³ at a rate of Φ_0 per cm² per second (fig. 7.2). The probability of a collision of type i of a photon with an atom of the substance is given by the following expression:

$$\Delta p_i = N \sigma_i \Delta x = -\frac{\Delta \Phi}{\Phi}, \tag{7.3}$$

where $\Phi(x)$ is the intensity of the photon stream at a depth x in the plate, $\Phi(d)$ is the intensity as it emerges from the plate, and $\Delta \Phi(x)$ is the reduction in intensity of the stream over an element Δx. Manipulation of eqn. (7.3) leads to:

$$\Phi = \Phi_0 \exp(-N\sigma_i d) = \Phi_0 \exp(-\mu d), \tag{7.4}$$

which describes the attenuation of a photon beam as a result of a single process i after it has passed through a plate of thickness d.

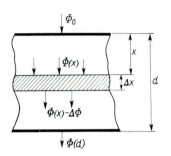

Fig. 7.2. Change in intensity of a photon beam Φ when passing through a plate of thickness d.

If more than one process occurs, σ_i should be substituted by a total summated active cross section $\sigma_S = \sum_i \sigma_i$. The unit of active cross section is the barn $= 10^{-24}$ cm². The quantity σ_S is known as the microscopic (atomic) active cross section, and $N\sigma_S$ is the macroscopic or total (summated) active cross section of a substance. The latter is often denoted by μ (as in §7.1) and called the linear absorption coefficient. There are also other quantities associated with the absorption phenomenon such as the mass absorption coefficient μ/ρ (where ρ is the density of the substance), the mean free path $1/\mu$, and the relaxation length l. This characterizes the rate of decrease in a given physical quantity (e.g. stream velocity, energy density, biological dose, etc.) according to the expression $q = q_0 \exp(-x/l)$. For γ-rays, its value is close to that of the mean free path.

Equations (7.1) and (7.4) are equally good for describing absorption phenomena, as there is a simple relationship between the values Φ and I, viz. $\Phi = I\omega$ (where ω is a solid angle). Plots of $\Phi = \Phi_0 \exp(-\mu d)$ and $I = I_0 \exp(-\mu d)$ are exponential curves (fig. 7.3 (a)); the value of I or Φ decreases asymptotically with d, the thickness of the plate, down to zero. As the thickness of the plate is increased from zero, there is always a value $D_{1/2}$ for which $\Phi_D = \Phi_0/2$. This is known as the half-thickness, and reduces the intensity of a photon beam by a factor of a half. It can be seen that the half-thickness is obtained from eqn. (7.4) as follows:

$$\Phi = \frac{\Phi_0}{2} = \Phi_0 \exp\left(-\mu D\right); \quad D_{1/2} = \frac{\ln 2}{\mu} = \frac{0\cdot693}{\mu}. \tag{7.5}$$

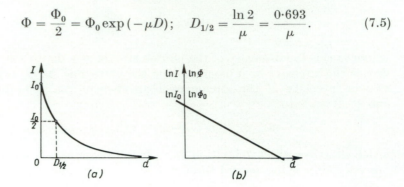

Fig. 7.3. The absorption law: (a) on a linear scale, and (b) on a semi-log scale.

Figure (7.3 (b)) shows the functions Φ and I plotted on a semi-log scale, e.g. $\ln \Phi = \ln \Phi_0 - \mu d$. As can be seen, it is a straight line intersecting the Φ axis (when $d = 0$) at $\ln \Phi_0$; the absorption coefficient μ for any substance, and hence also the active cross section σ_S, can be evaluated from the gradient of this line.

7.3. *Components of the Absorption Coefficient.* As mentioned previously, the total active cross section σ_S consists chiefly of three active cross sections for (a) the photoelectric effect σ_F, (b) the Compton effect σ_C, and (c) pair formation σ_P:

$$\sigma_S = \sigma_F + \sigma_C + \sigma_P. \tag{7.6}$$

Each component of the absorption coefficient μ corresponds to one of three active cross sections, viz. τ to the photoelectric effect, η to the Compton effect and κ to pair formation:

$$\mu = \tau + \eta + \kappa. \tag{7.7}$$

It is most usual to evaluate experimentally each of these three components for a given substance (e.g. lead or aluminium); the same components for other substances are then obtained by comparing their densities, atomic numbers and mass numbers with these values for the reference substance. The components of the absorption coefficient for different elements referred to the values for lead are:

$$\tau = \tau_{\mathrm{Pb}} \frac{\rho}{11\cdot3} \frac{207}{A} \left(\frac{Z}{82}\right)^4 \quad \text{for the photoelectric effect,}$$

$$\eta = \eta_{\mathrm{Pb}} \frac{\rho}{11\cdot3} \frac{207}{A} \frac{Z}{82} \quad \text{for the Compton effect,} \left.\vphantom{\begin{array}{c}a\\a\\a\\a\\a\\a\end{array}}\right\} \tag{7.8}$$

$$\kappa = \kappa_{\mathrm{Pb}} \frac{\rho}{11\cdot3} \frac{207}{A} \left(\frac{Z}{82}\right)^2 \quad \text{for pair formation.}$$

For low energy values the dependence is not continuous and sharp minima and maxima at values corresponding to the self-energy of the atom (electron shells K, L and M) are observed. Photons with these energy values are very strongly absorbed by the atom and cause K, L and M electrons to be ejected from the atomic shells. Other external electrons jump in to take their places and this results either in characteristic x-rays or fluorescent radiation. Between the discontinuities, the curve is monotonic in agreement with eqn. (7.8). Figure 7.4 shows the relationship

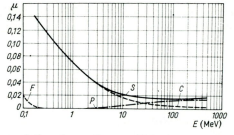

Fig. 7.4. Dependence of the absorption coefficient μ on energy, $E = h\nu$ photons, for x-rays in water. Ref.: Bacq and Alexander [19].

between the absorption coefficient μ in water and the photon energy for both light and heavy elements. It can be seen that for low photon energies the photoelectric effect F and the Compton effect C are of importance, while for high energies it is pair formation P which is most significant; the value of Z is also important here. The total effect is given by curve S.

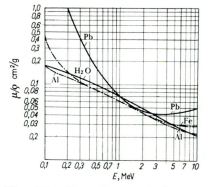

Fig. 7.5. Dependence of the mass absorption coefficient μ/ρ on the photon energy E for various substances. Ref.: Charlesby [8].

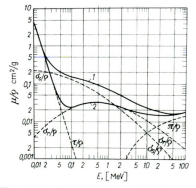

Fig. 7.6. Dependence of the mass absorption coefficient μ/ρ, and its components in water, on the photon energy E. Ref.: Jaeger [3].

In fig. 7.5, diagrams showing how the absorption coefficient changes for various substances depending on the initial photon energy are shown, while fig. 7.6 gives curves of the dependence of the mass absorption coefficient in water on the photon energy, and shows how the curves are divided into the individual components which make up the total effect.

As regards absorption of the energy of electromagnetic radiation (10–250 kev) with which we are mainly concerned here, the most important factor is τ or more specifically the ratio τ/ρ (where ρ is the density of the absorbing substance), known as the real mass absorption coefficient. The following equation relates the ratio τ/ρ to the atomic number of the absorbing element Z and the radiation wavelength λ:

$$\frac{\tau}{\rho} = CZ^m \lambda^n, \tag{7.9}$$

where C is a constant, and m and n are coefficients which vary for different substances and radiation energies. For $\lambda \leqslant \lambda_K$ (λ_K is the wavelength of K series) the expression becomes:

$$\frac{\tau}{\rho} = 0.0089 \frac{Z^{4.1}}{A} \lambda^n. \tag{7.10}$$

For the elements H, C and O, $n = 3.05$, while for elements where $11 < Z < 26$, $n = 2.85$. An approximate value:

$$\frac{\tau}{\rho} = CZ^4 \lambda^3, \tag{7.11}$$

is usually adopted, where C is a function of the wavelength.

Since the expressions

$$\mu_e = \frac{\mu}{\rho} \frac{A}{NZ}, \quad \text{and} \quad \tau_e = C_1(Z\lambda)^3 \tag{7.12}$$

relate the different absorption coefficients (where A is the atomic weight, N is Avogadro's number, μ_e (cm²/electron) is the beam attenuation coefficient per electron, ρ is the density and Z the atomic number), the following relationship can be written:

$$\frac{\tau}{\rho} = \frac{Z}{A} N(\tau_e)_c$$

or in a logarithmic form:

$$\log [N(\tau_e)_c] = 3 \log (Z\lambda) + \log (CN). \tag{7.13}$$

Now $C = 2.64 \times 10^{-26}$ and $N = 6.023 \times 10^{23}$, therefore the final form of eqn. (7.13) will be:

$$\log [N(\tau_e)_c] = 3 \log (Z\lambda) - 1.78. \tag{7.14}$$

From this equation and experimental data, Mohler [9] compiled a table of values of $N(\tau_e)_c$ for various $Z\lambda$ and plotted curves of these expressions which are helpful for quick calculation of the coefficient τ/ρ. Mohler's curves give good agreement for $100 < Z\lambda < 160$, while for other values, discrepancies of the order of 10–20% are observed. Over the range $5 < Z\lambda < 100$ better results can be obtained from the equation:

$$N(\tau_e)_c = 0.023(Z\lambda)^{2.78}. \tag{7.15}$$

Nomograms based on the above treatment, which enable values of the coefficient τ/ρ to be calculated quickly, have also been compiled.

Table 7.1. Mass absorption coefficients for a number of substances for wavelengths from 0·150–0·700 Å

Wavelength	Hydrogen	Carbon	Oxygen	Water		Hexane	Aluminium	Air
	Victoreen	Victoreen	Victoreen	Victoreen	(7.16)	Victoreen	Victoreen	Victoreen
0·150	0·308	0·159	0·166	0·182	0·172	0·183	0·199	0·164
0·200	0·326	0·174	0·190	0·205	0·192	0·199	0·274	0·186
0·250	0·337	0·188	0·219	0·232	0·214	0·213	0·388	0·212
0·300	0·350	0·209	0·261	0·271	0·242	0·232	0·557	0·250
0·350	0·355	0·231	0·317	0·320	0·280	0·251	0·786	—
0·400	0·360	0·259	0·383	0·380	0·330	0·275	1·087	0·356
0·450	0·363	0·295	0·470	0·458	0·393	0·306	1·472	—
0·500	0·366	0·336	0·577	0·553	0·473	0·341	1·950	0·525
0·550	0·369	0·389	0·715	0·672	0·571	0·386	2·532	—
0·600	0·374	0·449	0·865	0·810	0·691	0·437	3·226	0·775
0·650	0·376	0·519	1·405	0·980	0·834	0·496	4·043	—
0·700	0·379	0·605	1·26	1·180	1·004	0·568	4·993	—

Substance and equation used

For low-energy radiation, good agreement with experiment is obtained when Victoreen's [17] equations are used:

$$\left.\begin{array}{l} \dfrac{\tau}{\rho} = C\lambda^3 - D\lambda^4 \quad \text{for} \quad \lambda < \lambda_c, \\[4mm] \dfrac{\mu}{\rho} = C\lambda^3 - D\lambda^4 + \sigma_e N \dfrac{Z}{A} \quad \text{for} \quad \lambda < \lambda_c. \end{array}\right\} \tag{7.16}$$

The values for the coefficients using these equations are given in table 7.1.

7.4. *The Absorption Coefficient for Liquids.* When evaluating the mass absorption coefficient for compounds, and not for individual elements, use is made of the very important additive property of the coefficients for each atom in the molecule. It is sufficient to know the number p_i of atoms of each element in 1 g of the given substance and the values of μ_i/ρ_i for the individual atoms in order to evaluate the mass coefficient μ/ρ for the compound as a whole. Thus

$$\frac{\mu}{\rho} = \sum_i \frac{\mu_i}{\rho_i} p_i, \tag{7.17}$$

where

$$p_i = \frac{dN}{A_i}.$$

(d—the ratio of atomic to molecular mass.)

The ratio μ/ρ for hexane at $\lambda = 0.5$ Å can be evaluated as follows:

$$\begin{aligned} \left(\frac{\mu}{\rho}\right)_{C_6H_{14}} &= \left(\frac{\mu_1}{\rho_1}\right)_H \frac{14\cdot112}{86\cdot172} + \left(\frac{\mu_2}{\rho_2}\right)_C \frac{72\cdot06}{86\cdot172} \\[2mm] &= 0\cdot366 \times 0\cdot164 + 0\cdot336 \times 0\cdot836 \\[2mm] &= 0\cdot341 \text{ cm}^2\,\text{g}^{-1}. \end{aligned} \tag{7.18}$$

Coefficients for various wavelengths have been evaluated using this method and compiled in table 7.1; changes in radiation intensity after passing through a layer of a thickness d are shown in fig. 7.7.

When considering x-rays, it should be remembered that the general equation for absorption $I = I_0 e^{-\mu d}$ is true only for monoenergetic x-rays. The continuous x-ray spectrum has a blurred beam ranging from $\lambda_m = (12\,398/U)$ Å up to long wavelengths and displays a maximum. For this spectrum a correction to the absorption coefficient should be introduced in order to account for the thickness of the absorbent. An equation in a logarithmic form is then obtained:

$$\ln I = \ln I_0 - \mu_{\text{ef}} d, \tag{7.19}$$

where μ_{ef} is the effective absorption coefficient corresponding to the absorption coefficient μ for monochromatic radiation which gives the same attenuation in a given substance, and d is the absorbent layer thickness.

From both theoretical considerations and experimental work it has been found that μ_{ef} decreases continuously as the absorbent layer thickness increases, and approaches a certain limiting value for which the radiation beam is almost monochromatic. Under such conditions, eqn. (7.19) gives a straight line, i.e. indicates a constant value of μ_{ef}.

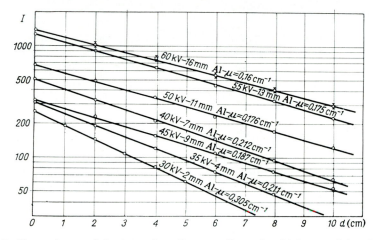

Fig. 7.7. Dependence of the intensity of x-rays, I, on the thickness of the absorbing layer d for hexane. Ref.: Kasztelan [18].

This can be explained by the fact that the long wavelength radiation is absorbed more rapidly than are short wavelengths; the use of filters in x-rays tubes is also based upon this phenomenon. For voltages from 6 to 200 kv, aluminium filters are most commonly used. Figure 7.8 shows the distribution of the x-ray spectrum, E_λ, using filters of various thicknesses.

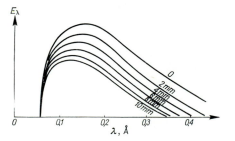

Fig. 7.8. Dependence of energy E_λ in the continuous x-ray spectrum on wavelength λ. The various curves correspond to the different thicknesses of aluminium filter used (0–10 mm).

A quantity λ_{ef}, defined as the wavelength of the monochromatic radiation wave that alters its intensity in the same way as polychromatic radiation, is connected with the coefficient μ_{ef}. Table 7.2 gives values of the mass absorption coefficient μ/ρ for water and hexane at various x-ray tube voltages and various filter thicknesses measured experimentally by

Kasztelan [18]. The data in the table indicate that the amount of energy absorbed is of very great importance as regards ionizing radiation dosimetry in liquids.

Table 7.2. Mass absorption coefficients μ/ρ for water and hexane at a number of x-ray energies

Tube voltage (kv)	Filter thickness (mm)	λ_{ef} (Å)	μ/ρ (cm² g⁻¹)	
			Water	Hexane
30	2	0·618	0·743	0·462
35	4	0·538	0·559	0·370
40	7	0·474	0·460	0·321
45	9	0·424	0·376	0·283
50	11	0·392	0·333	0·266
55	13	0·376	0·304	0·265
60	16	0·354	0·288	0·242

Chapter 8

Principles of Ionizing Radiation Dosimetry

8.1. *Dosimetry Units*. The effects of ionizing radiation on matter, described in the preceding sections, form the basis of dosimetry, which deals with the quantitative measurement of these radiation effects. The most important quantities encountered in dosimetry are: exposure dose or exposure (units are C/kg or the röntgen (abbreviated to R)), absorbed dose (unit: the rad), dose rate or rate of exposure (units: C/kg sec, R/h or μR/sec), and the absorbed dose rate (units: watts/kg, rad/h or μrad/sec).†

The unit of exposure dose, 1 R, corresponds to $2 \cdot 08 \times 10^9$ ion pairs produced in 1 sec from 1 cm³ of dry air (under standard conditions). The charge on this number of ion pairs is 1 e.s.u. When converted to 1 g of air, 1 R is equivalent to $1 \cdot 6 \times 10^{12}$ ion pairs (1 R corresponds to an energy of $0 \cdot 11$ erg). The unit of absorbed dose is joule/kg

$$(1 \text{ rad} = 100 \text{ ergs/g} = 10^{-2} \text{ J/kg}).$$

Assuming that the mean energy required to produce one ion pair in air is 34 ev, we obtain:

$$1 \text{ röntgen} = 0 \cdot 877 \text{ rad}. \tag{8.1}$$

C/kg, J/kg, röntgen and rad are the basic units admitted by the International Commission on Radiological Units (I.C.R.U.). Other units for specific quantities are also used, e.g. for corpuscular emission the unit is the rep (röntgen equivalent physical), and for tissue it is the rem (röntgen equivalent man). For x, γ and β-rays, 1 R = 1 rep = 1 rem, although this may not be true in the case of corpuscular emission (neutrons, protons, α particles, etc.).‡

The absorbed dose is measured in rads for all processes which occur when the energy of the ionizing radiation is absorbed by the substance. The exposure dose is only measured in röntgens when the energy absorbed

---

† 1 R = $2 \cdot 58 \times 10^{-5}$ C/kg; 1 rad = 10^{-2} J/kg; 1 rad/sec = 10^{-2} w/kg.

‡ In 1962, the I.C.R.U. suggested the introduction of a new quantity, the kerma (kinetic energy released in material) denoted by the letter κ. If ΔE_K is the sum of the initial kinetic energies of all the charged particles liberated by indirectly ionizing particles (neutrons, photons, etc.) in a volume element of the specified material, and Δm is the mass of the matter in that volume element, then kerma $\kappa = \Delta E_K / \Delta m$. It is only related to the 'first collision dose' (often used in work with neutrons) and not to the total absorbed dose, as it does not include the energy due to secondary particles. Only under equilibrium conditions (see § 17.1) are the first collision dose and total absorbed dose identical.

is considered to manifest itself as the number of ion pairs measured by a dosimeter. The total energy absorbed by air, if divided by the number of ion pairs measured in the substance, yields the average energy per ion pair equal to 34 ev; it is clear from table 8.1 that this energy considerably exceeds the true ionization energy (10–20 ev).

8.2. *Calculation of the Dose*.   Calculation of the dose in the neighbourhood of the radiation source is not usually easy since it is necessary to take into account the kind of radiation, its intensity distribution as a function of the energy of the quanta or particles, and the geometry of the radiation source. The extensive practical uses of dosimetry in physics, engineering and medicine, however, have made it possible to measure doses from a variety of sources with fair accuracy. General methods for dose calculation, with common examples, are described below. Calculations are usually performed for x, γ and β-ray sources.

Table 8.1. Ionization energy P and average energy W required to form one ion pair

Gas	W per ion pair (ev)	Ionization energy P (ev)	Difference $(W - P)$ (ev)
Argon (A)	36·4	15·68	20·72
Hydrogen (H_2)	36·3	15·6	20·7
Helium (He)	42·3	24·46	17·84
Krypton (Kr)		13·93	
Neon (Ne)		21·47	
Xenon (Xe)		12·08	
Nitrogen (N_2)	34·9	15·51	19·39
Oxygen (O_2)	30·9	12·5	18·4
Chlorine (Cl_2)		13·2	
Fluorine (F_2)		17·8	
Bromine (Br_2)		12·8	
Iodine (I_2)		9·7	
Ammonia (NH_3)		11·2	
Carbon dioxide (CO_2)		14·4	
Carbon disulphide (CS_2)		10·4	
Carbon monoxide (CO)		14·1	
Acetylene (C_2H_2)		11·6	
Ethylene (C_2H_4)		12·2	
Ethane (C_2H_6)	24·7	12·8	11·9
Methane (CH_4)	27·3	14·5	12·8
Mercury vapour (Hg)		10·39	
Water vapour H_2O	30·3	12·56	17·74
n-Pentane		10·55	
n-Hexane		10·43	
n-Decane		10·19	

The simplest cases are those involving a point source of γ-rays in air both with and without protective shielding. It is well known that γ-rays are extremely penetrating and interact with matter even at a large distance from the source. The distance factor is indeed very important, as the radiation intensity varies inversely as the square of the distance from the source. The inverse square law applies here since the total radiation energy (which can be expressed by the number of quanta emitted from the source during a period of 1 sec) passes uniformly through the whole surface of the sphere at whose centre the source is situated. If the source is of the point type and isotropic, the radiation energy density is distributed uniformly over the sphere and is proportional to $\frac{1}{4}\pi R^2$; it consequently decreases in inverse ratio to the square of the distance. If the source is anisotropic, the situation is similar, but for each of the individual solid angles.

The radiation or exposure dose at any given point is proportional to the activity of the source Q, the specific radiation constant or ionization constant Γ, and the irradiation time t; it is inversely proportional to the square of the distance from the source R:

$$D = \frac{Q\Gamma t}{R^2} \quad \text{(röntgens)}. \tag{8.2}$$

The units are as follows: Q is in millicuries, t in seconds or hours and R in centimetres or metres. The determination of D presents no great difficulties as regards measuring techniques. Determining the absolute value of the specific radiation constant Γ, on the other hand, is fairly difficult. It requires knowledge of the pattern of radioactive disintegration of the element, the energy and number of the quanta emitted by it, and the true absorption coefficient for each quantum, i.e. the number of electrons produced by the given quantum in air over a 1 cm path.

Tables and graphs are available for reference purposes which supply ionization constants for various radioactive elements. It follows from eqn. (8.2) that an ionization constant has units of (röntgen) (metres)²/ (hours) (curies), and the ionization constant is most usually defined as the dose (in röntgens) given out per hour by a source of activity 1 mCi at a distance of 1 m. A unit for ionization constant very frequently encountered in the literature is Rhm. For a radium preparation in a 0·5 mm platinum shield, the ionization constant is 8·4 R cm²/mg h (röntgens per mg of radium per hour), for ^{60}Co it is 13·2, for ^{198}Au it is 2·44 and for ^{125}Sb it is 2·75, etc.

8.3. *Dose Rate.* The ratio of the dose to the time for which the dose is received is known as the dose rate; if it does not vary with time, one may write:

$$M = \frac{D}{t} = \frac{Q\Gamma}{R^2} \quad \text{(röntgens/h)}. \tag{8.3}$$

If the half-life period of the element is comparable to the irradiation period, the quantity Q will be a function of time and M must be replaced by the mean dose rate $M(t)$:

$$M(t) = \frac{Q_0\Gamma}{R^2}\exp\left(-0.693t/T\right). \tag{8.4}$$

The total dose will then be:

$$D(t) = \frac{1 \cdot 44 Q_0 \Gamma T}{R^2} [1 - \exp(-0 \cdot 693 t/T)]. \tag{8.5}$$

For irradiation periods which are long compared with the half-life period $(t \gg T)$, eqn. (8.5) becomes:

$$D(\infty) = \frac{1 \cdot 44 Q_0 \Gamma T}{R^2}. \tag{8.6}$$

The following relationships exist between the various units for the dose rate:

$$1 \ \mu\text{R/sec} = 0 \cdot 06 \ \text{mR/min} = 3 \cdot 6 \ \text{mR/h} = 6 \times 10^{-5} \ \text{R/min} = 3 \cdot 6 \times 10^{-3} \ \text{R/h};$$

also

$$1 \ \text{R/h} = 10^3 \ \text{mR/h} = 16 \cdot 7 \ \text{mR/min} = 280 \ \mu\text{R/sec}.$$

The treatment so far has not considered the radiation absorbed in the shield of the radioactive preparation. It is accounted for by multiplying the activity of the source by a factor:

$$\exp(-\mu x), \tag{8.7}$$

where x is the thickness of the shield and μ is the absorption coefficient (strictly, attenuation coefficient) of the substance from which the shield is made. The equation for the dose rate from a continuous point source then becomes:

$$M = \frac{Q\Gamma}{R^2} \exp(-\mu x) \quad (\text{röntgens/h}). \tag{8.8}$$

8.4. *Other Dosimetric Methods.* Many dosimetric methods are based on the measurement of effects connected with ionization of the atoms and molecules of the detector, but a number of others are also in common use. These include methods based on colour changes, pH changes, the numbers of ferrous and ferric ions, optical density, absorption band shifts, light scintillation, heat generated in the detector, etc. They are described in detail in books on dosimetry; liquid dosimeters are dealt with in §17.2 of this book.

The accepted manner of expressing changes in the concentration of the various ions due to radiation is by the radiation yield G, which is the number of particles or ions formed during the absorption of 100 ev of radiation energy. Using chemical dosimeters, it has been found that the radiation yield is about 15·5 over the range 0·6–20 MeV. The radiation dose D is then given by:

$$D = 0 \cdot 964 \times 10^9 \frac{C}{G\rho} \quad (\text{rads}), \tag{8.9}$$

where C is the measured ion concentration (in mole/liter), and ρ is the density of the detector substance.

For large doses of the order of megarads, Rotblat *et al.* [20] stated that the relationship of the dose to the linear change in optical density ρ_0 of substances such as polyethylene terephthalate (Melinex), polyethylene methacrylate (Perspex), Lucite and Mylar has the following form:

$$D = a + b \frac{\rho_0}{h},\tag{8.10}$$

where h is the thickness of the detector plate, and a and b are constants. For Perspex, $a = 0 \cdot 045$ and $b = 3 \cdot 42$.

One of the most accurate ways of measuring the absolute value of absorbed dose is by the microcalorimetric method. This consists of measuring the quantity of heat generated by radiation in the detector substance. The quantities are very small (of the order of $0 \cdot 01$–$0 \cdot 1$ cal/h for preparations of the curie and millicurie order). For corpuscular emission with a mean energy of E Mev, it was found (Price [5]) that A(Ci) of the radioactive substance emits:

$$H \text{ (watts)} = (5 \cdot 94 \times 10^{-3}) \, A \text{ (Ci)} \, \bar{E} \text{ (Mev)}\tag{8.11}$$

Thermoluminescence methods (LiF, TLD) and methods based on the semi-conductor properties of solids (CdS, Si junction) have recently come to be widely applied. Table 8.2 shows the dose range covered by the various methods.

Table 8.2. Dose ranges covered by various types of dosimeter. Dose scale: rads $\times 10^x$ (Ref.: Boag [80])

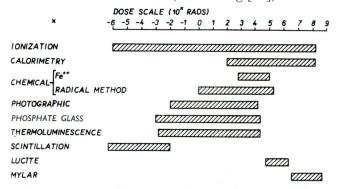

According to the 1966 I.C.I.P. directives, the following units are in general use in health physics: dose equivalent (DE), dose quality factor (QF), distribution factor (DF) and röntgen biological equivalent (RBE). This last unit is chiefly used in radiobiology, but cannot be measured directly. A quality factor, $\text{QF} = f(\text{LET}_\infty)$, is therefore used, which can be measured experimentally. A general correlation exists:

$$\text{DE} = \text{D(QF)(DF)} = \text{D(QF)} \, nk,$$

where n depends on the kind of radiation (e.g. for γ-rays, $n = 1$, for α particles, $n = 5$) and k depends on the organ undergoing irradiation. Section 17.2 of this book contains descriptions of certain special dosimetry methods.

Chapter 9

Mechanism of the Interaction of Ionizing Radiation in Liquids

9.1. *Types of Interaction Effect.* The following phenomena are able to yield information as to the mechanism of the action of ionizing radiation on liquids, particularly dielectric liquids:

(1) Changes in the physicochemical properties of irradiated liquids, e.g. density, viscosity, luminescence and absorption spectra; radiation yield of individual decay products, etc.

(2) Electrical conductivity in low electric fields, and particularly (*a*) current-voltage characteristics in ionized liquids; (*b*) measurement of the coefficients of mobility, recombination and diffusion of electrons and ions, and their dependence on molecular structure; (*c*) temperature dependence of the coefficients in (*b*), and especially their activation energy.

(3) Cathode-to-liquid photocurrents caused by ultraviolet radiation.

(4) Electrical conductivity in high electrical fields.

(5) Value of the breakdown stress in the liquid.

(6) Nuclear magnetic resonance and electron spin resonance in ionized liquids.

There is similarly a large amount of published work on research into the ionization of gases, but this must be approached with caution as the same reasoning cannot necessarily be applied to both gases and liquids. In many cases liquids display a greater similarity to solids rather than to gases, particularly when they are considered to have a close-ordered pseudocrystalline structure consisting of small areas composed of a large number of molecules.

This chapter is concerned with attempts to correlate experimental data collected for dielectric liquids with established theories of the physics of liquids, gases and solids. A number of new theories are put forward in the place of those which have been found unsatisfactory so far. Certain of them could form a suitable basis for further experimental work in this field.

The effects of x and γ-rays are of prime importance in work on liquid dielectrics. It is generally known that the phenomena which occur due to the action of primary and secondary photons, photoelectrons, Compton electrons, electron pairs and scattered electromagnetic radiation are as follows:

(1) Ionization of atoms and molecules (detachment of electrons from the atom or molecule).

(2) Excitation of atoms and molecules to higher energy states; electronic, oscillation, vibration and rotation states.

(3) Dissociation of molecules into neutral particles or ions.

(4) Formation of free radicals (i.e. electrically neutral molecular particles with unsaturated electron pairs) and a number of secondary processes using these radicals (e.g. oxidation and reduction).

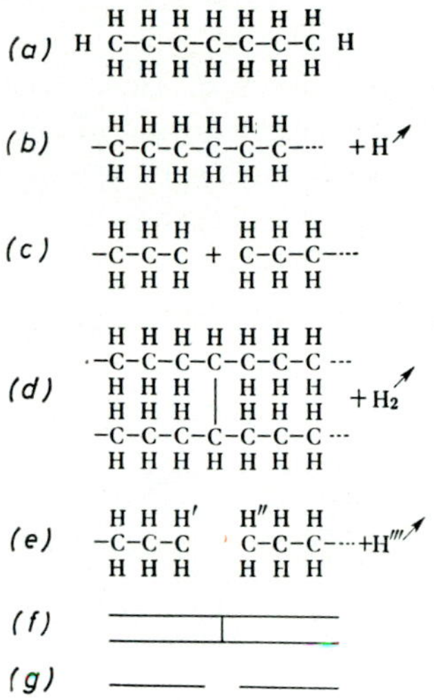

```
             H H H H H H H
(a)      H  C–C–C–C–C–C–C  H
             H H H H H H H

             H H H H H H
(b)        –C–C–C–C–C–C–---    +H
             H H H H H H

             H H H      H H H
(c)        –C–C–C  +  C–C–C–---
             H H H      H H H

             H H H H H H H
           ·–C–C–C–C–C–C–C ···
(d)          H H H | H H H      +H₂
             H H H | H H H
           –C–C–C–C–C–C–C ···
             H H H H H H H

             H H H'     H"H H
(e)        –C–C–C       C–C–C–----+H'''
             H H H       H H H

(f)        ═══════════╤═══════════

(g)        ─────────    ─────────
```

Fig. 9.1. Radiation effects on hydrocarbon molecules: (*a*) molecule in the normal state; (*b*) detachment of a hydrocarbon atom; (*c*) fracture of the molecule in the centre; (*d*) cross linking between two molecules with detachment of a hydrogen molecule; (*e*) degradation, (*f*) representation of cross-linking, (*g*) representation of chain fracture. Ref.: Charlesby [8].

Secondary effects include cross linking of whole molecules or parts of molecules, radical combination with atmospheric oxygen, oxygen in the liquid or other impurities in the liquid, polymerization, formation of free gases (e.g. hydrogen or methane), formation of acids, salts, alcohols, esters, etc. Detailed description and discussions of these phenomena have been published by Charlesby [8], Mohler [9], Lea [16], Chapiro [21], Laidler [22], Herzberg [23, 24], Massey and Burhop [25], Essex *et al.* [26, 27] and Haïssinsky *et al.* [28, 29]. Figure 9.1 shows some of the secondary reactions possible due to irradiation of a hydrocarbon molecule.

9.2. *Energy Levels in a Molecule. Excitation and Ionization.* An analysis of the energy relationships which occur in a molecule in its normal state and in excited states is of assistance towards understanding the phenomena which takes place when ionizing radiation interacts with matter. For simplicity, only a diatomic molecule will be considered here. The energy of a molecule can be divided into three parts connected with (*a*) the distribution of electrons, (*b*) the oscillatory motion of atomic nuclei about their equilibrium positions and (*c*) the rotatory motion of the whole molecule or part of the molecule. The three types of energy are of the order of 1, 0·1 and 0·01 ev, respectively.

An exact theoretical analysis of the problem is fairly difficult, but Born and Oppenheimer [30] rearranged the Schrödinger wave equation in the form (cf. Mott [31]):

$$\frac{\hbar^2}{2M_1}\nabla_1^2\psi + \frac{\hbar^2}{2M_2}\nabla_2^2\psi + [W - V(R)]\psi = 0, \qquad (9.1)$$

where M_1 and M_2 are the masses of two atomic nuclei, $\hbar$ is Planck's constant ($\hbar = h/2\pi$), and $V(R)$ is the potential energy of the molecule; for a state n, $W_n(R)$ defines energy values corresponding to the wave functions $\psi_n(R_1, R_2, q)$, and R_1 and R_2 are the coordinates of the positions of the atomic nuclei when at rest; further,

$$V(R) = Z_1 Z_2 e^2/R + W_n(R).$$

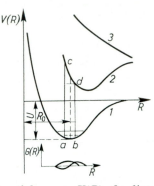

Fig. 9.2. Distribution of potential energy $V(R)$ of a diatomic molecule as a function of the distance R between the atoms (simplest forms): (1) in the ground state, (2) and (3) in excited states. $G(R)$ is a wave function of excited states. Ref.: Mott [31].

For values of R close to R_0, the function $V(R)$ can be written

$$V(R) = U\left[-1 + \frac{a(R-R_0)^2}{R_0^2}\right], \qquad (9.2)$$

where a is a non-dimensional constant and U is the dissociation energy of the molecule. (U is of the same order as atomic excitation potentials: 2–6 ev.)

Figure 9.2 shows the function $V(R)$ plotted against R. Each of the three curves is for a different state of the molecule: (1) the ground state, (2) the excited state of the stable molecule, and (3) an excited state in which the molecule cannot exist. The lower curves are of the wave functions of the excited states, which functions describe the oscillations of atoms in the molecule. With the molecule in the ground state, the electrons are on the quantized energy levels (horizontal lines) on curve 1; when in the excited state, they are on the corresponding levels on curve 2. External radiation energy can cause an electron to pass from one energy level a or b on curve 1 to a level c or d on curve 2, the energy difference in both cases being strictly determined by the Franck–Condon rule.

A molecule can revert from the excited state to the ground state either by emitting a light quantum (fluorescence) or by transferring the excitation energy to the oscillatory motion of the atoms. A third possibility is that it would undergo dissociation, but whichever of the three routes is taken depends largely on the state of oscillation at which the excitation takes place. The average duration of electron excitation in the molecule is of the order of 10^{-8} sec, i.e. long compared with the period of molecular vibrations (10^{-13} sec).

The solution of eqn. (9.1) yields the energy W in the form:

$$W = -U + \frac{\hbar^2 l(l+1)}{2MR_0^2} + h\nu(n + \tfrac{1}{2}), \qquad (9.3)$$

where n is an integer, and

$$\nu = \frac{1}{2\pi}\sqrt{\left(\frac{p}{M}\right)} = \frac{1}{2\pi}\sqrt{\left(\frac{2U_a}{MR_0^2}\right)}.$$

Also

$$M = M_1 M_2/(M_1 + M_2).$$

The term $\hbar^2 l(l+1)/2MR_0^2$ in eqn. (9.3) is the rotational energy of the molecule, where MR_0^2 is the moment of inertia and $l\hbar$ is the moment of momentum; $h\nu(n+\tfrac{1}{2})$ is the energy of the oscillator vibrations.

Various forms of molecular excitation and ways in which they are quenched (including dissociation) can occur depending on the type of molecule and the quantum energy of the incident radiation. The laws of radiation of excited molecules and transitions between the potential energy curves of the molecule in the excited and ground states are governed by the Franck–Condon rule and the Wigner rule. The former states that the most probable transitions are those which involve the smallest changes in distance between atomic nuclei in the molecule; the Wigner rule is connected with spin behaviour.

As a rule, photochemical excitation, quenching and dissociation in polyatomic molecules proceed similarly as in diatomic molecules, but they are more complicated. The two-dimensional potential energy curves referred to above become multi-dimensional potential surfaces which intersect each other, giving a high chance of predissociation. Further, the concentration of rotation and oscillation levels may bring about internal conversion of the excitation energy, increasing oscillatory and rotatory motion instead of electron excitation energy. It can even happen that the total energy is absorbed by only one bond. For this reason, the phenomena which occur during photochemical excitation of polyatomic molecules have not been studied in any great depth. A more detailed discussion, however, can be found in publications by Laidler [22], Herzberg [23, 24], etc.

According to Mohler [32], the phenomena occurring during the ionization and excitation of molecules can be represented diagrammatically as
follows:

(1) Excitation:

$$\rightsquigarrow AB \rightarrow AB^* \overset{h\nu}{\rightsquigarrow} AB \Big\rangle , \quad \text{fluorescence,}$$

$$\rightsquigarrow AB \rightarrow AB^* \Big\langle \begin{array}{l} \overset{h\nu}{\nearrow} AB \\ \searrow A' + B'' \end{array} \Big\rangle , \quad \text{dissociation,}$$

$$\rightsquigarrow AB \rightarrow AB^* \rightarrow A' + B'' \Big\rangle$$

$$\tag{9.4}$$

where AB^* denotes the molecule in the excited state, and A' and B'' are
the fragments obtained from it, possessing kinetic energy in an excited
electron state.

(2) Ionization:

$$\rightsquigarrow AB \rightarrow AB^+ + e, \quad \text{primary ionization,}$$

$$AB^+ \Big\langle \begin{array}{llll} \nearrow A^+ + B & \text{or} & A^+ + B'' \\ \searrow A + B^+ & \text{or} & A' + B^+ \end{array} \Big\rangle , \quad \text{dissociation.}$$

$$AB + e \rightarrow AB^- \Big\langle \begin{array}{llll} \nearrow A + B^- & \text{or} & A' + B^- \\ \searrow A^- + B & \text{or} & A^- + B'' \end{array} \Big\rangle , \quad \text{dissociation,}$$

$$\tag{9.5}$$

The products formed during these processes may change as a result of
mutual collisions (known as collisions of the second kind), undergoing
either exclusively physical changes, or chemical changes in which new
substances are formed. The former (physical) type of change can be
represented as:

$$\left. \begin{array}{ll} A' + e \rightarrow A + \overset{\uparrow}{e}, & \\ A' + B \rightarrow A + \overset{\uparrow}{B}, & \\ A' + B \rightarrow A + B'', & E_A > E_B, \\ A' + B \rightarrow A^* + B'', & E_A > (E_{A^*} + E_B), \\ A' + B \rightarrow A + B^+ + e, & E_A > E_B, \end{array} \right\}$$

$$\tag{9.6}$$

where E_A and E_B are the excitation energies of A' and B'', and E_{B^+} is the
ionization energy of B^+. The other (chemical) type of changes can be
represented as:

$$\left. \begin{array}{ll} A' + B \rightarrow AB^+ + e, & \\ A' + BC \rightarrow A + B + C, & E_A > E_d(BC), \\ A' + BC \rightarrow AB + C, & [E_A + E_d(AB)] > E_d(BC), \\ A' + B + M \rightarrow AB + \overset{\uparrow}{S} & \end{array} \right\}$$

$$\tag{9.7}$$

(S is a third substance which prevents the dissociation of AB).

These reactions between electrically neutral molecules are only a portion of the total number of processes which take place during ionization of a substance. Other important processes are those which occur between free electrons and positive ions, those between ions of opposite and identical sign (e.g. ion recombination), Coulomb repulsion, diffusion, and adhesion and detachment of electrons from neutral molecules. These have been discussed in detail by a number of authors, e.g. Laidler [22] and Massey and Burhop [25]. Sections 9.3 and 13.1 of this book also deal with the subject, but in this section it is only necessary to be acquainted with the simpler aspects, e.g. that as a result of an ionic recombination process (which has a pronouncedly exothermic character), excited molecules may be formed, which, in turn, may undergo the reactions described above. Besides the ionized fragments present in an irradiated substance, many electrically neutral fragments may be found, which nevertheless possess a high chemical activity due to the spins of the valency electrons being unsaturated. They are known as radicals and play an important rôle in the formation of new molecules.

The ionization processes outlined above are encountered in all states of matter, but have mainly been studied for the gaseous phase. In liquids, the phenomena become complicated insofar as, owing to the strong interaction of neighbouring molecules, each molecule which receives an energy quantum from the ionizing radiation may distribute it among the neighbouring molecules before undergoing dissociation and ionization. The time required for the excitation or ionization of a molecule is of the order of 10^{-15} sec, while the period of internal vibrations of a molecule is about 10^{-13} sec; the forces of intermolecular interaction, however, act far more rapidly. Neutral fragments of dissociated molecules and free radicals may either recombine to form the same molecules as previously, or else form completely new molecules of a different size; this could take place through the addition of molecules of other substances, e.g. oxygen from the atmosphere. Moreover, as a result of thermal motion of the molecules, electrons and ions may migrate from places where their volume concentration is higher to those where there is a lower concentration (ionic diffusion). The fate of free electrons produced by ionization has not yet been studied exhaustively.

It has been found that after a certain number of elastic collisions with molecules, electrons can combine with one of the molecules, forming a negative ion. During an elastic collision, the kinetic energy of the electron or molecule remains unchanged, and only the directions of velocities of the colliding particles change. During an inelastic collision, excitation or ionization of the atom or molecule occurs, and a sharp slowing down of the moving electron takes place. Dissociation of the molecule into electrically neutral parts, or even into ions, may also occur. Such processes are particularly common in a mixture of dissimilar atoms or molecules, when an inelastic collision between an excited atom of one substance and a molecule of another substance may cause the latter to ionize or dissociate.

Hart and Boag [33–36] have demonstrated the presence of 'hydrated electrons' in ionized water, which for other substances are known as 'solvated electrons'. These are secondary electrons liberated during ionization, which undergo a number of collisions with the surrounding molecules and then link up with a small group of molecules after polarizing it. Since the investigations were mainly concerned with the mechanism in water, they are discussed in more detail in § 9.3.

According to Magee [37], when electrons with an energy of 10^4–10^6 ev are formed in a substance, the number of elastic collisions is about 5% of the total number of collisions; the number of inelastic collisions leading to ionization is about 35%, and the number of collisions resulting in excitation is about 60%. As a result of additional collisions between electrons, ions or even molecules of neutral substances, the number of ionization processes may be increased by another 40%. Secondary electrons formed during the ionization processes usually have very low kinetic energies.

9.3. *The Radiolysis of Water.* The mechanism of the radiolysis of water has been studied with considerable care since it was discovered about 60 years ago, and in view of its importance in radiobiology, a large amount of research has been devoted to the subject. It should be stressed, however, that the mechanism has not yet been completely elucidated, and there are several hypotheses which attempt to explain the formation of the radiolysis products observed experimentally.

More comprehensive information connected with the problem can be found in publications by Ciborowski [38, 39], Allen [40], Charlesby [8, 41] and Kroh [42]. A comprehensive treatise on the radiolysis of water and aqueous solutions has also been published by Allen [43]; of equal importance are papers by Hart and Boag [33–36].

It is known that the water molecule contains a total of 10 electrons, 2 of which are in the K orbit of the oxygen atom; the remaining 8 are in internal shells, four of these forming two O—H bonds. Depending on which of these electrons becomes detached from the water molecule during the ionization process, either an H_2O^+ ion or H^+ or OH^+ ions will result. It is assumed that these are formed from the 8 external electrons in roughly equal proportions. The formation of H^+ or OH^+ ions indicates that the molecule dissociates. When subjected to ionizing radiation, water molecules may undergo excitation, ionization and decomposition. Ionization may be represented as follows:

$$\left. \begin{aligned} \rightsquigarrow H_2O \rightarrow H_2O^+ + e^-, \\ \rightsquigarrow H_2O \rightarrow OH^+ + H + e^-, \\ \rightsquigarrow H_2O \rightarrow H^+ + e^- + OH. \end{aligned} \right\} \qquad (9.8)$$

The free electrons arising during ionization may, after a number of collisions, recombine with the positive ion to give an electrically neutral molecule in an excited state:

$$e^- + H_2O^+ \rightarrow (H_2O)^*; \qquad (9.9)$$

otherwise they could link up with a neutral water molecule, yielding a negative ion:

$$e^- + H_2O \rightarrow H_2O^-. \qquad (9.9a)$$

This linking occurs most often in a resonance process, when the electron energy is 7·1 ev.

The positive and negative ions usually decay as follows:

$$\left.\begin{aligned}
H_2O^+ &\rightarrow H^+ + OH, \\
H_2O^- &\rightarrow H_2 + O^-, \\
H_2O^- &\rightarrow 2H + O^-, \\
H_2O^- &\rightarrow H + OH^-.
\end{aligned}\right\} \qquad (9.10)$$

There can also form even more complex ions (H_3O^+), for instance:

$$\left.\begin{aligned}
H_2O^+ + H_2O &\rightarrow H_3O^+ + OH, \\
H_2^+ + H_2O &\rightarrow H_3O^+ + H, \\
OH^+ + H_2O &\rightarrow H_3O^+ + O.
\end{aligned}\right\} \qquad (9.11)$$

Besides the positive and negative ions, excited molecules and radicals also arise during the radiolysis of water. Excited molecules are either formed directly in a primary process, mainly as a result of the action of δ electrons, or due to ion recombination. They usually decay into H and OH radicals:

$$(H_2O)^* \rightarrow H + OH.$$

The final stage of all the complex ionic changes in water is the formation of a number of products such as H, H_2, OH and OOH. These can react with one another in various ways, usually giving rise to the formation of H_2O_2 and O_2 according to the reactions:

$$\left.\begin{aligned}
&(1) \quad & H + OH &\rightarrow H_2O, \\
&(2) \quad & OH + OH &\rightarrow H_2O_2, \\
&(3) \quad & OH + OH &\rightarrow H_2O + O^*, \\
&(4) \quad & H + H &\rightarrow H_2, \\
&(5) \quad & H + O_2 &\rightarrow HO_2 \rightleftarrows H^+ + O_2^-, \\
&(6) \quad & HO_2 + HO_2 &\rightarrow H_2O_2 + O_2, \\
&(7) \quad & OH + HO_2 &\rightarrow H_2O_2 + O^*.
\end{aligned}\right\} \qquad (9.12)$$

The first of these is the regeneration of a water molecule. Reactions (2), (6) and (7) yield a hydrogen peroxide molecule, which is of great importance as far as biological effects are concerned. Reaction (2) may take place only when there is a considerable concentration of OH groups, for instance during ionization with α particles. Reaction (4) occurs when the hydrogen

7

atoms find no other combination partners. Reaction (5) takes place when the oxygen content of the solution is high, and reaction (6) occurs when the pH of the solution is low.

The hydrogen peroxide molecules can react with OH groups, or with hydrogen and oxygen ions, in the following manner:

$$H_2O_2 + OH \rightarrow H_2O + HO_2, \\ H_2O_2 + O^- + H^+ \rightarrow H_2O + OH + O. \qquad (9.13)$$

If the solution contains radicals of any dissolved substances, the following reactions may also take place:

$$H + R_1 \rightarrow H^+ + R_1^-, \\ OH + R_2 \rightarrow OH^- + R_2^+. \qquad (9.14)$$

The phenomenon of water radiolysis under different experimental conditions is studied quantitatively by determining the radiation yield $G(x)$ of each component (x). $G(x)$ denotes the radiation yield of component (x), i.e. the number of atoms, ions or molecules formed when 100 ev of radiation energy is absorbed. The radiolysis mechanism can then be written as:

$$G(H_2O) = G(H) + G(OH) + G(H_2O_2) + G(H_2). \qquad (9.15)$$

Kroh [42] and Allen [43] have determined the yield $G(x)$ for various fragments produced during the ionization of water as a function of the type and energy of radiation, e.g. for γ-rays with an energy of 0·5 Mev, for α particles with an energy of 5·3 Mev (from ^{210}Po), and for β particles with a maximum energy of 0·018 Mev (from tritium). The values obtained are shown in table 9.1. They satisfy the relationship:

$$G(H_2O) = 2G(H_2) + G(H) = 2G(H_2O_2) + G(OH). \qquad (9.15a)$$

Table 9.1.　Radiation yields $G(x)$ from the radiolysis of water

Type of radiation	$G(H)$	$G(OH)$	$G(H_2)$	$G(H_2O_2)$	$G(H_2O)$
γ-Rays (0·5 Mev)	3·64	2·86	0·48	0·87	4·6
β Particles (^{3}T)	2·90	2·10		1·0	—
α Particles (^{210}Po)	0·80	0·20	1·57	1·87	3·94

Kroh and Allen have pointed to the difficulties inherent in the determination of the amounts of certain radiolysis products, and have shown that there is a considerable dependence of $G(x)$ on LET values and that the ranges of individual ionizing particles are not of a particularly homogeneous character. They have also calculated similar yields for heavy water, D_2O.

Radicals of certain foreign substances in water have a much greater affinity for H and OH radicals, and therefore inhibit the reactions in (9.12), reducing the amounts of H_2, H_2O_2 and O_2 formed during the radiolytic process. Substances which behave in this manner are called radical scavengers. In water, radical scavengers include Fe^{++}, Br^-, Cu^{++}, NO_2^-, and Cl^- ions. The effect of these scavengers depends on their chemical properties and concentration. According to Dainton [44], the effect of radical scavengers during the radiolysis of water with γ-rays can be determined from the formula:

$$G'(m) = G(m) - kx^{1/3}, \qquad (9.16)$$

where $G'(m)$ is the radiation yield of products in the presence of the scavenger, $G(m)$ is this yield in the absence of the scavenger, x is the scavenger concentration and k is a constant characteristic of a particular scavenger and radical. It should be noted that the concentration of the individual radiolysis products is not established immediately, but that a certain period of time is required for equilibrium to be reached, expressed according to the formula:

$$S = A[1 + \exp(-kt)], \qquad (9.17)$$

where S is the final concentration, t is time and A and k are constants.

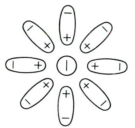

Fig. 9.3. Diagram of a negative polaron.

Elert and Boag [45] derived the following equation for the concentration $S(H_2O_2)$ in water irradiated with 1 MeV electrons:

$$S = 140[1 - \exp(-kx)] \quad (\mu mol/l), \qquad (9.18)$$

where x denotes the dose in erg/g and $k = 8.5 \times 10^{-8}$. The final value S calculated from this equation is 0.14 mmol/l, and the initial radiation yield is 0.38 particles/100 ev.

Another interesting theory of the mechanism of water ionization has been put forward by Weiss [46]. It is based on the polaron theory of Pekar [47] for explaining electron phenomena in ionic crystals. A polaron is a free electron produced by ionization and surrounded by the molecules of the medium, which molecules have been polarized by it (fig. 9.3). The radius of such a polaron is defined by the equation:

$$r \simeq \frac{10\hbar^2}{\kappa\mu e^2}, \qquad (9.19)$$

where $\kappa = 1/n^2 - 1/\varepsilon$, n is the refractive index, ε is the static dielectric constant, μ is the effective mass of the electron and $\hbar = h/2\pi$ (h is Planck's constant). For water $\kappa \approx 0.55$, and r is around 10 Å (if μ is equal to the rest mass). Weiss assumes that a polaron is relatively stable, and that it forms a trap for an electron. Besides negative polarons, positive polarons, whose centre is occupied by a positive ion, may also exist. Polarons can recombine with ions and polarons of opposite sign, just like normal ions, and can also form double systems known as bipolarons.

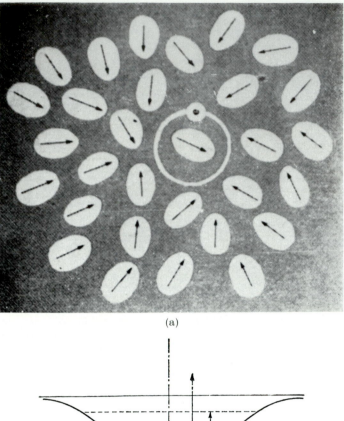

(a)

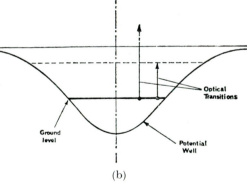

(b)

Fig. 9.4. Orientation of polar molecules around an electron. (*a*) Continuum model of the polaron; (*b*) energy distribution of the polaron. Ref.: Boag [56].

The lifetime of a polaron was estimated to be about 10^{-9}–10^{-11} sec; for this reason a polaron cannot be detected experimentally, and is a purely hypothetical entity. The orientations of polar molecules around an electron and the energy distribution of a polaron are shown diagrammatically in fig. 9.4.

Hart and Boag [33–36] and others were the first workers to demonstrate experimentally the existence of polarons in water and other dipolar substances. Their work enabled the properties and significance in radiolytic processes to be determined. The polarons were termed 'hydrated electrons'.

9.4. *Free and Solvated Electrons in Liquids.* A large amount of data on the lifetime of free electrons in liquids has recently been obtained by the pulsed ionization method (pulsed radiolysis) using times of the order of a few microseconds. This method permits the study of the mechanism of attachment of free electrons produced during ionization to various molecules, and detachment from these molecules. The method was chiefly developed by Boag *et al.* [48–53] and Keen [54, 55].

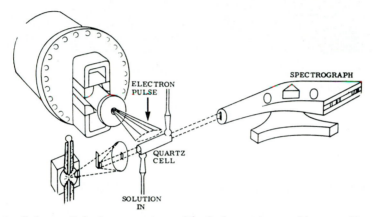

Fig. 9.5. Pulse radiolysis apparatus with flash spectrographic recording. Ref.: Boag [56].

The liquid to be examined is first irradiated with a beam of electrons from an accelerator with an energy of about 2 MeV in pulses lasting for about 2 μsec and, after a short time interval Δt ranging from a fraction of a μsec to several seconds, is illuminated with a light pulse from a flash lamp. The duration of the light pulse is around 5 μsec. After the light beam has passed through the ionized liquid, it is photographed on a spectrograph in order to study the absorption spectrum. Changes characteristic of the electrons solvated in the liquid appear in the absorption spectrum, which changes depend on the time Δt. Figure 9.5, taken from Boag's paper [56], shows the apparatus used in this type of work. Figures 9.6 and 9.7 show the time distribution of the electron and light pulses and the absorption spectrum in the liquid. The upper part of fig. 9.7 shows the spectrum prior to electron irradiation, and the lower part the spectra after irradiation at various times Δt. It can be seen from the centre portion that after a time $\Delta t = 0\cdot 5$ μsec, a broad absorption spectrum band caused by solvated electrons appears. This band disappears after some time (bottom photograph).

Another method of recording the same phenomenon is shown in fig. 9.8. Here, a light pulse from a spark passes through the liquid irradiated with an electron beam from the accelerator; the transmitted light is studied by means of a monochromator, photomultiplier and oscillograph. Figure 9.9 shows an example of an oscillogram obtained using this method.

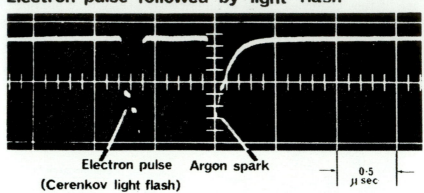

Fig. 9.6. Time relationship between pulse and flash. Ref.: Boag [56].

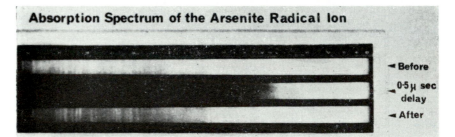

Fig. 9.7. Typical absorption spectrum obtained with the apparatus shown in fig. 9.5. The duration of the electron pulse was monitored by means of the Cerenkov light produced in the irradiated solution. The analysing light flash was a high-current spark in a jet of argon. Ref.: Boag [56].

From the distribution of the absorption spectrum and its changes with time, Hart and Boag were able to obtain results for the lifetime of a hydrated electron and other properties. The most important of these are: the charge on the particle is -1, its lifetime $\geqslant 300\ \mu\text{sec}$, the energy of hydration 1.7 ev, redox potential 2.7 ev, wavelength for maximum absorption 7200 Å $(1.72$ ev$)$, and coefficients of molar attenuation $9700\ \text{M}^{-1}\text{cm}^{-1}$ $(\lambda = 5780$ Å$)$ and $15\,000\ \text{M}^{-1}\text{cm}^{-1}$ $(\lambda = 7200$ Å$)$. The lifetime of a solvated or hydrated electron depends to a very large extent on the nature of the liquid, its purity, the amount of dissolved oxygen, etc., and can vary from 20 msec to 1 msec. An analysis of hydration under varying experimental conditions enables a new approach to be adopted for the study of chemical bonds in radiation chemistry, photo-chemistry, dosimetry, radiobiology, etc. A number of investigations in these fields have recently been conducted in some depth. They have included a series of studies on the oxygen effect in radiation chemistry.

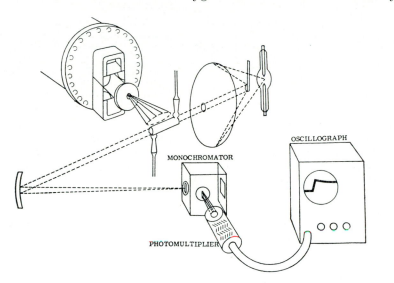

Fig. 9.8. Pulse radiolysis apparatus with kinetic spectroscopy at a single wavelength. Ref.: Boag [56].

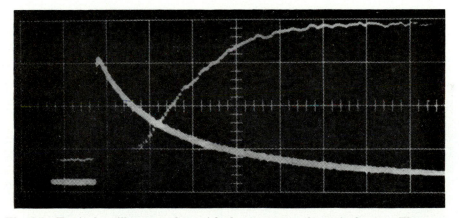

Fig. 9.9. Typical oscillogram taken with the apparatus shown in fig. 9.8. The upper curve shows the growth of absorption at 5000 Å in a neutral oxygenated solution of thiocyanate ion following a 200 nanosecond electron pulse; the abscissa scale is 1 μsec/cm and the ordinate is 1·49%/cm. The lower curve shows the decay of the same absorption on a time scale of 50 μsec/cm. Ref.: Adams *et al.* [53].

It is well known that dissolved oxygen can bring about wide variations in the effects due to ionizing radiation in solutions. It follows from research carried out on hydrated electrons that the dissolved oxygen readily captures electrons, thus producing O_2^- ions which, in a liquid of low pH, yield HO_2 radicals (fig. 9.10). When the liquid is saturated with air, not oxygen, and when the electron pulses are very short ($\sim 0.2\ \mu sec$),

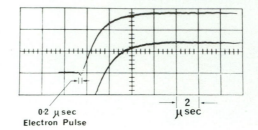

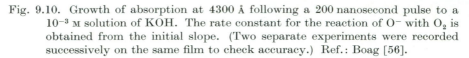

0·2 μsec
Electron Pulse

2 μsec

Fig. 9.10. Growth of absorption at 4300 Å following a 200 nanosecond pulse to a 10^{-3} M solution of KOH. The rate constant for the reaction of O^- with O_2 is obtained from the initial slope. (Two separate experiments were recorded successively on the same film to check accuracy.) Ref.: Boag [56].

O_3^- ions can be formed for very short periods of the μsec order. The following reactions are typical of this phenomenon, resulting in the disappearance of hydrated electrons (e_{aq}^-):

$$e_{aq}^- + H_2O \rightarrow H + OH^-,$$

$$e_{aq}^- + e_{aq}^- \rightarrow H_2 + 2OH^-,$$

$$e_{aq}^- + H \rightarrow H_2 + OH^-,$$

$$e_{aq}^- + H^+ \rightarrow (H^+)^- \rightarrow H.$$

Hart and Boag consider that the experimental discovery of hydrated and solvated electrons will be of considerable importance in radiation chemistry, radiobiology, etc.

Nitrous oxide, N_2O, displays a particularly strong reaction with hydrated electrons, viz:

$$N_2O + e_{aq}^- \rightarrow N_2 + O^-.$$

In this way, the numbers of O^- and O_3^- ions in a solution containing dissolved N_2O can be doubled.

More accurate quantitative research makes it possible to determine the yields from chemical reactions under various experimental conditions, i.e. with variations in temperature, pH, composition or concentration of the solution. The yield of OH groups formed in various alcohols irradiated with electrons from an accelerator has been measured. It was found that this yield increases systematically with the length of the alcohol chain; thus for methanol it is $4·8 \times 10^{-8}$ $M^{-1} sec^{-1}$, and for *n*-butanol it is $22·4 \times 10^{-8}$ $M^{-1} sec^{-1}$. It is expected that the methods for studying hydrated electrons described above will be improved by using even shorter irradiation times (of the order of nanoseconds) and light wavelengths below 2000 Å.

Much work has been carried out recently in order to determine the yield from the reaction of solvated electrons with molecules. Marcus [57–59] has developed a theory of electron transfer in liquids comparable with the previous polaron model and has calculated the thermodynamic functions for this process in various solutions.

As regards the conductivity of dielectric liquids (Part III of this book), certain radiolytic processes are of particular interest, e.g. those in which, even without external radiation, new electric charge centres (free electrons, positive and negative ions) may arise. Such processes can occur when electrons or ions collide with molecules possessing a high excitation energy. The excess energy of an excited molecule may exceed 10 ev, and can be greater than the ionization energy, provided it is not concentrated at one point in the molecule. If a highly excited molecule collides with electrons, ions, or even electrically neutral molecules, dissociation of the excited molecule into radicals or ions can take place; typical reactions are:

$$\left.\begin{aligned}
e^- + H_2O^* &\to H^+ + OH^- + e^-, \\
H_2O^+ + H_2O^* &\to H_2O^+ + H^+ + OH^-, \\
H_2O^* &\to H + OH, \\
H_2O^* + H_2O &\to H_2^+ + H_2O_2^-.
\end{aligned}\right\} \tag{9.20}$$

Forner and Hudson [60] have indicated the possibility of the reactions:

$$\left.\begin{aligned}
H_2O_2 + e^- &\to H_2O_2^+ + 2e^-, \\
H_2O_2 + e^- &\to H_2O^+ + O + 2e^-, \\
H_2O_2 + e^- &\to OH^+ + OH + 2e^-, \\
H_2O_2 + e^- &\to HO_2^+ + H + 2e^-,
\end{aligned}\right\} \tag{9.21}$$

which they observed during the radiolysis of water, where the ionization potential $I(H_2O_2)$ is about 11 ev. Further, owing to this high excess energy (which in certain excited states may be as high as 230 kcal/mole), excited molecules readily take part in numerous secondary reactions.

9.5. *Ionization Effects in Time and Space.* The distribution of electrons, ions, excited molecules and radicals in time and space is of great importance in studies of ionization effects in liquids. Quite a number of hypotheses for explaining these effects have been put forward; they are based mainly on experimental results obtained in Wilson chambers, bubble chambers, diffusion chambers and on nuclear plates. However, the various theories are not in agreement with each other, and the conclusions drawn from them lead to somewhat different results. The main work in this field is due to Lea [16], Gray [61], Magee [37] and Platzmann [62].

The fundamental problems here are the following: (1) How far are photoelectrons and Compton electrons able to move away from the parent ion? (2) What is the range of the electrical forces of the ion?

(3) What energy fraction is lost by electrons in elastic and inelastic collisions? (4) When can attachment of an electron to a neutral molecule to give a negative ion take place? (5) How are δ electrons distributed, and what rôle do they play in ionization processes? (6) What is the effect of various kinds of recombination reactions on these phenomena? (7) How are excited molecules and radicals formed, and in what manner do they give up their energy to the surroundings? (8) What is the possibility of secondary electrons and ions forming in the ionized liquid? A particularly important problem is the time distribution of these processes and its significance as far as the ability to detect and observe these phenomena is concerned.

In the initial period lasting for 10^{-18}–10^{-16} sec just after ionization, the phenomena which can be attributed to the ionizing particle in its path are as follows: over each element of the path (and in the track centre) there is a definite number of positive and negative ion pairs. Some ions are formed directly, while others arise due to secondary electrons (δ electrons) being knocked out during elastic collisions with atoms or molecules. These secondary electrons are arranged in small groups and possess very low energies of up to 100 ev. Various neutral atoms or molecules are also present in the path of electrons and photons, but are in higher excited states; radicals can also be detected. Both positive and negative electrons and ions take part in the thermal motion of molecules, and, in the absence of an electric field, can become neutralized as a result of collisions with ions of opposite sign.

Some workers, e.g. Magee [37], have assumed that a primary electron loses its energy mainly through inelastic collisions (about 4% of its energy per collision), and that if the action of electrostatic forces is disregarded, a 10 ev electron can find itself at a distance of 25 Å from the parent nucleus. However, as a result of the action of electrostatic forces, an electron can combine with a neutral molecule within 10^{-12} sec, cause it to dissociate and yield a heavy negative ion and a radical. The time necessary for an electron to slow down to a thermal velocity is of the order of 10^{-13} sec, and the collision cross section of an electron with water molecules is around 10^{-16} cm² per molecule. However, if the electron retardation to a thermal velocity takes place in the range over which the electrostatic forces of the parent ion are operative (up to 200 Å), the electron could be attracted by that ion, giving recombination. If electrons underwent elastic collisions only, many of them would be able to travel beyond the operative range of the ionic electrostatic forces; since they lose a great deal of energy in inelastic collisions however, the majority are confined to within that range.

One process which acts against the electron–ion recombination phenomenon is the attachment of a free electron to a neutral molecule. This leads to the formation of a heavy negative ion, which then dissociates into a smaller ion and a radical according to reaction (9.10). The cross section for the attachment of a free electron to a water molecule is of the order of

10^{-20} cm². Solvation of a free electron could also reduce the likelihood of recombination. The period of dielectric relaxation of dipoles in water is 10^{-11} sec, however, and it must therefore be assumed that the dipoles will not have sufficient time to arrange themselves round the electron prior to recombination.

A number of objections have been raised against Magee's theory, the most important of which are: (1) that certain quantities measured in the gas phase were assumed to be valid for liquids; (2) that the internal electric field present in water due to the polarity of molecules was not taken into account; and (3) that the laws of classical mechanics were applied in cases when quantum mechanics ought to have been used.

Lea [16], Gray [61], Platzmann [62] and Jortner and Stein [63] have assumed that an electron loses its energy as a result of an increase in the oscillatory and vibrational energy of the detector molecules. When an electron is able to travel a distance of about 50 Å and become surrounded by neutral molecules (electron solvation), the electron energy may fall to about 0·2 ev (the lowest level of vibrational energy) in about 10^{-12} sec. During the time 10^{-12}–10^{-7} sec after ionization, it is chiefly diffusion of ions and radicals which takes place; the diffusion is partly spherical and partly cylindrical. For an intermolecular distance in the liquid of about 3 Å, and at about 10^{11} collisions per second, the diffusion coefficient is around 10^{-5} cm² sec^{-1}. The initial non-homogeneous distribution of ions and radicals becomes more uniform due to the diffusion, and it becomes possible for two different radicals to collide and form a new molecule.

Within the time 10^{-7}–10^{-3} sec there is the likelihood of forming new secondary combinations of radicals, the decomposition of molecules and polymerization. According to Lea, an electron knocked out of a molecule may move 150 Å away from the parent ion, and there, having been retarded to a thermal energy level, combine with a water molecule to give a negative ion. Read [64] calculated that the electrostatic fields generated round the path of an ionizing particle with an ionization density of 1000 ion pairs per micron would be of the order of 10^5 v/cm. This would usually cause electron–ion recombination, if the effect of molecule polarity on the slowing down of the electron were not taken into account.

As a summary, the time and space distributions of the phenomena occurring in water radiolysis are shown in fig. 9.11 (a), (b). Figure 9.11 (b) illustrates the distribution of H and OH radicals according to Magee, and fig. 9.11 (a), (c) the distribution according to Lea. The latter author derived the following equation for the space and time distribution of the ions:

$$\frac{N_0}{N} = 1 + \frac{\alpha N_0}{8\pi D} \ln \frac{4Dt + r_0{}^2}{r_0{}^2}, \qquad (9.22)$$

where N_0 is the total number of ions produced per cm of the ionizing particle path, N is the number of ions of the same sign at any time t, α is the ionic recombination coefficient, D is the ionic diffusion coefficient, t is time and r_0 is the initial mean distance between ions.

Figure 9.11 (*c*) shows the distribution of H and OH ions round an α particle path in water at various times. The values shown were calculated using computers, but for certain cases only and with the aid of approximations. They have failed to give a definitive solution to the problem and indicate that certain theoretical assumptions are not entirely valid.

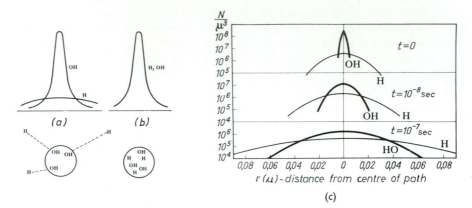

(c)

Fig. 9.11. Distribution of atoms and HOH radicals in tracks (*a*) according to the Lea–Gray–Platzmann theory, (*b*) according to the Samuel–Magee theory, (*c*) time and space distribution of H atoms and OH groups according to Lea. Ref.: Dainton [44].

Figure 9.12 summarizes the time span of the processes occurring in water radiolysed with α particles, as indicated below:

10^{-18}–10^{-17} sec	Traversal of the ionizing particle through the molecule; absorption of radiation energy; formation of δ radiation, primary ions, excited molecules and radicals
10^{-15} sec	Fluorescent vibrations; excitation of atoms; ionization
10^{-14}–10^{-12} sec	Excitation vibrations of the molecule, or instantaneous dissociation; recombination of electrons with positive ions
10^{-13} sec	Time between two consecutive collisions
10^{-12}–10^{-11} sec	Time required for the displacement of a small molecule, e.g. H_2O, through a distance equal to its diameter; radical recombination
10^{-10}–10^{-8} sec	Time required to slow down an electron to a thermal velocity
10^{-8}–10^{-7} sec	Excitation period for allowed states
10^{-5}–10^{-4} sec	Reaction time of ions and radicals in molar concentrations; lifetime of a hydrated electron
10^{-2} sec	Chemical equilibrium

9.6. *The Radiation Chemistry of Hydrocarbons.* In addition to investigations into the radiolysis of water, much work has been carried out on the effects of various types of ionizing radiation on hydrocarbons. Research in this field can be of great importance as regards industrial applications (§ 9.8), and special attention has been paid to the physics and chemistry of polymers, e.g. by Charlesby [8] and Chapiro [21]. A comprehensive summary of research work on this subject can also be found in books such as that by Topchiev and Polak [65].

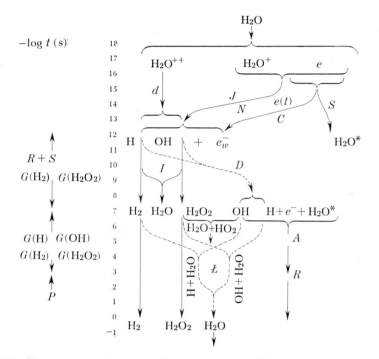

Fig. 9.12. Changes occurring over the time range 10^{-18}–10 sec in the phenomenon of water radiolysis: d denotes dissociation; D, diffusion; S, spin effect; $e(t)$ slowing-down of an electron to a thermal velocity; A, reaction, I, interaction within the track; J, ion–molecule reaction; N, neutralization reactions; C, electron capture. Ref.: Dainton in Charlesby [8].

Bach [66] has undertaken comprehensive research in the field of hydrocarbon radiochemistry. He expressed the chemical reactions which can occur as a result of irradiation in a general form as follows:

$$
\left.
\begin{array}{l}
RH \rightarrow R^{\cdot} + H, \\
R^{\cdot} + O_2 \rightarrow RO_2, \\
RO_2^{\cdot} + RH \rightarrow ROOH + R', \\
OR^{\cdot} + RH \rightarrow ROH + R^{\cdot}, \\
R'O_2^{\cdot} + R'' \rightarrow R'OOR'', \\
R'O_2^{\cdot} + R''O_2^{\cdot} \rightarrow R'OOR'' + O_2, \\
R'R''R'''CO_2^{\cdot} \rightarrow R'R''C^{\cdot} \rightarrow R'R''CO + OR^{\cdot\prime\prime\prime}, \\
\qquad\qquad \big\vert \\
\qquad\quad O\!-\!OR
\end{array}
\right\}
\quad (9.23)
$$

where R', R″ and R‴ are hydrocarbon radicals or hydrogen atoms, and dots by the chemical symbols denote radicals.

$$OR^{\cdot} + RH \rightarrow ROH + R^{\cdot}, \\ R'O_2^{\cdot} + R'' \rightarrow R'OOR'', \\ R'O_2^{\cdot} + R''O_2^{\cdot} \rightarrow R'OOR'' + O_2, \Biggr\} \quad (9.24)$$

An oxide radical RO_2 may react with a hydrocarbon molecule detaching a hydrogen atom and giving a hydroxide. For some of the reactions in (9.23) and (9.24) a certain activation energy is required, but this can be established experimentally by studying the radiation yield G as a function

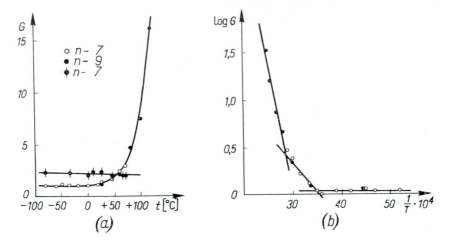

Fig. 9.13. (*a*) Radiation yield of peroxides and carbonyl compounds from the irradiation of *n*-heptane and *n*-nonane at various temperatures; (*b*) log G plotted against $1/T$. Ref.: Bach [66].

of temperature. Figure 9.13 (*a*) is a plot of G against t, while 9.13 (*b*) shows $\log G$ plotted against $1/T$. It can be seen that there are three distinct temperature ranges over which the activation energies for the oxidation of *n*-hexane and nonane are different. For the reaction $RO_2^{\cdot} + RH \rightarrow ROOH + R_2^{\cdot}$, this energy is about 12 kcal/mole. Values of G for the products from a number of hydrocarbons are shown in table 9.2 (taken from Davison [67]).

The yields in table 9.2 were measured using a gas chromatographic technique. It was found that they are independent of both the dose rate and the intensity of the radiation. The yields for ethane and propane are greater than those for methane, which points to the fact that the C—C bonds do not follow the statistical rule, i.e. that all bonds have an equal probability of rupture. The fragments obtained from the cleavage of a C—C bond are unlikely to travel a great distance and only combine with a molecule in the close vicinity, thus giving a number of possibilities of transverse, longitudinal and mixed bonds. The cleavage of a C—H bond

Table 9.2. Radiation yield G of various products from irradiated hydrocarbons

Paraffins	$G(H_2)$	$G(CH_4)$	$G(C_2H_4)$	$G(C_2H_6)$	$G(C_3H_6)$	$G(C_3H_8)$
n-Pentane	4·5	0·21	0·4	0·75	0·75	0·75
n-Hexane	4·0	0·12	0·22	0·36	0·2	0·36
n-Heptane	3·9	0·09	0·18	0·25	0·09	0·27
2,2-Dimethylbutane	2·2	0·09	0·04	0·14	0·01	0·02

G denotes the number of particles or ions formed during the absorption of 100 ev of radiation energy.

(which is stronger than a C—C bond) results in the liberation of a hydrogen atom which, due to diffusion, may travel much further than a molecular fragment. Table 9.3 shows the relative quantities of gaseous products liberated from paraffins irradiated with electrons. It can be seen from the table that the volume of gas liberated decreases as the length of the the normal paraffin molecule is increased, and the hydrogen content of the gas rises.

Table 9.3. Gaseous product yields from hydrocarbon irradiation

Paraffin	cm³ of gas at N.T.P.	Hydrogen (%)	Methane (%)	Gases non-volatile in liquid air (%)
n-Hexane	57·6	66·3	5·3	27·9
n-Heptane	51·4	76·9	3·9	19·2
n-Octane	48·3	78·8	2·8	18·3
n-Decane	41·6	78·9	2·1	18·2
n-Tetradecane	34·9	91·1	1·6	6·9
Cyclohexane	45·8	88·9	1·3	9·0
2,5-Dimethylhexane	49·8	42·1	11·6	46·2
2,2,4-Trimethyl-hexane	50·3	35·1	15·2	48·2
Methylcyclohexane	39·2	82·8	3·1	13·0

(From Schoepfle and Fellows [68] for an electron beam of 170 kev and 30 μA; the irradiation time was 30 min.)

Futrell [69] has accurately analysed the results obtained by various authors for the ionization of hexane in an electrostatic accelerator with an electron beam of 800 kev. The dose employed was of the order of 10^7 ergs/g of substance. The yields from both gaseous and liquid hexane were measured, and the effects of ethylene and oxygen impurities were studied. The results obtained are shown in table 9.4.

It can be seen from table 9.4 that the radiation yield of hydrogen is reduced by almost 50% due to the presence of impurities. The yields of

Table 9.4. Radiation yields G from pure hexane and from hexane with added impurities [69]

Product	Pure hexane		Additive	
	Vapour	Liquid	Ethylene	Oxygen
H_2	4·3	4·95	1·9	2·0
CH_4	0·5	0·13	0·5	—
C_2H_2	0·2	0·15	0·2	—
C_2H_4	1·1	0·63	—	1·1
C_2H_6	0·9	0·63	1·6	0·9
C_3H_6	0·3	1	0·5	0·4
C_3H_8	1·4	0·67	1·4	1·2
$i\text{-}C_4H_{10}$	0·1	0·005	0·1	0·05
$n\text{-}C_4H_{10}$	1·1	0·8	1·4	0·9
C_4H_8	0·1	0·46	0·1	0·2
$i\text{-}C_5H_{12}$	0·1	> 0·05	0·1	0·05
$n\text{-}C_5H_{12}$	0·1	> 0·05	0·1	0·05
$n\text{-}C_6$	9·5	—	—	—

other products remain unchanged. Futrell has suggested that the following reactions may occur between ions and molecules:

$$\left.\begin{aligned} R^+ + C_6H_{14} &\rightarrow RH + C_6H_{13}^*, \\ C_6H_{13}^+ + e &\rightarrow C_6H_{13}^* \rightarrow C_6H_{12} + H, \\ H + C_6H_{14} &\rightarrow C_6H_{13} + H_2. \end{aligned}\right\} \qquad (9.25)$$

After neutralization, the C_6H_{13} radical can be excited to such a high-energy state that it undergoes ionization or even dissociation; a hydrogen atom may be liberated and absorb most of the energy. Two hydrogen atoms may later form a hydrogen molecule.

Table 9.5 shows the results obtained by Hipple *et al.* [70] from the ionization of *n*-butane by electrons. The fragments were analysed using a mass spectrograph.

Table 9.6 contains the results obtained by Hustrilid *et al.* [71] for the irradiation of benzene with 72 ev electrons. It can be seen from tables 9.5 and 9.6 that there is a large number of fragments of different types, but that some of them occur in very small quantities. The occurrence of single carbon atoms should be noted, but they only have a very short lifetime of 10^{-6}–10^{-4} sec.

9.7. *Physical and Chemical Changes in Hydrocarbons due to Radiation.* The structural and physical changes which take place in paraffins due to ionizing radiation have been studied for many years. The substances investigated have been mainly paraffin waxes, which consist of a mixture of saturated hydrocarbons from $C_{22}H_{46}$ to $C_{29}H_{60}$, and unsaturated olefinic hydrocarbons of indefinite composition. In 1930, Pawłowski [72]

Table 9.5. The most frequently occurring fragments from the ionization of *n*-butane by electrons

Relative frequency	Ionized fragment	Probable cleavage of the molecule
40	$C_3H_7^+$	$H{-}\underset{\underset{H}{\|}}{\overset{\overset{H}{\|}}{C}}{-}\underset{\underset{H}{\|}}{\overset{\overset{H}{\|}}{C}}{-}\underset{\underset{H}{\|}}{\overset{\overset{H}{\|}}{C}}\ \Big|\ \underset{\underset{H}{\|}}{\overset{\overset{H}{\|}}{C}}{-}H$
14	$C_2H_5^+$	$H{-}\underset{\underset{H}{\|}}{\overset{\overset{H}{\|}}{C}}{-}\underset{\underset{H}{\|}}{\overset{\overset{H}{\|}}{C}}\ \Big|\ \underset{\underset{H}{\|}}{\overset{\overset{H}{\|}}{C}}{-}\underset{\underset{H}{\|}}{\overset{\overset{H}{\|}}{C}}{-}H$
13	$C_3H_5^+$	$H{-}\underset{\underset{H}{\|}}{\overset{\overset{H}{\|}}{C}}{-}\underset{\underset{H}{\|}}{\overset{\overset{H}{\|}}{C}}{-}\underset{\underset{H}{\|}}{\overset{\overset{H}{\|}}{C}}{-}\underset{}{\overset{\overset{H}{\|}}{C}}{-}H$
10·5	$C_2H_4^+$	$H{-}\underset{\underset{H}{\|}}{\overset{\overset{H}{\|}}{C}}{-}\underset{\underset{H}{\|}}{\overset{\overset{H}{\|}}{C}}{-}\underset{\underset{H}{\|}}{\overset{\overset{H}{\|}}{C}}{-}\underset{\underset{H}{\|}}{\overset{\overset{H}{\|}}{C}}{-}H$
9·5	$C_2H_3^+$	$H{-}\underset{\underset{H}{\|}}{\overset{\overset{H}{\|}}{C}}{-}\underset{\underset{H}{\|}}{\overset{\overset{H}{\|}}{C}}{-}\underset{\underset{H}{\|}}{\overset{\overset{H}{\|}}{C}}{-}\underset{\underset{H}{\|}}{\overset{\overset{H}{\|}}{C}}{-}H$
8	$C_3H_6^+$	$H{-}\underset{\underset{H}{\|}}{\overset{\overset{H}{\|}}{C}}{-}\underset{\underset{H}{\|}}{\overset{\overset{H}{\|}}{C}}{-}\underset{\underset{H}{\|}}{\overset{\overset{H}{\|}}{C}}{-}\underset{\underset{H}{\|}}{\overset{\overset{H}{\|}}{C}}{-}H$
5	$C_3H_3^+$	$H{-}\underset{\underset{H}{\|}}{\overset{\overset{H}{\|}}{C}}{-}\underset{\underset{H}{\|}}{\overset{\overset{H}{\|}}{C}}{-}\underset{\underset{H}{\|}}{\overset{\overset{H}{\|}}{C}}{-}\underset{\underset{H}{\|}}{\overset{\overset{H}{\|}}{C}}{-}H$

published his particularly interesting results of research into the structural changes caused by α particles in solid paraffin waxes, including photographs which show the effect of irradiation on the structure of the wax (see figs. 9.14 and 9.15). Later, Ścisłowski [73–75] described the ionization effects produced in solid paraffin wax by x-rays. His work was

8

chiefly concerned with the problem of initial currents which appear within milliseconds of a potential being applied to a paraffin wax capacitor. Particular care was taken with the degassing of the molten wax samples, and the effect of the amount of air present in the samples on the electrical conductivity was also studied.

Table 9.6. Relative yields of fragments from benzene irradiated with 72 ev electrons

Molecular weight	Ion	Relative frequency	Molecular weight	Ion	Relative frequency
78	$C_6H_6^+$	100	39	$C_3H_3^+$	6·6
77	$C_6H_5^+$	15·2	38	$C_3H_2^+$	3·6
76	$C_6H_4^+$	4·6	37	C_3H^+	2·5
75	$C_6H_3^+$	1·7	36	C_3^+	0·3
74	$C_6H_2^+$	4·0			
73	C_6H^+	1·3	28	$C_2H_4^+$	0·2
72	C_6^+	0·2	27	$C_2H_3^+$	1·3
			26	$C_2H_2^+$	1·1
63	$C_5H_3^+$	2·6	25	C_2H^+	0·14
62	$C_5H_2^+$	0·6	24	C_2^+	0·03
61	C_5H^+	0·5			
60	C_5^+	0·2	15	CH_3^+	0·01
			14	CH_2^+	0·02
52	$C_4H_4^+$	13·5	13	CH^+	0·03
51	$C_4H_3^+$	15·7	12	C^+	0·05
50	$C_4H_2^+$	13·3			
49	C_4H^+	2·1	78	$C_6H_6^{++}$	1·8
48	C_4^+	0·4	77	$C_6H_5^{++}$	0·18
			75	$C_6H_3^{++}$	0·42
			73	C_6H^{++}	0·015

Many papers on the subject have been published since 1945, and describe experimental work usually carried out using very powerful sources of radiation from atomic reactors and accelerators. Certain of the results obtained throw some light on the problem of ionization and electrical conductivity in liquid dielectrics, and these are consequently discussed in greater detail below.

It was found that the number of branchings caused by radiation does not depend on the type of hydrocarbon or the chain length; it averages a value approximately equal to $G(I) = 4$. For cyclic hydrocarbons, the quantity of gas evolved is much smaller (by one order of magnitude) than for aliphatic compounds. For example, the hydrogen yield from benzene is 0·036, from toluene it is 0·13 and from ethylbenzene 0·12; the corresponding methane yields are 0·0012, 0·008 and 0·03 respectively.

The structural changes which occur in hydrocarbon molecules due to radiation may, as mentioned previously, lead to gaseous products (hydrogen, methane) being liberated, the cleavage of C—C and C—H bonds, the combination of molecules to give different and more complicated substances (cross linking, lattice formation, dimerization), the formation of unsaturated hydrocarbons, the disturbance of the pseudo-crystalline structure of the liquid, colour changes and oxidation—particularly at the surface of the liquid. A strongly ionized substance can therefore even exhibit distinct changes in properties such as density, viscosity, colour,

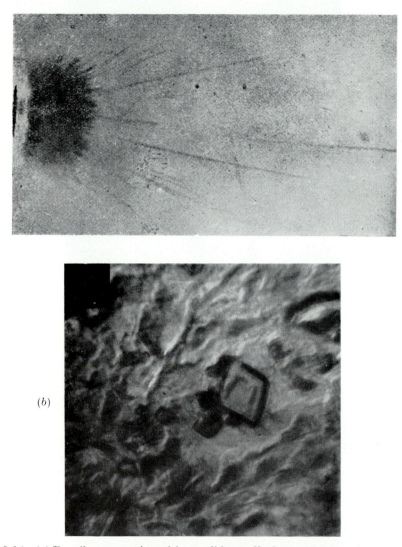

(a)

(b)

Fig. 9.14. (a) Recoil protons ejected in a solid paraffin by α particles; (b) crystalline forms in a paraffin irradiated by α particles. Ref.: Pawłowski [72].

mechanical properties, electrical conductivity, etc. Miller *et al.* [76] have found that for octacosane, $C_{28}H_{58}$, the variation of the molecular weight M_n of the liquid can be represented as a function of the radiation dose D

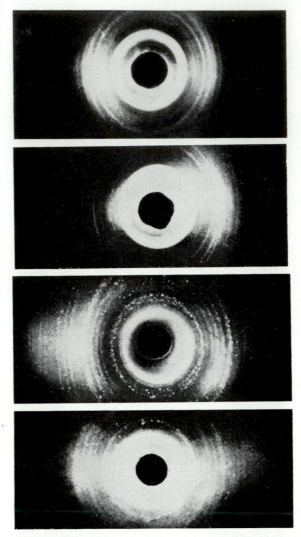

Fig. 9.15. Crystal structure of a solid paraffin. Photographs and x-rays of a thin
layer of the paraffin were taken using the Debye–Scherrer method. The
radiograms from top to bottom are of: a transparent paraffin, an opaque
paraffin, an irradiated transparent paraffin and an irradiated, heated transparent paraffin. Ref.: Pawłowski [72].

by the equation:

$$\frac{1}{M_n} = \frac{1}{M_{0n}} (1 - 1{\cdot}02 \times 10^{-3} D), \qquad (9.26)$$

where D is in megaröntgens (1 MR corresponds to an absorbed energy of
$83{\cdot}8 \times 10^6$ ergs per gram of octocosane). A plot of this function is shown
in fig. 9.16. A typical example of the variation of the melting point as a
function of the absorbed dose is illustrated in fig. 9.17 for several paraffin

hydrocarbons; the doses are defined in pile units† since they are large and usually only attainable in atomic reactors. It can be seen from the figure that melting point decreases as the dose increases, with the temperature

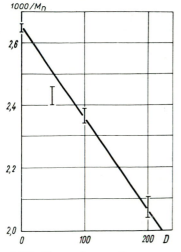

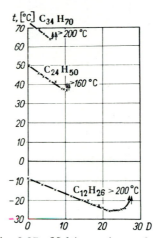

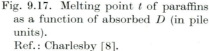

Fig. 9.16. Dependence of mean molecular weight M_n (determined by the cryoscopic method) on dose D according to the equation

$$10^3/M_n = (1/M_{0n})\,(1 - 1\cdot 02 \times 10^{-3}\,D).$$

Ref.: Charlesby [8] and Miller [76].

Fig. 9.17. Melting point t of paraffins as a function of absorbed D (in pile units).
Ref.: Charlesby [8].

being reduced by as much as 30% (if measured in °C) when the dose is large. Figures 9.18 and 9.19 show the variation of viscosity as a function of the absorbed dose (in Mrad). With very large doses (of the order of 10^9 rad), the viscosity is increased by as much as 600%.

† The pile unit has not been defined exactly, as its value depends on the reactor design, the nature and chemical composition of the irradiated substance, temperature, etc. The literature therefore contains a large number of values, but Charlesby's [8] data for the Oak Ridge reactor and BEPO may serve as an example. For the Oak Ridge reactor, the pile unit is a neutron flux of 10^{18} thermal neutrons per cm². It is written as $10^{18}\ nvt$, where n is the neutron flux density, v the neutron velocity and t the exposure period. For a 3·5 MW reactor, the neutron flux is $0\cdot 74 \times 10^{12}\ n/\text{sec}$, and the time required for absorption of a pile unit is about 16 days $(1\cdot 35 \times 10^6\ \text{sec})$. For polyethylene, it can be calculated that $10^{18}\ nvt$ is roughly equivalent to 10^9 R, and the daily dose of absorbed energy is 53·2 Mrad (the dose rate is then *circa* 615 rad/sec). For other hydrocarbons, the absorbed energy can be calculated by assuming the corresponding energies for carbon to be $4\cdot 6 \times 10^{-4}$ cal/g sec, and for hydrogen, 75×10^{-4} cal/g sec. For substances whose molecules contain other atoms besides carbon and hydrogen, calculations must be carried out using data on absorption coefficients to be found in the literature.

Charlesby [8] has also quoted the dose for the BEPO reactor under standard operating conditions. The pile unit was calculated for a thermal neutron flux of $10^{17}\ n/\text{cm}^2$ sec. For polyethylene, this corresponds to about 45 Mrads of γ radiation or electron radiation at 20°C.

In fig. 9.20, the specific volume (in cm³/g) is plotted against temperature with dose as parameter expressed in pile units (taken from ref. [77]).

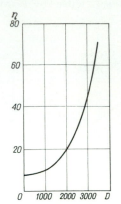

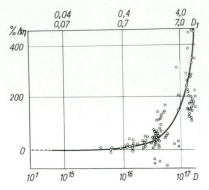

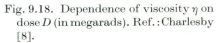

Fig. 9.18. Dependence of viscosity η on
dose D (in megarads). Ref.: Charlesby
[8].

Fig. 9.19.   Percentage variation of
viscosity η as a function of dose D
of fast neutrons; D is the number of
neutrons per cm², and D_1 is the dose
in rads $\times 10^8$. Ref.: Pomeroy from
Charlesby [8].

It can be seen that the specific volume is reduced, i.e. the density
increases, as the dose increases.   This behaviour is characteristic of long-
chain molecules and large doses, and is not observed for short hydrocarbons and small doses. It indicates the extensive changes in crystalline

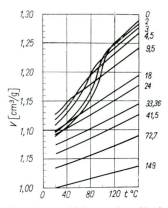

Fig. 9.20. Variation in specific volume V due to irradiation in a reactor as a function
of temperature t; the figures on the right denote the dose (in pile units).
Ref.: Charlesby [8].

structure which occur in irradiated polyethylenes.   Such changes have
been photographed; e.g. see fig. 9.21 showing the diffraction of x-rays in
polyethylene.   The diffraction rings on the photograph become fainter
as the radiation dose is increased.

There are similarly many other effects due to the changes in macroscopic physical properties (e.g. elasticity, extensibility, shape, mechanical strength, etc.) caused by irradiation. They occur most frequently, however, in substances in the solid state subjected to very high doses and are not dealt with in this book. Interested readers should refer to the comprehensive description of these phenomena by Charlesby [8].

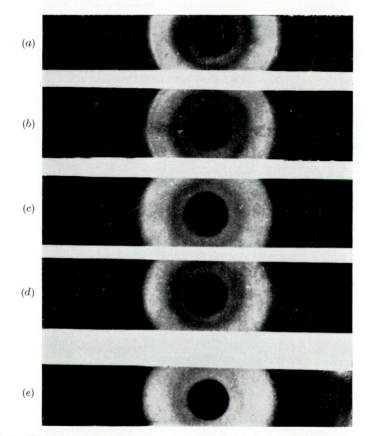

Fig. 9.21. Photographs of x-ray diffraction in polyethylene showing how the extent of destruction of the crystalline structure depends on the dose absorbed; (*a*) zero pile units, (*b*) 4 units, (*c*) 9 units, (*d*) 13 units, (*e*) 16 units. Ref.: Charlesby [8].

Of interest here, however, is an example of the variation of breakdown voltage on temperature for various radiation doses in polyethylene. It can be seen from fig. 9.22 that radiation alters the nature of the variation, viz. it exerts a somewhat stabilizing effect on the strength of a substance as regards electrical breakdown. Unfortunately, little information is available on the electrical properties of dielectric substances after being irradiated. Research in this field for lower dose levels is dealt with in the chapter on electrical conductivity in low fields (Chapter 11).

Attempts have been made to elucidate the radiation yields accompanying various molecular rearrangements which lead to the formation of new compounds. Figure 9.23 shows the yields for hydrogen (H_2), for the formation of new unsaturated compounds ($\times$) and for changes in initially unsaturated compounds ($+$); the doses used were high, around 20 мrad.

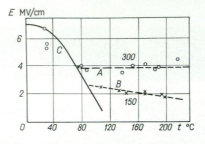

Fig. 9.22. Breakdown voltage E_p as a function of temperature t in polyethylene irradiated by 4 мev electrons; A, 300 мrads; B, 150 мrads; C, not irradiated. Ref.: Stark and Garton [79].

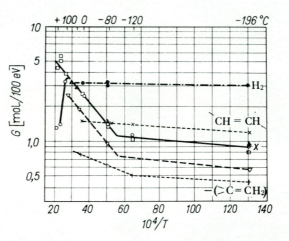

Fig. 9.23. Radiation yield G as a function of temperature T for various reactions in polyethylene; viz. formation of hydrogen (H_2), appearance of *trans*-unsaturation ($\times$) and disappearance of initial unsaturation ($+$). Dose, 20 мrads. Ref.: Charlesby [8].

9.8. *Industrial Applications of Hydrocarbon Radiochemistry.* Radiochemical research into the oxidation of hydrocarbons is not only of theoretical significance, but also has practical applications in the engineering and other industries. One particular aspect of this is based on the fact that the velocity v of chemical reactions is a function of the frequency of collisions n between molecules, the activation energy E for the given process and temperature T. This relationship can be expressed as:

$$v = n \exp(-E/RT). \tag{9.27}$$

In order to increase the rate of chemical reactions, therefore, it is necessary either to raise the temperature or to decrease the activation energy of a given process.

In industry, many chemical processes are conducted at highly elevated temperatures in order that they shall be economic, but some of the most recent work in radiation chemistry is concerned with attempts to increase the reaction rate by decreasing the activation energy. It is known that the activation energies for reactions of ions, excited molecules and radicals with molecules of the medium in the normal state are usually much lower than those for normal molecular collisions. Sometimes the energies even approach zero, implying that a reaction occurs at each collision. This explains why an ionized substance undergoes chemical changes much more rapidly than non-ionized substances, and why the whole process can be conducted at a temperature below that normally used. This fact can be of great importance for certain industrial manufacturing processes.

As an example let us consider some applications of radiation chemistry to the oxidation reactions of hydrocarbons. It is generally assumed that hydrocarbon oxidation reactions with molecular oxygen are radical reactions, i.e. that they take place according to:

$$\left.\begin{array}{l} \dot{R} + O_2 \rightarrow RO\dot{O}, \\[2mm] RO\dot{O} + RH \rightarrow ROOH + \dot{R}. \end{array}\right\} \tag{9.28}$$

The activation energy of the second of these reactions is around 5–15 kcal/mole, and it may therefore take place at higher temperatures (above 100°C), but the appearance of the same radical R at the end of the reaction enables a chain reaction to be induced. If the hydrocarbon contains impurities acting as inhibitors or scavengers and which trap the free radicals, the reaction is not able to develop into a chain reaction.

The conditions under which free radicals are formed and destroyed play an important rôle in determining the course of chemical reactions. This principally concerns those reactions which, under the action of ionizing radiation, may become chain reactions, causing a considerable increase in the radiation yield. Reactions of this type usually occur at higher temperatures (100–150°C) and in the presence of a large quantity of oxygen, but cases are also known of chain reactions taking place at room temperature with only the normal quantity of oxygen dissolved in the liquid. This happens with substances such as cumene and tetralin. For tetralin, the radiation yield of free radicals is usually $G = 0 \cdot 2$, but this increases rapidly with temperature; when the dose rate is 24 R/sec, it is $G = 65$ at 30°C, $G = 100$ at 40°C and $G = 236$ at 50°C. These figures point to an increase in length, as the temperature rises, of the reaction chain from 325 sections per radical at 30°C to 1180 sections per radical at 50°C. In such cases it is said that the ionizing radiation causes an action which initiates the chemical reaction.

Low molecular weight paraffins are generally resistant to oxidation at room temperature and under normal pressures. More interesting from the point of view of industrial applications is the radiation oxidation of aromatic hydrocarbons; benzene is probably the most important of these since its oxidation product is phenol, a valuable intermediate in many organic syntheses. On the whole, aromatic hydrocarbons are much more resistant to the action of oxygen than aliphatic and alicyclic hydrocarbons, but under certain experimental conditions, e.g. in aqueous solutions, the action of x-rays in the presence of air may increase the radiation yield of phenol by a factor of over 10. For this reason, the method is likely to find application in the industrial production of phenol.

It is evident from the examples quoted that studies of the mechanism of radiation reactions are directly related to the oxidation of organic compounds and may contribute to the development of new industrial processes. Further material in this same field can be found in the reports on the International Warsaw Conference held in 1959 [78], and summarized by Ciborowski [38, 39].

References to Part II

[1] Szczeniowski, S., 1961, *Experimental Physics*, Part 5 (Warsaw: Polish Scientific Publishers) (in Polish).

[2] Campbell, J., and O'Connor, D., 1956, *Principles of Use of Radioisotopes* (Warsaw: P.W.N.).

[3] Jaeger, R. G., 1959, *Dosimetrie und Strahlenschutz physikalische und technische Daten* (Stuttgart: Georg Thieme Verlag) (in German).

[4] Aglintsev, K. K., 1957, *The Dosimetry of Ionizing Radiation* (Moscow: Gostekhisdat) (in Russian).

[5] Price, W. J., 1964, *Nuclear Radiation Detection* (New York: McGraw Hill).

[6] Adamczewski, I., 1959, *Health Protection against Ionizing Radiation* (Warsaw: PZWL) (in Polish).

[7] Handbuch der Physik, vol. 45, 1958 (Berlin: Springer), (Glaser, D. A., *The Bubble Chamber*, p. 314).

[8] Charlesby, A., 1960, *Atomic Radiation and Polymers* (New York: Pergamon Press).

[9] Mohler, H., 1958, *Chemische Reaktionen ionisierender Strahlen* (Frankfurt am Main: Sauerländer) (in German).

[10] Glaser, D. A., 1954, *Nuovo Cim.*, **2**, Suppl. 2, 361.

[11] Adamczewski, I., 1951, *PostEpy Fizyki*, **1**, 210; 1951, *Ibid.*, **2**, 6.

[12] Bethe, H. A., 1938, *Z. Phys.*, **76**, 293.

[13] Möller, C., 1932, *Ann. Phys.*, **14**, 531.

[14] Miller, N., 1957, *Rev. pure appl. Chem.*, *R. Austr. chem. Inst.*, **7**, 123.

[15] Glasser, O. (Ed.), 1944–50, *Med. Phys.*

[16] Lea, D. E., 1955, *Action of Radiation on Living Cells* (Cambridge University Press).

[17] Victoreen, J. A., 1943, *J. appl. Phys.*, **14**, 95; 1948, *Ibid.*, **19**, 855.

[18] Kasztelan, S., 1961, M.Sc. Thesis, (Gdańsk: Pedagogical University).

[19] Bacq, Z. M., and Alexander, P., 1955, *Fundamentals of Radiobiology* (London: Butterworths).

[20] Rotblat, J., *et al.*, 1958, *Physics in Medicine and Biology*.

[21] Chapiro, A., 1962, *Radiation Chemistry of Polymeric Systems* (New York: Interscience).

[22] LAIDLER, K. J., 1955, *The Chemical Kinetics of Excited States* (Oxford: Clarendon Press).

[23] HERZBERG, G., 1950, *Spectra of Diatomic Molecules* (New York: van Nostrand).

[24] HERZBERG, G., 1950, *Molecular Spectra and Molecular Structure* (New York: van Nostrand).

[25] MASSEY, H. S. W., and BURHOP, E. H. S., 1952, *Electronic and Ionic Impact Phenomena* (Oxford: Clarendon Press).

[26] SMITH, C., and ESSEX, H., 1938, *J. chem. Phys.*, **6**, 188.

[27] WILLIAMS, N. T., and ESSEX, H., 1948, *J. chem. Phys.*, **16**, 1153; 1949, *Ibid.*, **17**, 995.

[28] HAÏSSINSKY, M., and MAGAT, M., 1954, *Compt. Rend.*, **232**, 954.

[29] HAÏSSINSKY, M., 1952, *Disc. Faraday Soc.*, **12**, 133.

[30] BORN, M., and OPPENHEIMER, J. R., 1927, *Annls Phys.*, **84**, 457.

[31] MOTT, N. F., and SNEDDON, I. N., 1948, *Wave Mechanics and its Applications* (Oxford: Clarendon Press).

[32] MOHLER, H., 1957, *Naturw. Rdsch.*, **10**, 177.

[33] HART, E. J., and BOAG, J. W., 1962, *J. Am. chem. Soc.*, **84**, 4090.

[34] BOAG, J. W., and HART, E. J., 1963, *Nature, Lond.*, **197**, 45.

[35] BOAG, J. W., 1963, *Am. J. Roentg. Rad. Ther. Nucl. Med.*, **40**, 896.

[36] HART, E. J., 1964, *Science*, **146**, 19.

[37] MAGEE, J. L., 1955, *J. chim. Phys.*, **52**, 508.

[38] CIBOROWSKI, S., 1962, *Wiad. Chem.*, **1**, 45.

[39] CIBOROWSKI, S., 1962, *The Radiation Chemistry of Inorganic Compounds* (Warsaw: Polish Scientific Publishers) (in Polish).

[40] ALLEN, A. O., 1954, *Radiat. Res.*, **1**, 85.

[41] CHARLESBY, A., 1954, *Proc. R. Soc.*, **A222**, 60.

[42] KROH, J., 1962, *Wiad. Chem.*, **16**, 135.

[43] ALLEN, A. O., 1961, *The Radiation Chemistry of Water and Aqueous Solutions* (Princeton, N.J.: van Nostrand).

[44] DAINTON, F. S., 1959, *Radiat. Res.*, Suppl. **1**, 1.

[45] ELERT, M., and BOAG, J. W., 1952, *Disc. Faraday Soc.*, **12**, 189.

[46] WEISS, J., 1960, *Nature, Lond.*, **186**, 751.

[47] PEKAR, S. J., 1951, *Investigation of the Theory of Crystals* (in Russian).

[48] BOAG, J. W., 1963, *Actions chim. biol. Radiat.*, **6**, 25 (Paris: Masson & Cie).

[49] BOAG, J. W., et al., 1965, *Radiation Effects in Physics, Chemistry and Biology* (Houston: Williams & Wilkins).

[50] BAXENDALE, J. H., 1962, *Radiat. Res.*, **17**, 312.

[51] HART, E. J., et al., 1964, *Radiat. Res.*, Suppl. **4**, 24.

[52] ADAMS, G. E., and BOAG, J. W., 1964, *Proc. chem. Soc.*, **112**.

[53] ADAMS, G. E., et al., 1965, *Trans. Faraday Soc.*, **61**, 492, 1417, 1674.

[54] KEENE, J. P., 1964, *Radiat. Res.*, **22**, 1.

[55] KEENE, J. P., 1966, *Pulse Radiolysis* (London: Academic Press).

[56] BOAG, J. W., 1965, *Physics in Medicine and Biology*, **10**, 457.

[57] MARCUS, R. A., 1963, *J. chem. Phys.*, **38**, 1858; 1963, *Ibid.*, **39**, 1734; 1965, *Ibid.*, **43**, 679, 3477.

[58] MARCUS, R. A., 1963, *J. chem. Phys.*, **37**, 853.

[59] MARCUS, R. A., 1965, *J. chem. Phys.*, **43**, 58.

[60] FORNER, S. N., and HUDSON, R. L., 1953, *J. chem. Phys.*, **21**, 1608.

[61] GRAY, L. H., 1951, *J. Chim. phys.*, **48**, 172.

[62] PLATZMANN, R. L., 1955, *Radiat. Res.*, **2**, 1.

[63] JORTNER, J., and STEIN, G., 1955, *Nature, Lond.*, **175**, 893.

[64] READ, J., 1949, *Br. J. Radiat.*, **22**, 366.

[65] TOPCHIEV, A. V., and POLAK, L. S., 1962, *Hydrocarbon Radiolysis* (Moscow: Academy of Science U.S.S.R.) p. 207 (in Russian).

[66] BACH, N. A., 1959, *Radiat. Res.*, **1**, 190.

[67] DAVISON, W. H. T., 1958, *Chem. Soc. Special Publ.*, **9**, 151.
[68] SCHOEPFLE, C. S., and FELLOWS, C. H., 1931, *Ind. Engng Chem.*, **23**, 1396.
[69] FUTRELL, J. H., 1959, *J. phys. Chem.*, **20**, 81, 5921.
[70] HIPPLE, J. A., *et al.*, 1937, *J. appl. Phys.*, **8**, 815.
[71] HUSTRILID, K., *et al.*, 1938, *Phys. Rev.*, **54**, 1037.
[72] PAWŁOWSKI, C., 1930, *J. Chim. phys.*, **27**, 266.
[73] ŚCISŁOWSKI, W., 1935, *Acta phys. pol.*, **4**, 123.
[74] ŚCISŁOWSKI, W., 1937, *Acta phys. pol.*, **6**, 403; 1938, *Ibid.*, **7**, 127.
[75] ŚCISŁOWSKI, W., 1939, *Acta phys. pol.*, **7**, 214.
[76] MILLER, A. A., *et al.*, 1956, *J. phys. Chem.*, **60**, 599.
[77] CHARLESBY, A., and CALLAGHAN, L., 1958, *J. phys. Chem. Solids*, **4**, 227, 306.
[78] 'The Industrial Applications of Powerful Radiation Sources, particularly for Chemical Processes', Paper presented at the Conference on Radiation, Warsaw, 1959.
[79] STARK, K. H., and GARTON, C. G., 1955, *Nature, Lond.*, **176**, 1225.
[80] BOAG, J. W., 1967, *Solid and Chemical Radiation Dosimetry in Medicine and Biology* (Vienna: I.A.E.A.).

ADDITIONAL BIBLIOGRAPHY TO PART II

ALLEN, A. O., 1952, *Disc. Faraday Soc.*, **12**, 79.
BECKEY, H. D., 1961, *Z. Naturf.*, **16a**, 505; 1962, *Ibid.*, **17a**, 1103; 1964, *Ibid.*, **19a**, 71.
BECKEY, H. D., and SCHULZE, P., 1965, *Z. Naturf.*, **20a**, 1329.
BECKEY, H. D., and WAGNER, G., 1965, *Z. Naturf.*, **20a**, 169.
BEYNON, J. H., 1960, *Mass Spectrometry and its Applications to Organic Chemistry* (Amsterdam: Elsevier).
BURTON, M., *et al.*, 1950, *J. chem. Phys.*, **20**, 760.
CHARLESBY, A., 1952, *Proc. R. Soc.*, **A215**, 187; 1955, *Radiation Research*, **2**, 96.
COLLISON, E., *et al.*, 1960, *Nature, Lond.*, **187**, 475; 1962, *Proc. R. Soc.*, **A265**, 407.
EBERT, M., 1966, *Pulse Radiolysis* (London: Academic Press).
FRANKIEVICH, E. L., and BARABANOV, E. N., 1965, *Fiz. twiord. tiela*, **7**, 1967 (in Russian).
JAKOVLEVA, B. S., and FRANKIEVICH, E. L., 1966, *Zh. fiz. khimii*, **6**, 40.
KUPER, C. G., and WHITFIELD, G. D., 1963, *Polarons and Excitations* (Edinburgh: Oliver & Boyd).
MAKOTO, N., and TSUTOMA, W., 1966, *J. phys. Soc., Japan*, **21**, 1573.
MOZUMBER, A., and MAGEE, J. L., 1966, *Radiat. Res.*, **28**, 203.
SCHULZE, P., 1964, Thesis, University of Bonn.
SMITH, D. E., 1959, *Radiat. Res.*, **1**, 190.
TALROZE, W. L., 1964, *Conf. chem. Rad. Polym.* (Moscow: Pribory Tekh. Eksper.) (in Russian).
Transactions of a Conference on Radiation Chemistry, U.S.S.R., 1958 (Moscow: Academy of Sciences) (in Russian).
WAGNER, G., 1964, Thesis, University of Bonn.

PART III

Electrical Conduction in Dielectric Liquids at Low Fields

Chapter 10

Measuring Apparatus

10.1. *Introduction*. This monograph is mainly concerned with research on electrical conduction in dielectric liquids at low and high electric fields and includes work on electrical breakdown. A great number of fundamental physical phenomena occur when an electric field exists in a dielectric liquid but the mechanisms are not yet fully understood. For this reason, it has been deemed necessary to discuss the most important results of experiments carried out by various scientists and in order to indicate the present state of knowledge in this subject, those theoretical concepts which are known have been gathered together. Although none of the theories have completely explained all the experimental facts, a critical presentation of both the experimental and the theoretical material may accelerate the solution of those difficulties which become apparent when discussing the mechanism of conduction and breakdown in dielectric liquids. The two previous parts of the monograph contain the general physicochemical foundation necessary for understanding the phenomena which occur in ionized liquids subjected to electric fields. This section includes the material concerned with conduction in low electric fields.

As noted in the introduction and in § 1.2, dielectric liquids may have quite different values of natural electrical conductivity. At the same time, the purity grade of the liquid, as determined by its conductivity, decides whether the results are reproducible and whether they offer an original contribution to the essential problems of the electrical conduction mechanism. The results of many experiments carried out with dielectric liquids not completely purified should therefore be accepted only with considerable reservations.

The value of natural or self-conductivity for the best insulating liquid dielectrics is of the order of 10^{-18}–$10^{-19} \, \Omega^{-1} \, cm^{-1}$. It is therefore necessary to prepare a well-purified sample of the liquid and to make careful measurements of electric currents in the range of 10^{-9}–$10^{-15} \, A$.

10.2. *Different Kinds of Liquid-filled Ionization Chambers.* A purified dielectric liquid is usually tested in a cell which is a special kind of ionization chamber (a type of capacitor) and subjected to a constant voltage in the range of 2–10 kv from a battery supply or from a special electronic circuit. Sensitive electrometers are needed in order to measure the current.

A basic diagram of the simplest measuring circuit, consisting of an ionization chamber K, an electrometer E and an accumulator battery B is shown in fig. 10.1. A capacitor which may be either parallel-plate,

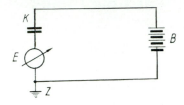

Fig. 10.1. Simplified circuit of an ionization chamber K, electrometer E and battery B; Z is earthed.

cylindrical or spherical is the main part of the ionization chamber. One electrode of the capacitor, set on a good insulator (e.g. ebonite), can be connected with the source of constant voltage, while the second electrode which collects a measured charge must be set on the best type of insulator such as amber or quartz and connected to the electrometer. The most frequently used solid insulators are listed in table 10.1.

Table 10.1. Solid insulators

Substance	Resistivity (Ω-cm)
Perspex	10^{13}–10^{20}
Quartz	10^{16}–10^{18}
Polyethylene	10^{21}
Sulphur	10^{22}
Silicon ceramics	10^{15}–10^{16}
Bakelite	10^{13}–10^{14}
Teflon	10^{20}
Amber	10^{19}
Polystyrene	10^{21}

Since amber, polystyrene, quartz and teflon are easy to machine, they are most frequently used for the collector electrode. The fact that some organic liquids have a strong chemical reaction with some insulating materials should also be taken into consideration. Hydrocarbons of the aromatic group (benzene, toluene), for example, dissolve amber, whereas saturated hydrocarbons (hexane, heptane) do not. Insulators such as ebonite, perspex or ceramics can be used for the second electrode providing that they are not exposed to harmful chemical effects.

In addition, an earthed metal ring must surround the collector electrode in order to shield it from stray currents. A diagram of this type of parallel-plate chamber is shown in fig. 10.2. The entire capacitor is placed in a closed vessel, the design of which will depend on whether the ionization chamber is to be filled with gas (under normal or increased pressure) or with a dielectric liquid.

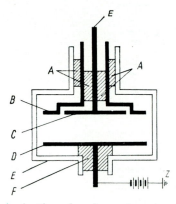

Fig. 10.2. Diagram of an ionization chamber. *A*, amber insulator supporting the collecting electrode; *B*, guard ring; *C*, collecting electrode; *D*, high-voltage electrode; *E*, chamber filled with liquid or gas; *F*, ebonite insulator.

If ionizing agents produce charges in a chamber with a capacitance C and a number n of ions with a charge e are brought to the collector electrode, the potential on the electrode is changed by:

$$dU = \frac{dQ}{C} = \frac{d(ne)}{C}. \tag{10.1}$$

If e is equal to the electronic charge ($e = 1 \cdot 6 \times 10^{-19}$ c), the capacity C is measured in picofarads and the voltage U in millivolts, the result is:

$$dU = 1 \cdot 6 \times 10^{-4} \frac{dn}{C} \text{ mv.} \tag{10.2}$$

If heavy particles (e.g. protons, deuterons, α particles, fission products) act as the ionizing agent the number of ions produced by each particle may be so great that they can be detected individually by means of a sensitive electronic apparatus and ionization chambers designed for such measurements are called pulse or counting chambers. However, if the ionizing agent consists of electrons (β particles from radioactive substances, photoelectrons or Compton electrons created by x or γ-radiation), their effect is measured in the form of the mean charge collected on the collector electrode per unit time (the so-called integrating or dosimetric chambers). The difference between the pulse and integrating ionization chambers is in the time constants RC of the input circuits (fig. 10.3). In pulse chambers resistances of the order of 10^8–10^{12} Ω are used, and with a capacitance of 10^{-12}–10^{-11} farads, a time constant of 10^{-3}–1 sec is obtained.

In integrating chambers the insulating resistance is about $10^{15}\,\Omega$ when measured by the loss-of-charge method and, as a result, the time constant is 10^4 sec and the capacitor discharges very slowly.

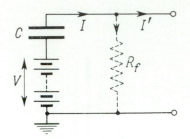

Fig. 10.3. Distribution of current I, I', resistance R_f and capacitance C during a measurement in an ionization chamber. I is the measured current, $I' = V/R_f$.

In measuring ionization currents in gases, the ionization effects in the chamber walls also play an important role because of the long range of photo- and Compton electrons. For this reason ionization chambers are divided into three groups:

(1) Standard chambers in which the whole ionization effect occurs exclusively in the effective volume of the measured gas.

(2) Chambers in which the measured ionization effect is caused not only by electrons produced in the effective volume but also by those electrons from the surrounding layers (so-called free gas chambers).

(3) Chambers in which the measured ionization effect is due to both electrons created in the effective volume and to electrons created in the walls of the chamber (so-called cavity chambers).

In dosimetry practice the cavity chambers are mainly used for relative measurements and chambers with air-equivalent and tissue-equivalent walls play an important part. The former have walls made of substance whose atomic number is equal to 7·64, i.e. equal to the value of the effective atomic number of air, while in the latter group the walls are made of a substance whose structure is similar to the chemical structure of tissue.

In order to eliminate random errors in the measuring of ionization currents a good ionization chamber must satisfy the following conditions:

(1) It must be free of vibrations which cause changes in both the capacitance and effective volume of the chamber, the electrodes should be well fixed.

(2) Insulators should be selected and purified in such a way as to prevent a leakage of charge and a chemical interaction with either gas or liquid in the chamber.

(3) The collector electrode should be surrounded by an earthed ring protecting it against stray currents.

(4) The distribution of the electric field in the chamber should be accurately defined and it should be possible to calculate its numerical value.

(5) The value of the applied voltage should be such as to enable the chamber to work in the region of saturation.

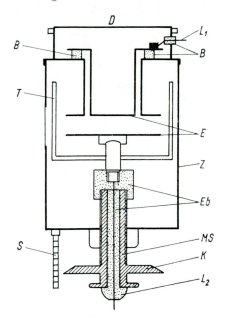

Fig. 10.4. Cross section of a liquid ionization chamber. B, amber insulator; Eb, ebonite insulator; E, chamber proper; T, glass vessel; Z, metal chamber with window; Ms, micrometer thread; K, disc with scale; S, scale; L_1, L_2, metal leads; D, metal cover. Ref.: Adamczewski [1].

(6) The side walls of the chamber should be of such a thickness, or alternatively they should be equipped with glass, quartz or mica windows, to prevent the absorption of too large a part of the measured radiation (this is particularly important for measurements of low-energy x-radiation, soft radiation and for measurements of α and β-particles).

(7) Electrodes and chamber walls should be made of substances which are neither easily contaminated nor form chemical compounds with the gas or liquid filling the chamber (they should be plated with chromium, gold, silver or nickel).

(8) The whole system connecting the collector electrode with the electrometer should be carefully screened electrically, placed in dry air and protected against scattered radiation.

The general theory of ionization chambers and the examples of special liquid-filled chambers are discussed in greater detail in §§ 16.2, 17.1 and 17.3. Examples of some different kinds of ionization chambers used by several research workers are given in this section (figs. 10.4, 10.5, 10.6, 10.7).

9

10.3. *Methods of Measuring very Low Electric Currents*. Electric currents in ionization chambers are measured by means of sensitive measuring devices, e.g. the galvanometers for the current range 10^{-8}–10^{-12} A and by electrometers for an even lower current range of about 10^{-16} A.

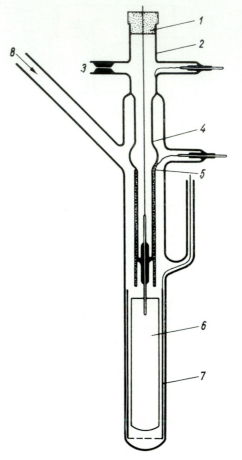

Fig. 10.5. Cross section of a liquid ionization chamber for studies of different temperature. 1, Amber insulator; 2, glass; 3, to diffusion pump; 4, platinum guard ring; 5, brass cylinder; 6, duraluminium rod; 7, copper cylinder; 8, glass connection. Ref.: Pao [3].

Since galvanometers are often used in many laboratories they will not be described, but the most important properties of those electrometers most frequently used should be discussed.

Electrometer measurements are most frequently made in two different ways: (*a*) the voltage drop is measured on a resistance or (*b*) the build-up time of the voltage is measured on a capacitor whose capacity is known.

In the first case the current is determined from the formula $I = U/R$; in the second case from the formula $I = C(dU/dt)$.

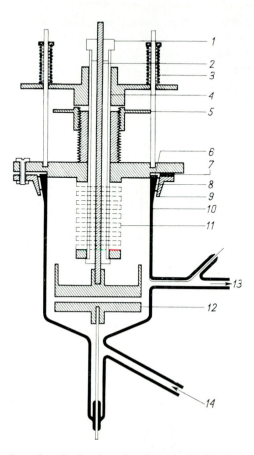

Fig. 10.6. Cross section of an ionization chamber. 1, Amber insulator; 2, brass rod; 3, spring; 4, brass cylinder; 5, brass disc with scale; 6, 20 A wire; 7, paraffin wax seal; 8, packing; 9, metal ring; 10, silvered glass vessel; 11, bellows; 12, brass electrode; 13, tube for emptying; 14, tube for filling. Ref.: Pao [3].

The method of measuring the charge collected on a capacitor:

$$dQ = CdU \qquad (10.3)$$

should be considered separately.

Electrometers can be generally divided in two different groups:

(1) Electrostatic electroscopes and electrometers.

(2) Vacuum tube electrometers (*a*) d.c. amplifiers and (*b*) dynamic amplifiers (conventional and vibrating capacitor electrometers with conversion of direct current into alternating current).

The following quantities serve to define the characteristic features of an electrometer:

(1) Voltage sensitivity $S(V)$—defined by a number of divisions per volt.

(2) Capacitance of the electrometer—C.

(3) Charge sensitivity $S(q)$ defined by the number of divisions per coulomb; $S(q) = S(V) C$.

(4) Range of measured voltage ΔU.

(5) Time for stabilization of the reading.

(6) Insulation resistance.

(7) Input capacitance.

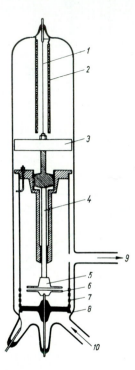

Fig. 10.7. Cross section of an ionization chamber. 1, Thin nickel wire; 2, brass cylinder; 3, soft iron dividing piece; 4, brass rod; 5, brass electrode; 6, copper electrode; 7, thin nickel screen; 8, guard ring; 9, tube for emptying; 10, tube for filling. Ref.: Pao [3].

A list of those electrometers most often in use and the numerical values of their characteristic features, is given in table 10.2.

Within the group of electrostatic electrometers the Lindemann and Lauritsen electrometers are still frequently used. Having a separate battery supply, they have the advantage of being independent of the mains supply and are not only resistant to shocks but are light and easily portable. The Hoffman electrometer which belongs to the same group of electrostatic devices deserves attention only in so far as it operates in a

Table 10.2. Characteristics of different electrometers

Electrometer	$S(V)$ (divisions/v)	C (pF)	$S(q)$ (divisions/c)	ΔU [v]	τ (sec)
The Wulf one-fibre electrometer	30–80	5	1.5×10^{13}	1·2	1
The Wulf two-fibre electrometer	1800	3·5	—	—	—
The Lauritsen electroscope		0·2	0.3×10^{14}	—	—
The Lindemann electrometer	3000 –5000	5	—	—	—
Torsional electrometer	7–60	5–6	10^{13}	1–10	0·1
Twin-plate electrometer	500	100	5×10^{12}	3	10
Quadrant electrometer	1000	20–100	0.4×10^{14}	1	60
The Compton electrometer	50 000– 100 000	30	—	—	60
Multicellular electrometer	—	100	—	150–1500	5–15
Vacuum-tube vibrating-capacitor electrometer	100 000	30	—	2×10^{-4}–10	1–30
Vacuum-tube direct-current electrometer	2000	20	10^{14}	2×10^{-4}–10	10^{-3}–20

vacuum and has a very light suspension. For this reason, its voltage and charge sensitivity can be maximal but it is not resistant to shocks. A diagram illustrating the construction of a Lindemann electrometer is

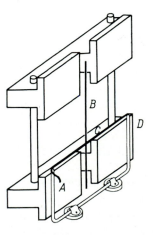

Fig. 10.8. Basic component of a Lindemann electrometer (one pair of quadrants not shown). *A*, input to the ionization chamber; *B*, indicator torsion wire; *C*, suspension; *D*, quartz frame. Ref.: Campbell and O'Connor [6].

given in fig. 10.8. This electrometer is based on the quadrant electrometer principle. A tiny quartz needle is placed between the quadrants and is fixed to a 6 μm thick thread. When the measured potential is applied to the needle, the latter then turns inside the quadrants, one pair of which carries a positive charge and the other a negative charge.

Figure 10.9 shows a diagram of a circuit which includes an ionization chamber K and an electrometer E with its associated resistance and voltage values. The voltage sensitivity of this electrometer amounts to about 4000 divisions per volt and the current sensitivity may be as high as 10^{-16} A per division. Deflexion of the indicator on the scale is observed by means of a microscope.

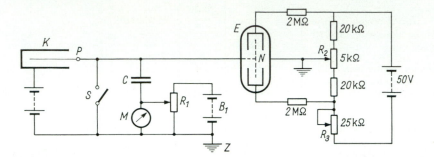

Fig. 10.9. Circuit diagram for a Lindemann electrometer. K, ionization chamber; P, electrode; E, electrometer; S, switch; C, capacitor; R_1, R_2, R_3, resistances; B_1, cell; M, microammeter. Ref.: Campbell and O'Connor [6].

The increasingly often-used vacuum tube electrometers are based on the action of special electron tubes, called 'electrometer tubes', and on high ohmic resistors of about 10^9–10^{13} Ω. The main features of electrometer tubes are their very high input resistance—$R_{in} = 10^{14}$–10^{16} Ω—and a very low grid current—$I_g = 10^{-14}$–10^{-16} A—whereas in ordinary radio valves these parameters are in the region of $R_{in} = 10^8$–10^9 Ω, $I_g = 10^{-8}$–10^{-10} A. The input capacity of the electrometer tubes is about 3–10 pF. There are more than twenty types of known electrometer tubes, e.g. FP-54, T-113, T-114, 4060, XE2, XE3 or the substitutes 6Z1Z, 955, etc. The most common are triodes or single and double tetrodes. The plate current in a tetrode is in the region of tenths of a milliampere, while the screen grid current is several times higher. The amplification factor is almost 1.

The highohmic resistors of about 10^{10}–10^{13} Ω present a special problem, because the common radio resistances are not higher than about 10^9 Ω. Highohmic resistors are produced by several methods including a bakelite composition mixed with soot, glass rods platinum-coated and placed in a vacuum pipe, glass rods with a carbon layer, the liquid resistors of defined chemical compounds (frequent polarization is a disadvantage) and ionization resistors based on the principle of a gaseous ionization chamber containing a constant α or β radioactive source, which can be partially shielded (so-called Bronson resistors).

Vacuum tube electrometers can work in different connections: as common circuits with an anode or cathode meter, as bridge circuits (Du Bridge, Barth, Caldwell) or as balanced circuits with two tubes or one double tube. The circuit must be adequately balanced in order to function in stable conditions and, as far as possible, should be independent of ambient temperature fluctuations and of the cathode filament current.

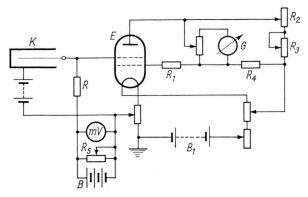

Fig. 10.10. Circuit diagram for a Du Bridge and Brown electrometer. B, B_1, cells; R, R_1, R_2, R_3, R_4, R_5, resistors; G, galvanometer; K, ionization chamber; E, electrometer. Ref.: Campbell and O'Connor [6].

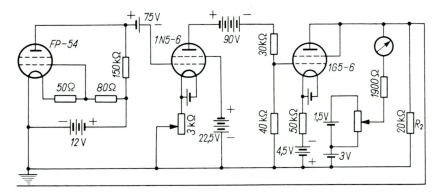

Fig. 10.11. Electrometer circuit using an FP–54 valve.

Further details concerning the construction of vacuum tube electrometers can be found in the works of Bierleyev [8], Bonch-Bruyevich [9], Schintlmeister [10], Du Bridge [11], Campbell and O'Connor [6], Price [2] and Aglintsev [7]. In figs. 10.10 and 10.11 diagrams of the two vacuum tube electrometer circuits illustrate the frequently used Du Bridge and Brown circuit [11] and a circuit which was used by Gzowski and Terlecki [12] in their research described in § 12.4.

In recent years the vibrating capacitor electrometer is more frequently applied to the measuring of extremely low currents in the order of 10^{-15}–10^{-17} A. With this electrometer the measured charge q is collected on a capacitor whose capacitance C periodically varies with time owing to the vibrations of one of the electrodes. Let the frequency of the vibrations be ν, then

$$C_t = C(1 + k\sin 2\pi\nu t), \qquad (10.4)$$

where C is the mean capacitance and k is a constant depending on the amplitude of the vibration. Therefore, the measured potential difference between the electrodes of the condenser will be:

$$V_t = \frac{q}{C(1 + k \sin 2\pi\nu t)}. \tag{10.5}$$

This is a periodically alternating voltage and it can therefore be amplified far more easily than a steady voltage.

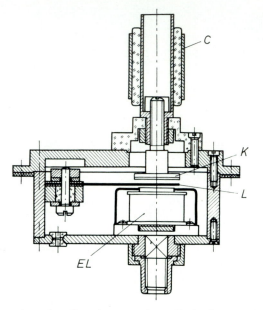

Fig. 10.12. Cross section of a vibrating capacitor. Ref.: Campbell and O'Connor [6].

In fig. 10.12 a cross section of a vibrating capacitor is shown. The capacitor consists of two highly polished steel plates, one of which (the upper one, called an anvil) is placed on a high-quality insulator (e.g. polystyrene) and can be connected to an ionization chamber. An elastic plate is the lower electrode of the capacitor and can be set into vibration by means of an electromagnet. A block diagram of the whole circuit with the ionization chamber K, vibrating capacitor C, alternating current amplifier W, oscillator O and phase-sensitive detector D is shown in fig. 10.13.

The resistor R (fig. 10.13) is only used when the current method is applied and it is removed when the voltage is used. Such a circuit normally works with a negative feed-back in order to eliminate the fluctuations due to variations of the amplitude k. Detailed descriptions of the vibrating capacitor electrometers can be found in the works of Thomas and Finch [13] and in the book by Campbell and O'Connor [6]. The following parameters of an electrometer of this type are typical: vibration frequency of capacitor—550 c/s, vibration amplitude—0·5 mm, electrode gap (minimum)—0·13 mm, coupling capacitance—8 pF, input resistance—$10^{15}\,\Omega$ and the measured current—10^{-15} A for $R = 10^{12}\,\Omega$. Circuits based on the principle of charging a capacitor by the measured current are also used and currents as low as 10^{-17} A and charges of 4×10^{-16} c can be measured with these circuits with about 2% error.

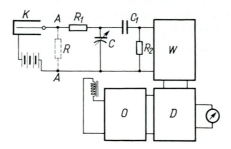

Fig. 10.13. Circuit diagram for a vibrating reed electrometer (vibrating capacitor). W, a.c. amplifier; O, oscillator; D, phase-sensitive detector; K, ionization chamber; R, R_1, R_2, resistors; C, variable capacitor; C_1, capacitor. Ref.: Campbell and O'Connor [6].

Table 10.3 shows selected examples of several different types of vibration electrometers as well as their parameters.

Table 10.3. Parameters of some vibration electrometers

Type	Input resistance (Ω)	Resistance (Ω)	Current sensitivity (A)
Ekco Electronics (England)	10^{15}	10^8–10^{12}	3×10^{-15}
S.A.I.P. (France)	10^{15}	10^8–10^{12}	5×10^{-16}
Frieseke (Germany)	10^{15}	10^9–10^{11}	3×10^{-15}
Intertechnique (France)	2×10^{15}	10^7–10^{12}	1×10^{-16}
Victoreen 474 (U.S.A.)	10^{15}	3×10^8–3×10^{11}	3×10^{-15}
Victoreen 475 (U.S.A.)	10^{16}	10^8–10^{12}	3×10^{-16}

In some cases the current measured in the ionization chamber can vary greatly (e.g. by several orders of magnitude) and logarithmic electrometers are then used, with their response proportional to the logarithm of the current. (This type of electrometer is shown in the simple diagram in fig. 17.20.) The main difference between such a measuring circuit and a common direct current amplifier is that in place of an electrometer tube with an input resistance R_{in} an electronic tube is used. This latter tube can be the 954 pentode which gives the following relationship between the external voltage U and the current I:

$$U = A \log BI, \tag{10.6}$$

where A and B are constants.

Such electrometers can be used to measure currents in the range of 10^{-7}–10^{-12} A. The resistance of the ionization chamber should be high in relation to the input resistance of the tube, if the electrometer is to function correctly.

This electrometer has one drawback, namely that its response depends to a large extent on the characteristics of the tube and is not always proportional to the measured current and for this reason it is necessary either to work only in the linear section of the characteristic or to apply additional electronic circuits to compensate for the distortion.

Chapter 11

Electrical Conduction in Dielectric Liquids at low Fields

11.1. *Natural Conduction* (*Self-conduction*). When a liquid sample which has been carefully purified by physical and chemical methods described in § 4.1 is tested in an ionization chamber which has been carefully cleaned and dried, it is found that after applying the voltage, the electric current usually decreases rapidly with time to a certain minimum value. Such a process is often called 'electrical purification', because it is evident that a liquid, subjected to the effect of the electric field, is characterized by a reduction in its electrical conductivity even to the extent of several orders of magnitude (sometimes in a period of less than 100 hours). After such a process of purification the liquid does not regain its initial conductivity value, unless it has been contaminated again. There are no signs of polarization and this means that on the whole there is a symmetry of charges (positive and negative) collected on the electrodes of the ionization chamber. Both the rapidity with which the electric current in the chamber decreases immediately after the electric field has been applied, and the residual value of the current indicate whether the liquid is well purified and suitable for careful measurements.

The mechanism of electrical purification has not yet been fully clarified. It is generally thought that a sample of liquid immediately after distillation may still include traces of electrolytic impurities, air and oxygen bubbles or vapour bubbles of the same liquid and other impurities may also be present if the chamber has not been thoroughly cleaned. For this reason it is vital that the chamber (vessel, electrodes, insulators) should be properly cleaned and the final stage of this process is the rinsing of the chamber with the liquid under investigation. Purification by means of the electric field may include the following processes:

(1) Removal of traces of electrolytic impurities from the effective volume of the chamber.

(2) Disappearance of gas bubbles accumulated in the liquid (mainly air and oxygen).

(3) Disappearance of gas bubbles adsorbed on the surface of the electrodes.

(4) Removal of the accidental electric charge carriers (free electrons, ions) accumulated in the liquid before it is poured into the ionization chamber.

(5) Possible rearrangement of molecules in the liquid into a quasi-crystalline lattice and a slow transition of the liquid sample from an n-type semiconductor into an insulator.

With imperfectly purified liquids the first process is generally very slow and rarely yields useful information. The result of electrical purification is relatively small and after the voltage to the chamber has been removed, the chamber polarizes and occasionally the liquid regains its initial conductivity value. In such a case it is usual to change the liquid sample and in the event of a similar occurrence, the whole distillation process should be repeated.

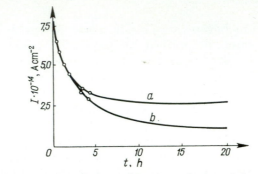

Fig. 11.1. Intrinsic current I as a function of time (electrical purification); *a*, experimental curve; *b*, theoretical curve. Ref.: Blanc *et al.* [35].

The processes (2), (3) and (4) normally lead to a rapid and irreversible disappearance of conductivity immediately after switching on the voltage and then proceed to change very slowly until the final value of conductivity is reached. The last process (5) is as yet hypothetical and special measurements are required to confirm its existence. An example of decay curves of the measured electric current I in the time t when the electric field is active is shown in fig. 11.1. The curve *a* corresponds to a liquid with electrolytic impurities and the curve *b* (theoretical) corresponds to a pure liquid, in which the final value of the current I should be two to three orders of magnitude lower than the initial value.

Although the phenomena occurring during the electrical purification of a liquid described above have been generally recognized by almost all those who have investigated the electrical conductivity of liquids, the essence of the problem has not yet been explained. It is the present writer's opinion that the first purification process which takes place very rapidly has an exponential form of the type:

$$I_1 = I_{01} \exp(-\lambda t), \tag{11.1}$$

since the amount of removed impurities (gas or vapour bubbles) is proportional to the total number of impurities or bubbles. I_{01} denotes the initial current and λ is a constant. The half-life time of such a process $\tau = 0 \cdot 693/\lambda$ is estimated to be about 15 to 30 min depending on the liquid. Further electrical purification should be attributed to more subtle changes in the liquid structure, caused by the rearrangement of certain regions of the quasi-crystalline molecular lattice; a similar rearrangement can be noted in some electronic semiconductors in solids. The difference between solids and liquids lies in the fact that in liquids such distortions, deformations and impurities are mobile and can be removed by an electric current flowing through the liquid. It is possible to describe such a phenomenon in a way similar to that adopted by Fowler [104] for polyethylene:

$$I_2 = \frac{I_{02}}{1 + (\alpha/kT) I_{02}(W - W_0)t},$$ (11.2)

where W_0 denotes the lowest trap level, W—other trap levels, k—Boltzmann's constant, T—absolute temperature, I_{02}—the initial current of that type, and α is a constant. This phenomenon causes in a number of hours a slow but systematic decrease in the electrical conductivity of the liquid under the influence of an electric current. It has been observed in work done by the present author, Gzowski and Terlecki [12], Pao [3] and Blanc *et al.* [35] that these two different processes were occurring in the natural conduction of the liquid. A system of two intersecting straight lines was usually obtained when plotting the logarithm of current against time. Blanc *et al.* [35] have given some attention to the electric purification of liquids and found the relationship:

$$I = \frac{I_0}{1 + bt},$$ (11.3)

where b is a constant and I_0 denotes the initial value of the current. The results obtained, however, were not entirely consistent since their experimental curves corresponded more closely to the two different processes described in the eqns. (11.1) and (11.2).

Much attention was given to the work on the degassing of dielectric liquids in the recent investigations of such English scientists as Sletten and Lewis [155], Kahan and Morant [121, 154, 348]. Their experiments with the whole cycle of the physical and chemical purification of the liquid, fractional distillation, filling the ionization chamber and the measuring of the electric current in the chamber took place in a hermetically sealed glass apparatus, which could be completely de-aerated and filled with another gas, e.g. nitrogen. These investigations have not—so far—produced uniform results. In some cases the electrical conductivity of the liquid has exhibited a strong dependence on the degree of its degassing, while in other cases this dependence has been rather slight. In the author's opinion the process of electrical purification, being also a process of degassing, might in many cases be a substitute for very tedious investigations carried out in a completely sealed glass apparatus whose purpose is to degas the liquid.

Gzowski, Terlecki [12, 307] have made preliminary measurements of the dependence of initial currents on time and temperature. They established the existence of two processes in this phenomenon, the first of which lasts a number of minutes while the second lasts a number of hours. The $\ln I = f(1/T)$ plot showed two intersecting straight lines and the author believes that the second of these lines corresponds to the rearrangement of the inner structure of the liquid. In order to avoid the possibility of contaminating the liquid while it is being transferred from the distillation column to the ionization chamber, those conducting the experiments placed the chamber in the closed system of the distillation column and frequently rinsed it with the liquid under investigation before taking measurements. This procedure was adopted by Nikuradse [23] and several specialists from Queen Mary College in London. Such a method, however, did not influence the reduction of the residual value of self-conductivity in the case of many saturated hydrocarbons. For this reason, this rather troublesome experimental method is not in general use.

Apart from the question of electrical purification of liquids, attention should be given to those phenomena which result from the molecular structure of liquid and influence the conduction mechanism of the electric current. There is a very marked difference in this respect between saturated and aromatic hydrocarbons in which appear π-electrons delocalized in the molecule. Saturated hydrocarbons, when in a state of the highest chemical purity and after careful electrical purification, show an electrical conductivity of the order of 10^{-19}–10^{-20} $\Omega\,cm^{-1}$, whereas the natural conductivity in aromatic hydrocarbons, after the same purification treatment, is greater by about 10^4.

In recent years a great number of measurements have been taken in the aromatic hydrocarbon group, mainly in benzene (Nowak [122], Forster [119, 147–9], as well as Chong [157] and Zając [213]).† Nowak was chiefly concerned with ionization conduction in benzene and he measured currents, mobilities and ion recombination. Forster's investigations were confined to self-conduction, but he included a large group of liquids in his research. He performed a number of experiments on the electric field distribution in the liquid under investigation and arrived at some interesting conclusions concerning the mechanism of electric conduction in those liquids.

According to Forster's theory, self-conduction consists of two different processes: electronic conduction σ_e and conduction σ_d caused by excited molecules. In the former process a free electron jumps from one molecule to a neighbouring molecule where it is held up for a short time (this is called the trap conduction model). The latter process is based on the hypothesis that a liquid at room temperature contains a small but strictly defined number of molecules, which are excited to the lowest degree of excitation. Rice and Choi [150] maintain that when such molecules are in collision they might produce both a positive and a negative ion. Forster has shown that it is possible in many cases to combine the activation energy of conduction with the excitation energy of the molecule. Many other scientists have concerned themselves with self-conductivity in liquids but mainly from the point of view of temperature-dependence and for this reason they are discussed below (§ 11.6).

† See also Forster and Langer [342] and Sharbaugh and Barker [356].

Apart from the phenomena discussed above, the following causes of self-conduction in dielectric liquids have also been suggested:

(1) Electron emission from the cathode which takes place under the influence of an applied electric field—this is especially true in high fields.

(2) The influence of traces of electrolytic impurities in the liquid.

(3) Dissociation of molecular impurities still present in the liquid.

(4) Dissociation of the molecules of the investigated liquid as a result of an electric field.

(5) Ionizing influence of radioactive impurities present in both the walls of the vessel and in the air.

(6) Ionizing effect of cosmic rays.

(7) Temperature processes similar to those in semiconductors.

Under adequate experimental conditions each of the causes given above may play some part: at high electric field stress (in the range of hundreds of kv/cm) for example, one can easily observe the influence of electron emission from the cathode. This phenomenon will be discussed in detail in Part IV, § 16.7. The influence of electrolytic impurities has been studied by Standhammer and Seyer [189] who determined the number of ion pairs produced in cyclohexane when gradually mixed with water. They established that the number of ion pairs produced per cm^3/sec— and therefore the electric conduction in such a liquid—depended on the concentration of water. At 50°c the maximum number of ion pairs per cm^3 sec in cyclohexane fully saturated with water was $1 \cdot 6 \times 10^5$, whereas the minimum number (only obtained by an extrapolation to completely pure cyclohexane) was 200 ion pairs per cm^3/sec, which is of the same order of magnitude as the value obtained by the author's direct measurements [18] (cf. § 11.2). It is worth noting that in his first measurements of currents in dielectric liquids, Jaffe [14] estimated the number of ions to be about 150 ion pairs per cm^3/sec.

Both Onsager [64] and Reiss [22] discussed the possibility of the dissociation of impurity molecules in the liquid. The possibility of the dissociation of the molecules of the liquid itself taking place under the influence of high electric fields was examined by Plumley [21], who developed a complete theory of this process which will be presented in § 16.7. This theory, however, was not generally well received.

11.2. *The Influence of Cosmic Rays.*   Some attention should be devoted to the investigations of different scientists concerned with the influence of cosmic rays on the electrical conductivity of highly purified dielectric liquids. According to Jaffe [16], Adamczewski [18], Białobrzeski [17] and Rogoziński [19], this effect can be observed only in the highest purity grade liquids whose natural conductivity is of the order of 10^{-19}–$10^{-20}\ \Omega^{-1}\,cm^{-1}$. After the ionization chamber has been shielded with a thick layer of lead (up to 15 cm), this effect is dominant when the measured ionization current decreases in proportion to the thickness of the lead layer. The influence of cosmic rays can account for the fact that different investigators (Jaffe, Adamczewski, Nikuradse and Rogoziński), although working under different experimental conditions, achieved identical boundary values of the self-conductivity in the liquid

$$(\sigma_n \approx 10^{-19}\text{--}10^{-20}\,\Omega^{-1}\,\text{cm}^{-1})$$

despite differing methods of purification. While the initial self-conductivity of the order of $10^{-12}\text{--}10^{-13}\,\Omega^{-1}\,\text{cm}^{-1}$ in some pure dielectric liquids can be relatively easily lowered by several orders of magnitude, the lower limit has not been exceeded by any of the investigators except those who shielded the ionization chambers with thick layers of lead.

As early as 1936 the author [18] measured self-conduction currents in hexane in order to estimate the influence of cosmic rays. He experimented with a large ionization chamber consisting of a number of flat capacitors with the total area of collecting electrodes being 540 cm² and the effective volume of the liquid being 810 cm³. The liquid was carefully cleaned ($\sigma_n = 10^{-19}\,\Omega^{-1}\,\text{cm}^{-1}$) and the whole chamber surrounded with a lead shielding screen of 5 cm thickness. Under such conditions the current density at a stress of 2000 v cm⁻¹ was 10^{-16} A cm⁻². The author has calculated that under such conditions about 420 ion pairs per cm³/sec are produced in hexane; after a simple density calculation this result would correspond to a value of approximately 0·7 ion pairs per cm³/sec in air under normal pressure. Having assumed that the average flux of incident cosmic particles on the earth's surface is 1·48 particles per cm²/min (according to Millikan), the author concluded that in hexane, cosmic particles would create about $1·7 \times 10^4$ ion pairs per 1 cm track, which would correspond to 28·6 ion pairs per 1 cm of air. Without accepting the validity of these detailed calculations, it can be generally stated that the lowest value of the electric conductivity of the purest dielectric liquids ($10^{-19}\text{--}10^{-20}\,\Omega^{-1}\,\text{cm}^{-1}$), as established by different scientists, is influenced by factors including the effect of the cosmic rays penetrating every substance on the earth.

On the basis of those observations concerning the influence of cosmic rays on the conductivity of dielectric liquids, Białobrzeski and Adamczewski [24] as early as 1935 used liquid-filled ionization chambers to investigate the secondary effects of cosmic rays—namely, the so-called ionization 'bursts' of Hoffmann (*Hoffmannstösse*). These have been observed in gas-filled ionization chambers by Hoffmann [25], Compton *et al.* [26], Steinke [29] and others. During the observation of the ionization current in the chamber it was noted that, from time to time, rapid increases of the current took place under the influence of such phenomena as 'cascades' and 'showers' of ionizing particles which were caused by high-energy cosmic particles. On the whole, this phenomenon occurred very rarely (only once in many hours of experimenting) and its frequency increased with growing pressure and, therefore, with density. For this reason some research workers used long ionization chambers filled with gas (argon, helium) subjected to pressures of 40–60 atm.

Large chambers filled with hexane were used for this purpose by Białobrzeski and Adamczewski [17]. The frequency of registered pulses was as high as 1·5 per hour. The highest pulses registered on photographic paper approximately corresponded to the number of $5·9 \times 10^7$ ion pairs, the most frequent pulses being about 2×10^7 ion pairs. The ionization chamber consisted of a cylindrical capacitor with a liquid capacity of about 800 cm³ and with a lead or aluminium cylinder placed inside. The chamber was surrounded by a thick layer of lead. The self-conduction current in the liquid was balanced with a Bronson resistance. In later investigations of a similar nature, the author [63] used a liquid-filled ionization chamber connected with a system of four Geiger–Müller counters which were placed both above and below the chamber and worked in the coincidence régime for the purpose of detecting 'showers'. Readings of individual 'showers' on the counter and ionization 'bursts' in the chamber were registered simultaneously. A great number of 'showers' was observed and a small number of 'bursts' (1 or 2 per hour) which frequently coincided with one of the 'showers'. Unfortunately, a large proportion of the collected statistical data was destroyed during the last World War.

11.3. *Ionization or Induced Conduction in Gases.* Research on ionization conduction in dielectric liquids does not involve so many difficulties as that on natural or self-conductivity. In a very pure liquid the current may increase hundreds of thousands of times (when ionizing with x-rays, for example) and may vanish immediately after ionization ceases and when the remainder of the ions are removed from between the ionization chamber (in a fraction of a second). The measured effects can be reproduced with great precision and there is general agreement between the results obtained by various research workers. Despite this, there are, in this field, basic phenomena which so far cannot be explained in any detail.

Because of a strong resemblance between the mechanisms of electric conduction in ionized gases and dielectric liquids, the most important features of conduction in gases will now be recapitulated. More detailed information on this subject can be found in the following textbooks: *Ionized Gases* by von Engel [224], *Basic Data of Plasma Physics* by Brown [67] and *Handbuch der Physik* [134] and *Fundamental Processes of Gaseous Electronics* by Loeb [135]. A number of special problems concerning electric conduction and break down in gases and vapours can be found in the following works: Crowe [298], Heylen [221–23], Craggs *et al.* [195, 234, 299] and Kuffel [229].

Under normal conditions gases are not conductors of electricity but they can conduct an electric current, which may be large, if a great number of ions are produced in the gases by means of an ionizing agent such as heating with a flame, irradiation with x-rays, γ-rays, α and β-particles, protons and deuterons, etc. The activity of ionizing agents

10

can be explained in the following manner: at the expense of the energy of incident radiation, single electrons are separated from neutral atoms or gas molecules (ionization process) and the electric charge is distributed; the positive charge remains with the rest of the atom or molecule thus forming a positive ion, while the negative charge initially moves in the form of a free electron which, after a number of collisions with the neutral molecules of the gas, becomes attached to a neutral atom, molecule or even a whole complex of molecules and forms a negative ion. As a result, after each act of ionization one pair of ions, both a positive and a negative

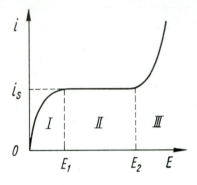

Fig. 11.2. The dependence of ionization current in gases on the electric field E; i_s, the saturation current.

ion, is produced and the energy used for this purpose is more or less constant and its average value is for air about 34 ev. Once the energy of the incident particle is known, it is possible to calculate the number of ions this particle will produce after having been completely stopped in the gas. For example, an α-particle of 3·2 мev energy will form 10^5 ion pairs on its track. If the ionized gas fills a capacitor and an electric field is applied across the electrodes, the ions produced will drift towards electrodes of opposite signs and in the circuit of such a capacitor it is possible to detect the electric current by means of a sensitive electrometer. This current is known as an ionization current. The intensity i of the ionization current depends to a large extent on the voltage U applied across the electrodes, i.e. on the strength E of the electric field in the inter-electrode region. If the field is uniform (e.g. in a parallel-plate capacitor) the field strength at the electrode separation d will be $E = U/d$. The relationship $i = f(E)$ for ionized gases is shown in outline in fig. 11.2. It can be seen from the diagram that at the beginning of the curve for low fields (in region I), the intensity of the current i increases (non-linearly) with the growing strength of the field to a certain saturation value i_s. At the field E_1 the current reaches a constant value (the saturation current i_s) and up to the value $E = E_2$ a further increase in the field does not cause the current to increase. At the value $E = E_2$, the current starts increasing up to the spark discharge (region III).

Region II, in which there is a saturation current, is the easiest to explain. In this region of the electric field all the ions produced in the gas are carried away by the field to the electrodes of the capacitor and are registered on the electrometer. In region I the phenomenon of ion recombination is of great importance. At low values of the electric field, when the transit time of the ions between the electrodes is sufficiently long, recombination causes the neutralization of a certain fraction of the ions and the lower the voltage, the more ions are neutralized.

The value of the saturation current i_s is calculated on the basis of the number of ions q produced by the ionizing agent in 1 cm³/sec, charge e of the ions, the electrode area S and the distance d between the electrodes:

$$i_s = qeSd. \tag{11.4}$$

Above the saturation region there again occurs an increase in the current with an increase in the strength of the field which can be explained as follows: the ions of the gas acquire such a high velocity $\bar{v} = uE$ (with u denoting the mobility of the ions) that their kinetic energy $mv^2/2$ is sufficient to ionize other neutral atoms or molecules in the gas. It is found that the logarithm of the current is proportional to the electrode separation for a given field strength and thus showing an avalanche process in the growth of the electron and ion population. This means that every electron or ion produced acquires sufficient energy in the field to produce a larger number of ions. The final region ends with a spark discharge, i.e. with a current so high that the gas is intensively heated along the narrow channels connecting the electrodes.

The above-mentioned problems concerning the electrical conduction in ionized gases will be discussed in more detail from a mainly theoretical point of view in Chapter 16. Analogies will be drawn between problems concerning liquids and gases. Detailed information is already available for the latter. A general description of those phenomena which take place in ionized gas under normal pressure becomes very complicated when the pressure of the gas is increased. As a rule, the ionization current then increases in proportion to the density of the gas but this phenomenon is dependent on a number of factors, such as the type of ionizing radiation and its energy, the range of electrons produced in the gas, uniformity of ionization, the kind of gas, etc.

In figs. 11.3, 11.4 and 11.5 diagrams are given showing the relationship between the ionization current and both the density and pressure of several gases. It can be seen from the diagrams that the ionization current grows with the increasing density and pressure of the gas. Therefore, for example, in the analysis of phenomena connected with cosmic rays, Steinke [29] and others used ionization chambers of about 100 cm in length, filled with heavy gases (argon) at a pressure of 40 atm. It is more difficult to obtain a saturation current in gases under high pressure, but a further gradual increase in the current is possible and can be expressed by the general formula:

$$i_s = i_0 + f(E). \tag{11.5}$$

This can be generally explained by the fact that in a medium of high density some additional types of recombination occur (columnar, electron, group

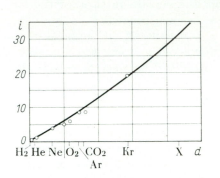

Fig. 11.3. The dependence of ionization current i on density d for various gases.
Ref.: Telegdi and Zünti [27].

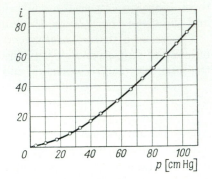

Fig. 11.4. The dependence of ionization current i in gases on pressure p (at low pressures). Ref.: Telegdi and Eder [28].

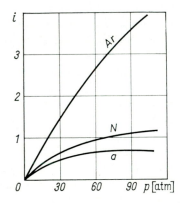

Fig. 11.5. The dependence of ionization current i on pressure p for various gases (at high pressures). Ref.: Compton *et al.* [26].

recombination, etc.—cf. §16.5). These distort the uniform space distribution of the ions, such as would exist under normal conditions, and, consequently, higher field strengths are necessary to protect the electrons and ions from recombination.

Undoubtedly, however, other phenomena exist which may be of importance in the region of the electric fields above the so-called saturation level. These include secondary electron ionization, ionization as a result of the collision of excited atoms, molecular dissociation, etc.

11.4. *Ionization or Induced Conduction in Dielectric Liquids.* In dielectric liquids the phenomenon of the lack of saturation of ionization currents is clearly apparent. In such a case, there is a linear increase in

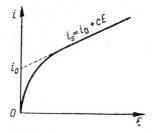

Fig. 11.6. The dependence of ionization current i on electric field E in dielectric liquids.

the current with electric field strength as shown in fig. 11.6. The ionization current i_s can be expressed as a function of the field strength in the following way:

$$i_s = i_0 + cE. \tag{11.6}$$

The coefficient c depends on the number of ions produced in the liquid by the ionizing agent, electrode separation, the energy of ionizing radiation and also depends on the physical and chemical properties of the liquid. Curie [20], Jaffe [15], Białobrzeski [17], Adamczewski [1], Pao [3], followed by the contemporary works of Terlecki [33], Jachym [46], Blanc *et al.* [35], Nowak [122], Gazda [126], Vermande [166] and others who have done research on conduction in ionized dielectric liquids, have all arrived at the general form of this equation.

In experiments with liquids not completely purified, however, ionization was superimposed upon the electrolytic effect of the impurity traces and the process of ionization was distorted. This fundamental error in research work on conduction in dielectric liquids is, unfortunately, made quite frequently even in contemporary investigations and the results obtained are therefore of limited value or are completely distorted.

During the years 1932–39 the author carried out a large number of experiments [1, 17, 18, 41, 43] on the characteristic features of electrical conduction in dielectric liquids, mainly in the group of saturated hydrocarbons subjected to processes of careful physical and chemical purification. The methods of purification were based on specialist literature and were put into practice under the supervision of the most eminent Polish physicists and chemists, including Professor Świętosławski. Using these methods, it was possible to obtain self-conductivity limits in liquids (cf. §§ 11.1 and 11.2) and to remove all traces of electrolytic impurities from the liquids investigated. This facilitated extensive research on physical magnitudes which characterize conduction both in dielectric liquids and in gases, and in particular research was made possible on the coefficients of mobility and the recombination of ions produced in the liquid exposed to x or γ-radiation. At a later date, Pao [3], Reiss [22], Gzowski [32], Terlecki [34], Blanc *et al.* [35], Gibaud [36], Ivanov [37] and Vermande [166], together with a large group of research workers in our Institute, carried out a series of experiments in this field. These will be discussed below. It should be stated that in recent years considerable activity has been observed in this branch of physics. This is a result of the great importance of this research for the theory of the electric strength of liquid dielectrics (to be discussed in Part IV), and of the fact that while both the ionization mechanism and the mechanism of electrical conduction in dense media of the best insulators (gases under high pressure,

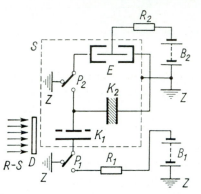

Fig. 11.7. Diagram of apparatus for measuring the electrical conductivity of dielectric liquids. K_1, ionization chamber filled with liquid; K_2, supplementary capacitor; E, electrometer; B_1, B_2, accumulators giving 2000 v and 50 v; R_1, R_2, resistors; R–S, x-ray beam; D, screen; S, earthed electrostatic screen; P_1, P_2, platinum switches. Ref.: Adamczewski [1].

dielectric liquids and, to some extent, solids) play an important part in many theoretical and practical fields (e.g. in plasma physics, dosimetry of ionizing radiation, radiation chemistry, radiobiology, etc.), they are not yet completely understood. There are several different theoretical interpretations of ionization and electric conduction mechanisms in liquids. Such interpretations are based on the phenomenon of preferential recombination (Lea [80], Kramers [184], Jaffe [15]), electron 'hopping' (Crowe [79], Stacey [38], Le Blanc [57]) and on the process of ionization or secondary dissociation of excited molecules or on the change in the activation energy (Adamczewski [94, 270]). Also Frenkel's [101] theory, or perhaps the theory of electronic semi-conductors, might contribute to the interpretation of this phenomenon. These theories will be discussed in Chapter 16, following the presentation of those experimental results which have so far been obtained.

    The main results of the author's work on current voltage characteristics in liquids ionized by x and γ-rays under varying experimental conditions will be discussed first. A simplified diagram of the measuring apparatus used by the author [1, 40, 41] is given in fig. 11.7. The measuring circuit was enclosed in an electrostatic screen made of zinc sheet while the air inside was dried using calcium chloride. The ionization chamber most frequently used by the author is shown in fig. 10.4. For various experiments, however, the author used several chambers differing in shape with their volume varying from several millilitres to 10 l. A Dolezalek quadrant electrometer with a capacitance of about 80 pF and a sensitivity

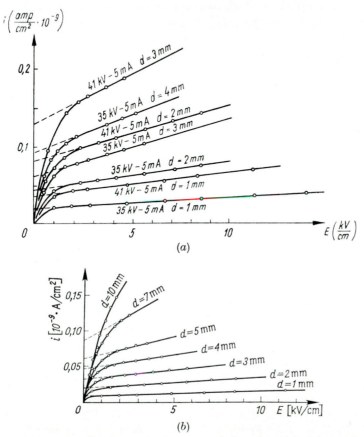

Fig. 11.8. (a) Current density as a function of electric field strength in ionized hexane using x-rays under various experimental conditions; the numerical values denote voltage on the x-ray tube, electron current in the tube and electrode separation d. Ref.: Adamczewski [1]. (b) i as a function of E for various electrode separations d with constant ionization (30 kv, 5 mA). Ref.: Adamczewski [1].

between 300 and 1500 mm/v was most frequently used. This instrument was suitable for most current and charge measurements, while for other measurements a Lindemann and Hoffmann (vacuum) electrometer of a very high-charge sensitivity was used.

Some examples taken from the author's work [1] on current–voltage plots for different compounds of the alkane group are given in fig. 11.8(a), (b). X-rays came from a Seemann apparatus which was demountable and had a copper anode and was connected to a system of vacuum pumps giving vacuum of the order of 10^{-5} mm Hg.

The x-ray equipment operated in an autoregulation circuit within the current and voltage limits of 5 mA, 25 kv to 25 mA, 41 kv. The maximum load of the generator was 30 mA, 125 kv. The incident beam of radiation, filtered through an aluminium window, passed through 3 cm air, two glass panels, a 0·19 mm thick aluminium foil and a 3 cm layer of liquid. The experimental conditions are described here in detail because the dose in the liquid has not been accurately specified and the data may be needed for comparison with the results obtained by other research workers.

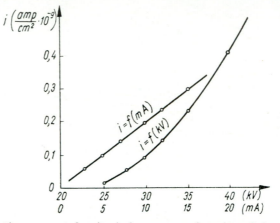

Fig. 11.9. Ionization current density in hexane as a function of the electron current in the x-ray tube ($i = f$(mA)) and the voltage applied to the tube ($i = f$(kv)). Ref.: Adamczewski [1].

The voltage up to 2000 v was taken from a low-capacity accumulator battery. The whole volume between the electrodes of the ionization chamber was ionized, while the separation of the electrodes was altered from 0 to 20 mm. Figure 11.9 shows diagrams of the ionization current i in the ionization chamber as a function of the voltage on the x-ray tube $i = f$ (kv) and of current in the tube $i = f$ (mA). As can be seen from the diagram, the function $i = f$ (mA) is linear while the curve $i = f$ (kv) is almost parabolic. The course of both curves is consistent with theoretical data. Under these conditions, the x-ray beam consists of the characteristic spectrum of copper (excitation potential of K series is 8·86 kv) and a continuous spectrum limited by the short-wave boundary

$$\lambda_m = 0\text{·}309 \text{ Å (at 40 kv)} \quad \text{and} \quad \lambda_m = 0\text{·}494 \text{ Å (at 25 kv).}$$

It is well known that under such experimental conditions, x-ray intensity can be expressed by the formula:

$$i = \text{const.}\, U^{3/2} IZ, \tag{11.7}$$

where U denotes the constant voltage applied to the tube, I—current in the tube and Z—atomic number of anode material.

Ionization currents measured during irradiation by x-rays oscillated from 6.8×10^{-12} A cm^{-2} at 25 kv, 5 mA to 166×10^{-12} A cm^{-2} at 41 kv, 5 mA. A radium source (11·95 mg), placed at the same distance as the x-ray tube, induced a current of 0.865×10^{-12} A cm^{-2} (at the electrode separation $d = 0.3$ cm and the voltage across the ionization chamber $U = 722$ v).

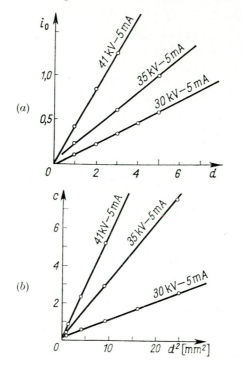

Fig. 11.10. (a) i_0 [10^{-10} Acm^{-2}] as a function of d [mm] under various ionization conditions. (b) c as a function of d^2. Ref.: Adamczewski [1].

Under such conditions (about 0·1 röntgen/hour) 1 mg of radium induced a current greater by a factor of 28·3 than a self-conduction current while x-ray-induced currents were greater by about 10^4–10^5. At 50 kv, 20 mA, currents greater than those of self-conduction by a factor of 10^6 would be expected.

In fig. 11.10 (a), (b) several relationships have been shown between the electrode separation and the values i_0 and c as given in the eqn. (11.6). These values have been measured by the author and, as seen from the diagrams, the general form of current–voltage characteristics is highly regular and the values i_0 and c vary systematically when there is a change in the experimental conditions. The $i_0 = f(d)$ relationship shows that the phenomena of ionization in dielectric liquids are bulk processes and that surface processes (e.g. diffusion) have little significance (straight lines run very near to the origin of the coordinates). This work indicated an interesting relationship between the coefficients c and d, namely that $c = \text{const.} \times d^2$, i.e. that the current–voltage characteristic can be expressed as $i_s = i_0 (1 + \gamma U)$, where U denotes voltage applied across the electrodes of the ionization chamber.

The author has pointed out that all the current–voltage characteristics he measured in hexane can be expressed by one formula:

$$i = 1\cdot09 \times 10^{-10} \, Sd\dot{D} \, (1 + 3\cdot21 \times 10^{-4} dE)$$
$$= 1\cdot09 \times 10^{-10} \, Sd\dot{D} \, (1 + 3\cdot21 \times 10^{-4} U) \quad (\text{A cm}^{-2}), \quad (11.8)$$

where S denotes the area of the collecting electrode, d—electrode separation, $\dot{D}$—dose rate and E—electric field strength in v cm^{-1}. It follows from this formula that straight-line plots of the current–voltage characteristics $i_s = i_0 (1 + \gamma U)$ should intersect at one point on the axis of the abscissae (for $U = -3090$ v). Such diagrams facilitate the drawing of current–voltage characteristics especially if the method which allows the localization of the origin of the saturation region (cf. § 16.3) has been taken into consideration.

Having extended these measurements to a series of other saturated hydrocarbons, the author [94] arrived at the following generalized formula:

$$i = C\dot{D}(\rho + 0\cdot4) \, Sd(1 + \gamma U), \quad (11.8a)$$

where $C = 3\cdot43 \times 10^{-13} \, (\text{A cm}^{-3})/(\text{R/h})$, $\dot{D}$ denotes the dose rate in the ionization chamber and is expressed in R/h and ρ denotes the density of the liquid. This formula can in practice be widely applied in the dosimetry of ionizing radiation using dielectric liquids. This subject will be developed in Chapter 17. Recent research indicates that this formula is a special example, which applies at relatively low fields (up to about 50 kv/cm), of a more general expression, and the author's new formula as presented in § 16.9 for the electrical conduction in ionized liquids also reflects the dependence of the current on both temperature and very high fields.†

Table 11.1. Ionization effects in hexane
($d = 0\cdot3$ cm, $U = 722$ v)

Ionization source	Current densities (A cm^{-2})	Conductivity ($\Omega^{-1} \text{cm}^{-1}$)	Relative values of conductivity
—	$2\cdot6 \ \times 10^{-15}$	$4\cdot5 \ \times 10^{-19}$	1
Radium, 2·19 mg	$1\cdot61 \times 10^{-13}$	$2\cdot8 \ \times 10^{-17}$	62
Radium, 9·76 mg	$7\cdot2 \ \times 10^{-13}$	$1\cdot25 \times 10^{-16}$	277
Radium, 11·95 mg	$8\cdot65 \times 10^{-13}$	$1\cdot5 \ \times 10^{-16}$	332
x-rays:			
25 kv, 5 mA	$6\cdot8 \ \times 10^{-12}$	$2\cdot78 \times 10^{-15}$	6200
30 kv, 5 mA	$3\cdot96 \times 10^{-11}$	$1\cdot64 \times 10^{-14}$	36 600
41 kv, 5 mA	$1\cdot66 \times 10^{-10}$	$6\cdot9 \ \times 10^{-14}$	154 000

Tables 11.1 and 11.2 contain a summary of the results obtained by the author, with reference to the ionization of hydrocarbons by x and γ-rays. A comparison between the ionization effects in the conductivity of liquids under the influence of different sources of radiation is given in Table 11.1.

† See also: Adamczewski [332, 335].

Table 11.2. Physicochemical properties of a group of saturated hydro-carbons

Liquid	ρ	M	$N \times 10^{-21}$	$N_0 \times 10^{-8}$	$\gamma \times 10^4$	$q \times 10^{16}$
Pentane	0·63	72·14	5·28	6·66	3·19	1·27
Hexane	0·663	86·17	4·64	6·86	3·21	1·45
Heptane	0·709	100·19	4·26	7·04	3·14	1·54
Octane	0·725	114·22	3·84	7·14	2·88	1·57
Nonane	0·75	128·25	3·52	7·35	2·97	1·77

It can be seen from table 11.2 that the number of ions produced increases systematically with, but not in proportion to, the density of the liquid and that the coefficient γ decreases slowly with the density for the liquids investigated. Figure 11.11 shows the dependences of the current

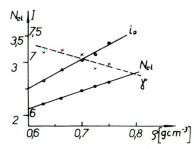

Fig. 11.11. Liquid density dependence on electron density (N_{el}), extrapolated initial current (i_0) and the coefficient (γ). Ref.: Adamczewski [1].

i_0 and coefficient γ on liquid density as well as on the number of electrons N_{el} in 1 cm³ of the liquid.

The last column of the table gives the cross sections for ionization or ion dissociation of the liquid molecules caused by electrons accelerated in the field. This represents the author's attempt [94] to explain the increase in current in the liquid at the region of saturation. A great part of the author's experimental results was later confirmed by other research workers using different experimental conditions (e.g. Pao [3] using a radium source and Vermande [166] who used cobalt 60). From 1962 to 1963 Vermande found a number of current–voltage characteristics in hexane which satisfied the relationship:

$$i_s = 2 \times 10^{-3} d\varphi(1 + 2 \times 10^{-4} dE), \tag{11.9}$$

where i_s denotes the density of the saturation current in A cm⁻², φ is the number of photons per cm² sec and the other symbols retain the same meaning as before.

Unacquainted with the author's formula, Vermande interpreted his experimental results in a different way but it should be emphasized that Vermande's experimental results are consistent with the author's formula.

Special attention should be drawn to the dependence of the current on the distance d between the electrodes as well as to the change in the numerical value of the coefficient γ. Slight deviations in the values of the numerical coefficients for small inter-electrode distances ($d = 0.05$, 0.1 and 0.2 mm) in Vermande's work can be explained by surface effects such as the photoeffects on the electrode surface and ion diffusion effects in the liquid. As an ionizing agent γ-rays are less convenient because their penetrating power is too high and the absorption effect in the liquid is low and they may cause a disturbance in the whole measuring apparatus. On the other hand, the effect of an x-ray beam can easily be localized inside the sample of liquid.

In recent years further investigations in this field have generally confirmed the described phenomena. Many of these experiments, however, were carried out with liquids not carefully purified but in many cases a number of important new details were discovered for the phenomenon described above. Experiments with very pure liquids or with liquids deliberately contaminated with specified admixtures have been especially significant. These experiments have been conducted by a group of research workers in Great Britain (Lewis, Tropper, Coe, Kao, Secker, Angerer, Krasucki, Morant, Jessup, Ward, Watson, Calderwood, Hughes and others), in France (Blanc, Mathieu, Vermande, Guizonnier and Felici), in Poland (Adamczewski, Gzowski, Terlecki, Jachym, Gazda, Januszajtis, Kalinowski, Nowak, Zając and others) and in the United States (Watson, Sharbaugh and Crowe).

Plumley's [21] work, published in 1941, dealt with research on the self-conductivity of heptane in high fields (up to 600 kv/cm) at temperatures varying from $-190°$c to $20°$c. The purpose of Plumley's experiments was to support his theory which was based on the assumption that the high-field conductivity of dielectric liquids is caused by the dissociation of the molecules of the liquid into ions under the influence of an applied electric field and to explain the further increase in the current together with increasing field stress. This theory is described in § 16.7.

A large number of experiments on self-conduction and induced conduction in *iso*-octane and liquid oxygen were reported by Pao [3] in 1943. The value of self-conductivity in *iso*-octane after careful purification was very low (of the order of $3 \times 10^{-20} \, \Omega^{-1} \text{cm}^{-1}$). Pao ionized the liquid with γ-rays from weak radium sources in the range of 10–4000 μg. Almost a linear increase in the current was obtained with the increased activity of the source (fig. 11.12). The current–voltage characteristics for various intensities of ionizing radiation were similar to those given in the author's work [1]. Pao conducted a series of measurements on the dependence of the ionization current on the temperature of the liquid and these results are described in § 11.6.

Guizonnier's group [173–9, 185–7] published many papers between 1954 and 1963 concerning research on conduction in dielectric liquids either unpurified or deliberately mixed with various impurities. The papers produced by this group mainly concentrated on electric field distribution inside the ionization chamber as well as on the influence of admixtures and temperature on electrical conduction.† Attention should be paid to the non-uniform field distribution inside the ionization chamber in the case of liquids not carefully purified (Guizonnier [173]) and also to investigations on the activation energy of ionization currents in those liquids (Guizonnier and others [177, 178, 185–7]). Gibaud's [36, 179] research, which belongs to the same group, was concerned with investigating ionization currents in petroleum ether and his analysis of the

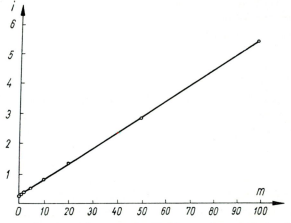

Fig. 11.12. Current density i (in A cm^{-2} × 10^{-14}) as a function of radium activity m (in mg); field strength $E = 5830$ v cm^{-1}. Ref.: Pao [3].

results was based on Jaffe's theory. Gibaud used a liquid which was not carefully purified and which had a high measure of self-conduction (of the order of 10^{-13}–10^{-14} Ω^{-1} cm^{-1}). The liquid was ionized by weak radium sources (2–6 mg Ra). He measured the ionization currents subtracting the self-conduction current and discovered a linear dependence between this difference in the currents and the activity of the radioactive source and the separation of the electrodes. Applying Jaffe's theory, he found the values of the saturation current (for $E = \infty$) and estimated the number of ion pairs produced by the γ-radiation from 6 mg Ra in 1 cm^3 of petroleum ether to be as high as $3 \cdot 8 \times 10^7$ ion pairs/sec. In 1959 and 1960 Ivanov [37] reported a number of experiments in the field of ionization and conduction in the dielectric liquids. He made measurements in toluene, heptane, *iso*-octane and benzene and ionized them with x-rays perpendicular to the electrode surface and applied very low voltages between 20 to 50 v (at the distance $\delta = 0 \cdot 4$ cm) to the electrodes.

Ivanov mainly investigated current–voltage characteristics in low fields (for $i/i_0 < 0 \cdot 5$) and determined the dependence of the resistance R of the ionized liquid on the voltage U which was applied to the electrodes.

† What concern the potential distribution in dielectric liquids see also the new papers: Guizonnier [343], Bloor and Morant [362], Pollard and House [363], Huck [354] and Hill and House [355].

He found the following relationship to be the most accurate in the voltage region 0–70 v:

$$R = R_0 + \frac{U}{\kappa i_0}, \tag{11.10}$$

where κ is a constant characteristic for the given chamber. In Ivanov's opinion this proved that in the observed cases volume recombination and not columnar or group recombination plays the main part.

Ivanov compared the ionization in liquid with that in air and found the current several times greater in liquids. He calculated that 10^{11} ion pairs/cm^3 sec were produced in the chamber and he believed that this corresponded to 10^6–10^7 columns. He was able to estimate from the time taken for the current to decrease (after irradiation had ceased) that the mobility of the ions was as high as $(0\cdot5–1\cdot6) \times 10^{-4}$ cm^2 v^{-1} sec^{-1}, the possible error of the measurements being about 100%. On the basis of Jaffe's, Lea's, Magee's and Boag's theories, which have been partially described in § 9.3, Ivanov discussed the mechanism of ionization in liquids. Sections 16.4 and 16.5 are devoted to a more detailed analysis of Ivanov's theory. The liquids investigated by Ivanov were not carefully purified and had a high degree of self-conduction and, in addition, he applied very low electric fields to the liquids. For these reasons the results of his experimental work can only be regarded as having a qualitative value. They can have no major significance in deciding the validity of any one of the above-mentioned theories of ionization and conduction. This, however, is not to deny the value of Ivanov's reflections on the mechanism of ionization.

Between 1961 and 1966 Blanc, Mathieu, Boyer and Vermande [35, 159, 166] published the results of their experiments with ionization chambers filled with liquid hexane at room temperature and ionized by γ-rays (^{60}Co–12 mci), α particles (^{210}Po) and neutrons.† The hexane was carefully cleaned ($\sigma_n \approx 10^{-18}$ Ω^{-1} cm^{-1}), and its self-conduction current was about 2×10^{-14} A/cm^2, the ionization current being about 100 times greater. The electric fields applied in the course of this experiment were as high as 65 kv/cm. The above-mentioned scientists performed a number of preliminary measurements confirming the following experimental data:

(*a*) A slow decrease taking anything up to 20 hours in the self-conduction current.

(*b*) An exponential increase in the self-conduction current at fields of the order of 20 kv/cm.

(*c*) An increase in the current with increasing temperature (from $-10°$c to $10°$c).

(*d*) Typical current–voltage characteristics in the region up to 10 kv/cm ($i_s = i_0 + cE$) with a clearly linear dependence on E.

They obtained characteristics which can be expressed in the form of the relationship $i_s = i_0(1 + \gamma U)$ which was established by the author [1] as early as 1934. Attention should be drawn to some aspects of the original results of their work and especially to the data obtained with a pulse

† See also the new papers: Mathieu, Blanc and Patau [319, 322, 329, 333,, 338, 358].

chamber. The diagram of the circuit used for this purpose is shown in fig. 11.13. The currents were measured by means of a vibrating capacitor electrometer and were registered on a recording tape at a speed of 16 cm/hour. The voltage was taken from a 5000 v dry battery. Figure 11.14

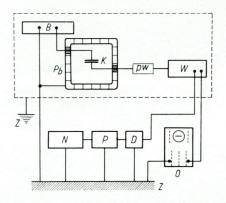

Fig. 11.13. Apparatus for measuring electrical conductivity and ionization pulses in liquids. *K*, ionization chamber; *B*, high voltage; *Pb*, lead screen 5 cm thick; *pw*, *W*, amplifiers; *N*, counter; *P*, scaler; *D*, discriminator; *O*, oscillograph. Ref.: Blanc *et al.* [35].

shows the photographs of two pulses induced by α-particles, one pulse having an amplitude of 160 μv and a duration of 20 μsec while the other had an amplitude of 104 μv and a duration of 18 μsec. The top section of fig. 11.14 shows the statistical fluctuations of these pulses induced in hexane by α-particles. Figures 11.15 and 11.16 show variations of the current I and the pulse heights V as a function of the electric field stress E in the ionization chamber. These scientists, therefore, have discovered the existence of the effects of ion multiplication in liquids which is similar to those effects observed in ionized gases and p–n semiconductors. All the other results of their work concerning the dosimetry of ionizing radiation are given in §17.3. It should be noted that their investigation of α-induced pulses is reminiscent of the research conducted by Białobrzeski and Adamczewski [17, 24] on the detection of ionization bursts which was published in 1936.

In recent years our Institute has been responsible for a great deal of research directed towards the explanation of current–voltage characteristics in dielectric liquids ionized by x, γ and β-rays of various energies and under different experimental conditions as well as of the photocurrents from the cathode immersed in the liquid. This work was generally planned to form an introduction to certain complex scientific problems such as mobility, recombination and diffusion of ions, the application of dielectric liquids in dosimetry, etc. In the course of this research, a great deal of interesting experimental material has been added to the investigations on current–voltage characteristics and to the theoretical study of the mechanisms of ionization and electrical conduction in liquids.

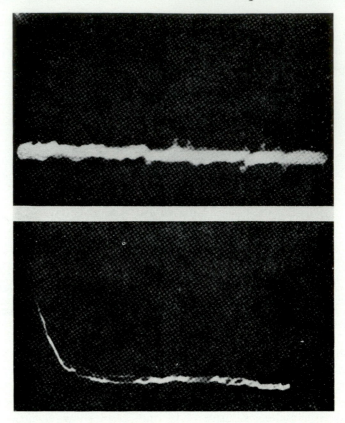

Fig. 11.14. (*a*) Number of current pulses in hexane generated by α particles; field strength $E = 33000$ v cm^{-1}, velocity of motion of the film $V = 0.5$ cm msec^{-1}. (*b*) Two following pulses generated by ^{210}P α particles; $E = 44\,000$ v cm^{-1}, $V = 100$ cm msec^{-1}. Ref.: Blanc *et al.* [35].

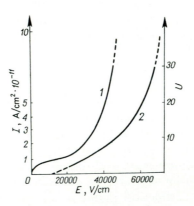

Fig. 11.15. Curve 1, $i = f(E)$ characteristic for radium γ-ray ionization. Curve 2, pulses generated by α ^{210}Po particles; U in v × 10^5. Ref.: Blanc *et al.* [35].

From among the more important results the following deserve particular attention:

(1) The confirmation of the general relationship $i = f(E)$ in the form given in the formula (11.6) for x and γ-radiation. In some investigations, however (Jachym [193], Januszajtis [47], Nowak [122], Zając [213]), the relationship between the coefficients c and d was somewhat different ($c = Ad^{5/3}$ or $c = Ad$).

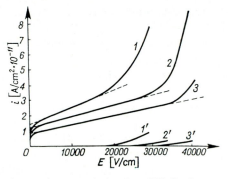

Fig. 11.16. Current–voltage characteristics $i = f(E)$ for hexane of various purities; $d = 0 \cdot 1$ cm. Curves 1, 2 and 3 for ionization current, of electrical purification ($E = 15 \, \text{kv cm}^{-1}$). Curves 1', 2', 3' were obtained under the same conditions but without ionization. Ref. Blanc *et al.* [35].

(2) The establishment of the fact that the further increase of the current in the region above saturation is strictly linear up to the fields of about 300 kv/cm, this being valid both for x and γ-radiation (Jachym [190], Terlecki [33]) and for photocurrents (Terlecki [33]) where the relationship $i = AE^n$ is fulfilled when $n = 1 \cdot 12$.

(3) The determination of the working conditions of a standard ionization chamber filled with liquid (see § 17) and of the possibility of applying liquid-filled chambers in dosimetry (Adamczewski [94–96], Januszajtis [47], Terlecki [307]).†

11.5. *Research on Emission and the Injection of Electrons into Dielectric Liquids.* The introduction of free electrons to liquids is an interesting method of testing conduction in dielectric liquids. This can be done in four different ways: (*a*) by means of a photoelectric effect, (*b*) by injecting electrons from a glowing cathode, (*c*) by the emission of electrons from cathodes at high electric fields and (*d*) by superimposing a β-radiating substance on one of the electrodes.

The most developed of the above-mentioned methods is that based on the photoelectric effect, i.e. research on photocurrents induced by irradiating the cathode with ultra-violet rays. Jaffe [15] was the first to take measurements in 1913. Further work was done by Dornte [90], Reiss [22],

† See chapter 17.

11

Morant [49], Swan [51], Chong and Inuishi [112], and more recently by Terlecki and Gzowski [50], Vermeil *et al.* [188], Kalinowski [238, 286, 347] and Schmidt [283].

Research work in our Institute will be discussed first of all, because many of the results obtained illustrate the basic features of the phenomenon. An ionization chamber filled with either hexane or decane was used by Terlecki and Gzowski. The lower electrode (anode) was made of glass (silibor) and covered with a thin transparent layer of tin oxide which has a low electric resistance (of several hundred ohms) and is 90% transparent to ultra-violet light (up to $\lambda = 2600$ Å). The cathode was made of aluminium and placed on a micrometer screw which enabled the separation between the electrodes to be controlled with great accuracy. Some of the measurements were carried out with the cathode placed in the lower part of the chamber while the glass plate was coated with a very thin layer of aluminium. A quartz lamp, having a Q-400 Hanau burner, was the source of ultra-violet radiation. Measurements were taken by means of a vacuum tube electrometer which is described by the same authors [12]. They measured the photocurrent I as a function of the applied electric field E in the region from 10–200 kv cm^{-1} and in the course of their experiments found the following relationships:

$$I = AE^{1 \cdot 09} \quad \text{for hexane,}$$

$$I = AE^{1 \cdot 12} \quad \text{for decane,}$$

where A denotes a constant which depends on the kind of cathode as well as on the liquid, time of irradiation and E is expressed in kv cm^{-1}. The relationships therefore, obey the general formula:

$$I = AE^n. \tag{11.11}$$

Previous workers have arrived at this formula but the value of n has differed according to the author. For example, Jaffe found that $n = 1 \cdot 38$ for hexane and Swan that $n = 1 \cdot 21$. Terlecki and Gzowski have found an explanation for the differences in the value of n. They put forward the theory that in some cases a photoeffect could occur in the liquid. This photoelectric effect depends on the lower limit of the wavelength of ultra-violet radiation acting on the cathode as well as on the purity grade of the liquid. It should be added that Jaffe observed a distinct volume effect which was not noticed by other research workers.

In fig. 11.17 the relationship $I = f(E)$ for hexane is shown at various electrode distances (1·5, 2·0, 2·5 mm) and, as can be seen, the current does not change with electrode distance. Figure 11.18 shows the results of various workers to the dependence of the photocurrent on the electric field strength. Figure 11.19 (*a*) shows the relationship between the photocurrent and the field strength in the case of a chamber filled with air under normal pressure and the dependence of the photocurrent on the intensity of light supplied by a quartz lamp is shown in fig. 11.19 (*b*).

Terlecki and Gzowski devoted a great deal of attention to devising a method of preparing the photocathode. They investigated the sensitivity of the cathode which had been treated in different ways and also the rate

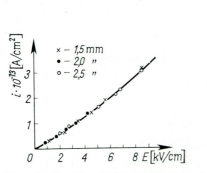

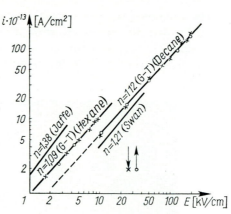

Fig. 11.17. $I = f(E)$ relationship for photocurrent in hexane at various electrode separations. Ref.: Terlecki and Gzowski [50].

Fig. 11.18. Dependence of photocurrent intensity i in hexane and decane on the field strength E. Ref.: Terlecki and Gzowski [50].

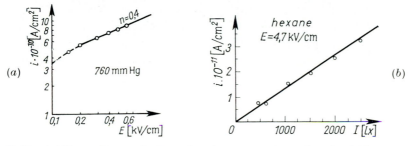

Fig. 11.19. (a) Dependence of photoelectric current i on electric field strength. (b) Dependence of photocurrent intensity i on the intensity I of ultra-violet light. Ref.: Terlecki and Gzowski [50].

of its fatigue during irradiation. One of these fatigue curves is shown in fig. 11.20. Kalinowski used a brass chamber in which the first of the two electrodes had the form of a silver grid (0·1 mm dia. mesh) and the second was a sliding nickel-plated electrode equipped with a protective ring. By using a grid electrode which was placed against the quartz window, it was possible to obtain from nickel a photoeffect (the amount of work necessary to extract the electron from the surface was 4·84 ev) which required a light whose wavelength was $\lambda \leqslant 2500$ Å. Ultra-violet rays in both cases were supplied by a quartz lamp with a Q-400 Hanau burner.

According to Kalinowski n equals 1 in eqn. (11.11) within the bounds of experimental error. He maintained that this relationship was dependent on a 'hopping' mechanism of the electron's movements in the liquid.

This relationship is expressed in the formula:

$$I = \frac{2\delta e^2 M n_0 E}{\xi^2 \rho \tau_0 kT} \exp\left(-\frac{W}{kT}\right), \tag{11.12}$$

where δ denotes the mean width of the barrier of the potential which has been formed by an electron-induced molecular dipole; e signifies the electronic charge, M—the molecular mass, n_0—the number of electrons in a unit volume, ξ—the mean linear dimension of the molecule in the

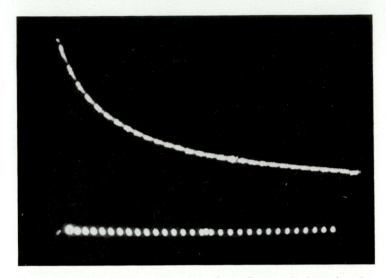

Fig. 11.20. Oscillogram showing the time dependence of photoelectric current.
Ref.: Terlecki and Gzowski [50].

liquid, ρ—the liquid density and τ_0—the shortest time of the electron's presence in the vicinity of the molecule. W is the activation energy of the process concerned with the migration of the electron in the electric field in contrast to the energy of the inductive influence of the electron on the molecule ($W \approx \alpha e^2/2r$), where α is the mean polarizability of the molecule, r is the mean distance between the critical radius r_k, below which the electron follows a spiral track in the area of the molecule, and the limiting radius r_0 where the forces of repulsion begin to dominate. The linear dependence of I on the intensity of the light has been discovered by experimentation and its confirmation in eqn. (11.12) was based on the obvious assumption that n_0 is in direct proportion to the flow of photons. In order to give a quantitative verification of the formula (11.12) it is necessary to devise a method enabling the absolute determination of n_0 and to determine with precision a physically justified criterion for the selection of δ, τ_0 and r_0.

In the years (1960–61) Morant [49] and Swan [146] have carried out investigations on photocurrents in liquid hexane. The results are given in figs. 11.21 and 11.22 which show the current–voltage characteristics in various regions. The results obtained by both scientists generally confirm the description of the course of the phenomenon given above. They also contribute a number of new ideas to the interpretation of this phenomenon and to the assessment of its significance to the breakdown mechanism in dielectric liquids.

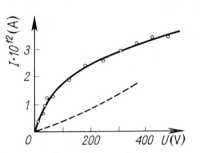

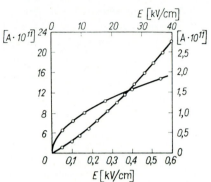

Fig. 11.21. Dependence of photocurrent *i* in hexane on applied voltage *U*. Ref.: Morant [49].

Fig. 11.22. Dependence of photocurrent *i* in hexane on electric field strength *E* for constant flux of ultra-violet light. Ref.: Swan [51].

At the Durham Conference [180, 181], Vermeil *et al.* [188] gave a report of their work concerning the action of ultra-violet radiation on hydrocarbon liquids. In their experiments *iso*pentane, propane and *iso*butylene were exposed to the radiation of a well-defined quantum of energy ($hv = 6\cdot7$, $8\cdot4$ and 10 ev) at low temperatures and current–voltage characteristics were measured in the ionization chamber. The molecules of the investigated hydrocarbons have considerable powers to absorb radiation and for this reason it was necessary to maintain the ionization chamber at low temperatures (in order to reduce vapour density). It was found that ionization occurs even when the energies of the photons are lower than the values of the first adiabatic potential of the liquid molecules. No ionization effects were observed for quanta with $6\cdot7$ and $8\cdot4$ ev energies and the characteristics were similar to those of self-conduction currents. On the other hand, for the quantum energy $hv = 10$ ev, Vermeil *et al.* found both an ionization effect and a characteristic linear increase of the current in the saturation region, typical for ionization currents. This effect is considered to be the result of a reduction in the ionization energy of the molecules in the liquid caused by subtracting the electron affinity energy and the solvation energy. The results of the electrical measurements were compared with those of the chemical measurements (photopolymerization) both being carried out under the same experimental conditions. An identical ionic mechanism was observed in both cases. The photoionization yield was estimated at 10^{-2} ion pairs per one quantum of radiation.

The effect of volume ionization by means of ultra-violet radiation ($\lambda \geqslant \sim 1900$ Å) also occurs in such typical dielectric liquids as saturated hydrocarbons to which small admixtures of substances, having relatively low ionization potentials, have been added. A great deal of work has been done on photoconductivity in anthracene solutions. Reiss [22] used this phenomenon in the method he devised for measuring the mobility of ions which were produced by means of ultra-violet radiation in an anthracene solution in hexane. Figure 11.23 illustrates the dependence established by Reiss of a photocurrent on the separation of the electrodes.

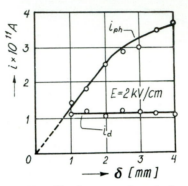

Fig. 11.23. The dependence of residual current i_d and photocurrent i_{ph} on electrode separation. Ref.: Reiss [22].

It is interesting to note that in many works a relationship has been found between photocurrent and the electric field strength. This relationship is similar to current–field relationships for currents induced by means of x or nuclear radiation. A linear dependence $I = f(E)$ or a dependence not precisely defined has been found in other research work. There are certain differences of opinion about the character of the dependence of the photocurrent on the intensity of light. Aukerman [285] maintains that this dependence in a hexane solution of anthracene can be expressed by the formula:

$$i = a + b\sqrt{I}, \tag{11.13}$$

where a and b are constant.

Pitts *et al.* [237] investigated triphenylamine solutions in hexane and benzene and found a distinct linear dependence within the limits of concentration from 10^{-6} to 10^{-3} M/l. Some scientists have stated that the spectral maximum of the photocurrent is almost equal to the optical maximum of the absorption spectrum. Photoionization of the admixture has been explained by others as a result of the short-wave extremity of the irradiation band (approx. 2000 Å).

In order to describe the initial products of the irradiation of triphenylamine and some other substances in hexane, Rüppel and Witt [297] devised a method of taking pulse measurements of photoconduction. A diagram of their measuring apparatus is given in fig. 11.24. The measuring chamber K consists of a rectangular quartz tray with internal electrodes. A well-fitted closure keeps the chamber air-tight and maintains its contents in a desired atmosphere. One of the electrodes of the measuring chamber is directly connected to the grid of the pre-amplifier tube while the other is connected to the positive potential of a high-quality, high-voltage $0·25 \, \mu\mathrm{F}$ capacitor. A pulse of ultra-violet light produced by the discharge of a $100 \, \mu\mathrm{F}$ capacitor through a flash mercury lamp was directed between the electrodes of the tray. While the pulse lasted ($T \sim 10^{-4}$ sec)

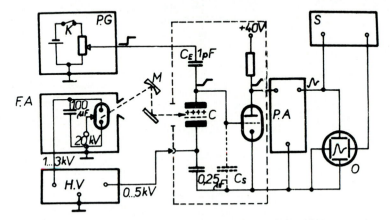

Fig. 11.24. Diagram of Rüppel and Witt's [297] apparatus for measuring photo-conductivity in a liquid using a pulse method. For further explanation see text.

the charge carriers produced in the chamber gave a short current pulse in the electric field, the pulse being determined by the lifetime of the charge carriers. This pulse was transmitted on to an oscillograph after amplification and synchronization. With the type of amplifier used, the magnitudes of the charge produced by radiation in the chamber can be read and compared from the peak of the oscillograph. The sensitivity and rating of the apparatus can be established both by means of a standard capacitor C_E and by a controlled pulse generator.

The sensitivity has been determined as 8×10^{-16} coulomb which was roughly equal to 5000 monovalent ions. If a narrow beam of light produces in the layer Δx at a distance x from one of the electrodes N_0 of monovalent ion pairs with a mean lifetime τ and mobilities u_+ and u_-, the ion pairs will move accordingly within that lifetime to a distance $\delta_+ = u_+ E\tau$ and $\delta_- = u_- E\tau$. At $\delta_+ \ll l - x$ and $\delta_- \ll x$ this movement will produce the following charge on the electrodes:

$$Q = eN_0 \frac{\delta_+ + \delta_-}{l} = \frac{eN_0(u_+ + u_-)E\tau}{l}, \tag{11.14}$$

where l is the electrode separation. The lifetime of the ions τ in the layer Δx depended on the mechanism of the decay of the ions.

Kavada and Jarnagin [235] investigated anthracene and phenanthrene solutions in hexane and tetrahydrofuran. Using ultra-violet radiation with a wavelength $\lambda \approx 3000$ Å they detected ionization in tetrahydrofuran solutions but not in hexane solutions. A number of works published in the 1960's testified to the existence of photoconductivity in pure organic liquids which represent the borderline between dielectrics and semi-conductors. Kallmann *et al.* [236] discovered and investigated photo-conductivity in methylnaphthalene (conductivity $\sigma \approx 10^{-14}\,\Omega^{-1}\,\text{cm}^{-1}$).

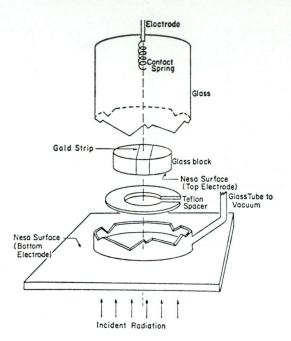

Fig. 11.25. Diagram of the conductivity cell used by Kallmann *et al.* [236]. For further explanation see text.

A sample of α-methylnaphthalene, purified and chromatographically checked, was placed in a vacuum chamber (fig. 11.25) and exposed to ultra-violet radiation of $\lambda < 3200$ Å. Radiation was directed through $10^{-4}\,\text{m}^2$ of the tin-oxide-coated lower electrode which transmitted up to about 3% of the radiation $\lambda = 2800$ Å. The second electrode consisted of a glass disc also coated with tin oxide and having a strip of golden foil, the purpose of which was to ensure a better connection to the contact spring. The separation of the electrodes was altered within the limits $(2\cdot5$ to $25) \times 10^{-4}\,\text{m}$ by means of the exchange of teflon spacers insulating the electrodes. The most important result of this research was the discovery of an essential difference in the values of the so-called positive and negative photocurrents. These photocurrents were measured when the illuminated lower electrode was first an anode and then a cathode (fig. 11.26).

Under the same experimental conditions, the positive current can be fifteen times greater than the negative current and the mobility of the positive and negative ions was found to be $1{\cdot}2 \times 10^{-4}$ and $0{\cdot}5 \times 10^{-8}$ m^2/vs. respectively. Figure 11.27 illustrates the dependence between viscosity

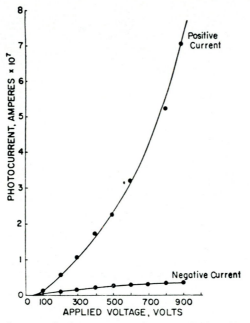

Fig. 11.26. Dependence of photocurrent in α-methylnaphthalene on applied voltage, area of specimen: 1 cm^2, thickness of specimen: $0{\cdot}025$ cm, exposure time: $0{\cdot}1$ sec. Ref.: Kallmann *et al.* [236].

(η) and the mobility of the positive ions u_+ and $u_+ \times \eta$ on the reciprocal of the temperature. Stokes's law was applied in order to determine u_+ for different values of η ($u_+ = e/6\pi\eta \times r$ where e is the charge and r is the radius of a spherical ion). It is clear from fig. 11.27 that $\eta \times u_+ = f(10^3/T) \neq$ const. The authors, however, did not pay much attention to this fact, assuming that in this case the inconsistency with Walden's law was negligible and that the photocurrent depended on the movements of ions. It is worth noting that when the absorption of photons was $3 \times 10^{15}/s$ and their range in α-methylnaphthalene δ was 3×10^{-6} m at $\lambda = 3130$ Å, the maximum yield of the photoions should be 10^{-2}. It was also established that there existed a linear dependence of the photocurrent on the intensity of the light and that the admixture of 9,10-diphenylanthracene decreased photo-conductivity by 25% (the full spectrum of a mercury lamp was used for radiation). The results of measurements taken by Kallmann and his associates [236] were supplemented and discussed by Baessler [217–19, 220].

In benzene, naphthalene, diphenyl, pyrene and benzophenone photo-conductivity was independent of self-conductivity. Photoconductivity only occurred when the electrodes of the measuring chamber were metallic.

When the electrodes were coated with quartz it was seen that the effect of photoconductivity was reduced.

It follows from the above experiments briefly described that the mechanism of photoconductivity cannot be interpreted in a uniform manner and that it can be more or less complex depending on the type of

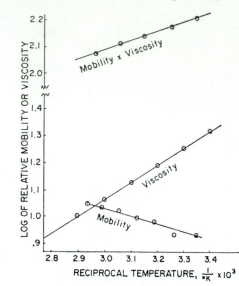

Fig. 11.27. Dependence of η, u and the product $u\eta$ on $1/T$.
Ref.: Kallmann *et al.* [236].

liquid and solution and the wavelength of the light. The following energetic condition must be fulfilled before ionization can take place:

$$\sum_{i=1}^{n} E_i \geqslant I_D - A_A - 2P, \tag{11.15}$$

where $\sum\limits_{i=1}^{n} E_i$ is the total of the energies of all the excited states (n) of the molecule or the molecular complex, I_D—the ionization potential, A_A—the electron affinity of an insulated acceptor of the electron and P— the polarization energy of the surrounding medium.

Kavada and Jarnagin listed the values of various types of energy for anthracene and phenanthrene in hexane and tetrahydrofuran—THF (see table 11.3). In the way they indicated that delayed ionization of those molecules in THF is possible as a result of their collision in excited triplet states. On the other hand, delayed ionization should not occur in hexane or other liquids with a small dielectric constant when the solution is irradiated with a light having a wavelength $\lambda \approx 3000$ Å. Their experiments confirmed this supposition. The mechanism of ionization suggested by Kavada and Jarnagin competes with the process of delayed fluorescence. The spectrum of this fluorescence can have two forms depending on whether emission takes place from the singlet excited state of the molecule

Table 11.3.

Admixture	Type of energy $2E_t$ (triplet) (ev)	I_D (ev)	A_A	2P (ev)		Surplus of excited energy (ev)	
				THF	Hexane	THF	Hexane
Phenanthrene	$2 \times 2 \cdot 69$	$8 \cdot 03$	$0 \cdot 25$	$3 \cdot 6$	$1 \cdot 9$	$1 \cdot 2$	$-0 \cdot 5$
Anthracene	$2 \times 1 \cdot 8$	$7 \cdot 55$	$0 \cdot 61$	$3 \cdot 6$	$1 \cdot 9$	$0 \cdot 3$	$-1 \cdot 4$

After Kavada and Jarnagin [235].

(delayed fluorescence—DF) or from the excited state of the bimolecular complex of the so-called excimer (delayed excimer fluorescence—DXF).

The following diagrams illustrate all these processes:

$$(I) \ 2 \, ^3X \rightarrow [X+X]^* \rightarrow \begin{cases} 2X + DXF \\ X^* + X \rightarrow 2X + DF \\ X^+ + X^- \quad \text{or} \quad X^+ + X + e^- \\ X + X^* \rightarrow X^+ + e^- + X \end{cases} \qquad (11.16)$$

$$(II) \ 2 \, ^3X \rightarrow X^* + X \rightarrow \begin{cases} 2X + DF \\ [X+X]^* \rightarrow 2X + DXF \\ X^+ + e^- + X \\ [X+X]^* \rightarrow X^+ + X^- \quad \text{or} \quad X^+ + X + e^- \end{cases} \qquad (11.17)$$

Both types of mechanism for delayed luminescence were observed in liquid solutions but it was found that the first mechanism was more apparent in solutions of a lesser viscosity. The mechanism of producing ions by the annihilation of excited states leads, however, to a linear increase in the photocurrent as a function of the concentration of the admixture. On the other hand, the measurements of Pitts *et al.* [237] show that the photo-current is dependent on the square root of concentration. An analysis of those mechanisms described above led Pitts *et al.* [237] and Kalinowski [286]† to the conclusion that ionization in two stages resulting from the absorption of two photons in the same molecule is the mechanism most likely to be correct for photoionization. This conclusion is in agreement with experiments showing the dependence of a linear increase in the photocurrent on the field and with those illustrating that the photocurrent is proportional to the root of concentration:

$$\begin{rcases} h\nu \leadsto X \rightarrow X^* \rightarrow {}^3X, \\ h\nu \leadsto {}^3X \rightarrow X^{**} \rightarrow X^+ + e^-. \end{rcases} \qquad (11.18)$$

Kalinowski maintained that if the energy $2h\nu$ was not sufficient to ionize the molecule there still existed a possibility of obtaining the photo-ionization of an excimer which is formed as a result of the diffusion of molecules in the first excited triplet states:

$$2 \, ^3X \rightarrow [X+X]^*, h\nu \leadsto [X+X]^* \rightarrow X^+ + X^- \quad \text{or} \quad X^+ + e^- + X. \quad (11.19)$$

† See also Kalinowski [347].

Excited states of the molecules play an essential part in photoconductivity in such homogeneous liquids as benzene, α-methylnaphthalene (at room temperatures), naphthalene, pyrene and diphenyl (above their melting point).

In order to explain these experimental results Kallmann and his associates suggested the following mechanism of photoionization in α-methylnaphthalene. The ion is produced on the electrode by the exchange of a charge between an excited molecule and the electrode.

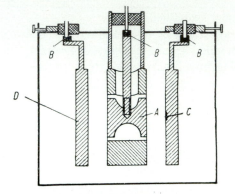

Fig. 11.28. Diagram of the ionization chamber of Gzowski and Chybicki [130].
A, collecting electrode; *B*, amber insulators; *C, D*, voltage electrodes. On electrode *C* the β-radioactive source is marked.

The fact that the positive current has a far greater value can be explained by the probability of the electron moving from the cathode to the excited molecule. According to Kallmann and his colleagues, the relationships presented in fig. 11.26 show that the photocurrent is either a result of the movement of ions or holes. They compared the mobility ($\eta = 0.024$ P at 20°C), $u = 0.71 \times 10^{-8}$ m²/vs calculated on the basis of Stokes's law, with the value obtained in the course of measuring the time t' taken by the ions to travel between the electrodes ($u = a^2/(U \times t')$; U—voltage, a—distance). The mobility of negative ions was less than that of positive ions ($u_+ = 1.2 \times 10^{-8}$ m²/vs; $u_- = 0.5 \times 10^{-8}$ m²/vs). Both values, however, are of the same order of magnitude as those derived from Stokes's law. These results are analogous to those obtained by Adamczewski [40, 41] for dielectric liquids ionized by x-rays.

Secker *et al.* [227] devised a technique for injecting electrons into transformer oil and silicone fluids. They investigated the characteristics $I = f(U)$ as a function of the potential of the grid placed in a two-phase triode (cf. fig. 18.22). Begun in 1966, this research concentrated on finding a technical application to this kind of triode.

In 1962 Gzowski and Chybicki [130] applied a different method of investigating electron emission from the surface of a metal into a dielectric liquid. They coated one of the electrodes of the ionization chamber, placed opposite the collector electrode, with a β-ray source (0.02 ci of tritium) (fig. 11.28) and investigated current–voltage characteristics in relation to temperature at various inter-electrode separations and for different polarities of the voltage. When the electrode covered with the radioactive source was negative the currents obtained were greater by a constant value and independent of electric field strength (in the so-called saturation region) (cf. fig. 11.29) than those currents obtained when the

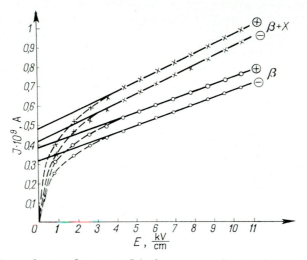

Fig. 11.29. Dependence of current I in hexane on electric field strength E with tritium β-ray layer ionization (0·02 ci) and with β and x-ray ionization ($\beta+$x). Ref.: Gzowski and Chybicki [130].

potential of the electrode was positive. The curves $i_s = i_0 + cE$ are in both cases parallel up to the field of 10·5 kv cm^{-1} with $\Delta i_s/i_s$ amounting to about 21% and the value of c being equal to $3·8 \times 10^{-14}$ A cm v^{-1}. In the investigated range of temperatures the value of c remains constant and i_0 increases linearly with the temperature (in the region of 20 to 50°c) within the limits of experimental error.

The authors have explained the difference in the current values when the voltage polarity was changed as being the result of the influence of the difference in track length of β-electrons while drifting in the electric field. That is, when the collector electrode is positive the electrons travel the whole distance between the electrodes. When the electrode is negative, however, the electrons quickly return to the electrode from which they originated. The return journey is twice as far as their range in hexane. In both cases the basic value of the current is determined by the positive and negative ions produced in equal numbers in a small amount of the liquid which is close to the electrode coated with tritium. It should be added that Gzowski and Chybicki have not completed their investigations, but special attention should be paid to their work on the mechanism of electric currents induced in the liquid when the effects of β-rays in a thin layer of liquid, and the effects of x-rays in the whole volume of the liquid, are superimposed on one another. Their main purpose was to examine the possibility of inducing secondary ionization in the liquid or molecular dissociation of the ions when β-particles move through the ionized liquid in which there are many excited molecules. The current–voltage characteristics obtained in both cases are shown in fig. 11.29. As can be seen from the diagram, the slope of the straight lines increased when additional ionization was produced by x-rays. The difference in the distance between the straight lines when using the positive and negative potential has not been observed under given experimental conditions. This problem requires further research.

In his later works Chybicki† maintained that the mechanism of the collecting of charges in the electric field is based on well-known physical phenomena, such as columnar ionization by α and β-particles and the formation of not only slow but also fast electrons δ. Owing to the existence of a strong radial field in the column, slow electrons rapidly disappear in an initial and marked recombination process. The radial field also disappears at the same time. A small number of positive ions and electrons δ emerge from the column. The electrons can avoid recombination if the external forces are greater than the attractive Coulomb force operating among the electrons. That is, if $eE \geqslant (e^2/\varepsilon r^2)f(\zeta, \theta)$ where $f(\zeta, \theta)$ is a certain function which takes into consideration both the orientations of the column towards the external field and the direction of the electron, which has been pushed out, in relation to the axis of the column. In order to simplify the analysis it is assumed that $f(\zeta, \theta) = 1$. Therefore these ions (electrons) whose distance from the original ion of the opposite sign is not less than the value $r = (e/\varepsilon E)^{1/2}$, are collected at the field E. For the electron δ to be initially, at least at the distance r from the original ion, it must have an energy W which is determined by the relationship $r = CW^\gamma$ in which C and γ are certain constants. For this reason, only those electrons δ having the energy

$$W \geqslant \left[\frac{1}{C} \left(\frac{e}{\varepsilon E} \right)^{1/2} \right]^{1/\gamma}$$

are collected at the field E. The number of electrons δ produced by an ionizing particle whose energy is E_k, charge is ze and which travels along a track dx whose electron density N_e is higher than the energy W, can be expressed by the formula:

$$dN = \Phi(W, E_k, ze) N_e \, dx, \qquad (11.20)$$

where $\Phi(W, E_k, ze)$ is a cross section for the formation of the electrons δ whose energy is either W or more than W. If $[1/C(e/\varepsilon E)^{1/2}]^{1/\gamma}$ is substituted for W and the range of ionizing particles in a given centre is substituted for dx, the number of electrons δ collected by the applied electric field E is :

$$N_\delta = \Phi \left\{ \left[\frac{1}{C} \left(\frac{e}{\varepsilon E} \right)^{1/2} \right]^{1/\gamma}, E_k, ze \right\} N_e x. \qquad (11.21)$$

† See Chybicki [337].

In 1962 Gaeta [131] carried out similar measurements with β-rays but he used a different liquid and worked under different experimental conditions. He irradiated helium II at temperatures of $0\cdot83°$K and $1\cdot925°$K with β [147]Pm rays ($E = 0\cdot229$ мev). The most important results of Gaeta's work are shown in fig. 11.30. It can be seen from the diagram that at a

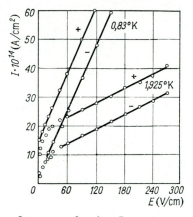

Fig. 11.30. Dependence of current density I on electric field strength E at two different temperatures in liquid helium II. The signs $+$ and $-$ indicate voltage polarities of the collecting electrode in each case. Ref.: Gaeta [131].

temperature of $1\cdot925°$K, the current–voltage characteristics are similar in shape to those found by Gzowski and Chybicki, whereas at a temperature of $0\cdot83°$K they tend to agree with Ohm's law. In 1956 Sato *et al.* [56] devised a different method of injecting electrons into a liquid. They produced electrons from a glowing cathode in a vessel which contained oil and from which all air had been removed and injected them into the liquid by means of an electric field. The main purpose of their work was to investigate the mobility of negative carriers in the liquid. For this reason, their work is described in greater detail in the following chapter. Watson and Clancy [181, 206] used a similar method when measuring conductivity in liquid siloxane. A diagram of their apparatus is shown in the fig. 11.31. They applied the first electron beam in order to introduce negative charge carriers into the liquid and a second beam in order to determine the potential of the liquid. The layer of the liquid used was of the order of 10^{-2} cm in thickness and the resistivity of the liquid was calculated from the formula:

$$\rho = \frac{E}{J} \approx \left[\frac{10D}{ukJ}\right]^{1/2}, \tag{11.22}$$

where D denotes the thickness of the layer of the liquid, J—density of the current, u—the mobility of the charge carriers and k—the dielectric constant.

11.6. *The Influence of Temperature on the Conductivity of Dielectric Liquids.* Research on self-conductivity and ionization conductivity in dielectric liquids has been done by many scientists, including Jaffe [14, 15], Adamczewski [42–45], Plumley [21], Pao [3], Gzowski [32], Jachym [193], Guizonnier and his colleagues [185], Gibaud [179], Forster [147–9],

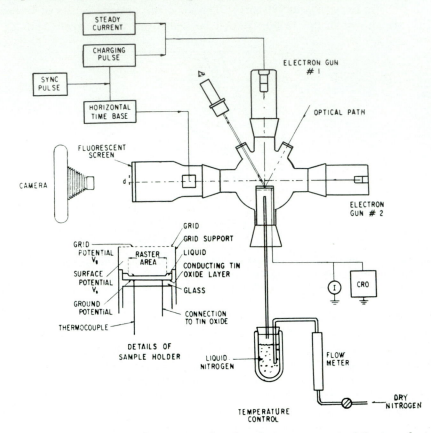

Fig. 11.31. General view of apparatus showing the arrangement of the two electron guns and the associated circuits. The optical arrangement for thickness measurement is indicated and also the method of temperature control. Ref.: Watson and Clancy [206].

Bässler and his collaborators [214–17, 220], Freeman [240–2], Nowak [122], Silver [245] and others. Those results concerned with an explanation of the mechanism of electrical conduction in dielectric liquids will be discussed below, whereas the results of other research, especially that connected with the dependence of mobility, recombination and diffusion of ions on temperature, will be analysed in chapter 15.

Pao [3] conducted a long series of measurements of current–voltage characteristics under different temperatures in *iso*-octane and liquid air, both of which were ionized by γ-rays from a radium source. Figure 11.32 is a diagram, based on the work of Pao [3], of current–voltage characteristics in *iso*-octane ionized by γ-rays. As can be seen, the current increases systematically with an increase in temperature but the slope of the straight lines remains constant in the saturation region and no differences were found as a result of changing the direction of the applied voltage. Having examined Jaffe's theory of columnar ionization, Pao arrived at the following relationship:

$$\frac{1}{I} = \frac{1}{I_\infty}\left(1 + S\,\frac{1}{E}\right). \tag{11.23}$$

One of the plots drawn by Pao for various temperatures is given in fig. 11.33. It is an interesting fact that all the straight lines intersect at

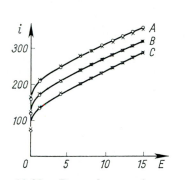

Fig. 11.32. Dependence of current density i (in A cm$^{-2}\times 10^{14}$) on electric field strength at three temperatures: A, 315°K; B, 260·6°K; C, 210°K. Ref.: Pao [3].

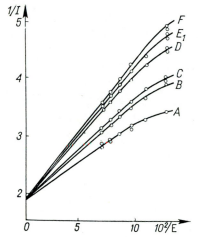

Fig. 11.33. $1/I = f(1/E)$ dependence in *iso*-octane with radium (4 mg) γ-ray ionization at different temperatures: A, 313·3°K; B, 273°K; C, 260·6°K; D, 230°K; F, 191°K. The open circles are experimental points for two voltage signs, positive and negative. Ref.: Pao [3].

one point (the same I_∞ value). This indicates that the initial number of ions produced by the ionizing agent is independent of the temperature of the liquid. This is an obvious conclusion, but although some changes in ionization might be expected when the density of the liquid is varied and the radiation is absorbed, these changes were apparently too small to be noticed. On the basis of the experiment, $S = \text{const.}/T^{1.28}$. Pao also investigated variations resulting from changes in temperature in induced and natural conductivity in *iso*-octane and liquid oxygen.

The difference in shape of both characteristics is shown in fig. 11.34. The results of Pao's research on natural conductivity depending on temperature in *iso*-octane are given in fig. 11.35. The plot of the relationship $\log i = f(1/T)$ (fig. 11.35) indicates that this relationship consists of two intersecting straight lines whose slopes decrease in proportion to the reduction in the electric field. Plumley's theory (§ 16.7) did not account

12

for the fact that the line bends at a certain point. Pao explained this phenomenon as a result of the action of a different mechanism. According to Pao, the field-induced molecular dissociation is active in the first case, while in the second case it is the result of the ionization effect which has been caused by cosmic rays.

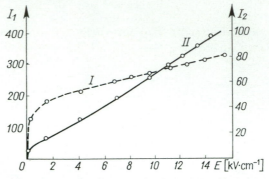

Fig. 11.34. *I* as a function of *E*. Curve I, in *iso*-octane ionized by radium (4 mg) γ-radiation. Curve II, in liquid oxygen ionized by radium (4·5 mg) radiation at 90°ᴋ. Ref.: Pao [3].

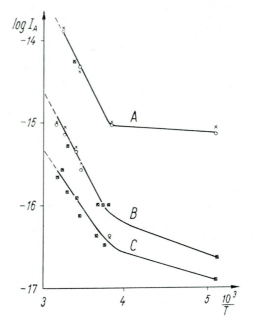

Fig. 11.35. The relationship $\log I = f(10^3/T)$ in *iso*-octane for various electric field strengths: *A*, 14·850 v cm⁻¹; *B*, 4·010 v cm⁻¹; *C*, 1·215 v cm⁻¹ (intrinsic current for two voltage signs). Ref.: Pao [3].

In 1962 Forster [148, 149] published the results of his research on electric conduction in benzene. Under normal conditions—i.e. when the liquid was saturated with air—the self-conductivity in benzene was of the order of $10^{-14}\,\Omega^{-1}\,cm^{-1}$, while after the de-aeration of the freshly purified liquid, it was about $10^{-16}\,\Omega^{-1}\,cm^{-1}$. Forster found that the distribution of the field was unsymmetrical in the area between the electrodes of the ionization chamber. He obtained the following relationship for the dependence of conductivity on temperature at the various separations of the electrode:

$$\sigma = \sigma_0 \exp\left(-W/kT\right), \tag{11.24}$$

where $W = 0.42$ ev.

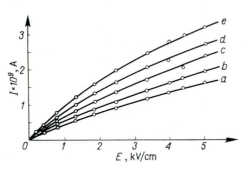

Fig. 11.36. Dependence of current i on electric field strength E in cyclohexane ionized by x-ray at a number of temperatures: a, 7·5°c; b, 17°c; c, 30°c; d, 40°c; e, 50°c. Electrode separation $d = 0.8$ cm. Ref.: Jachym [193].

The values of activation energy obtained by Forster for aromatic hydrocarbons were about 0·4–1 ev, and for benzene, toluene and xylenes about 0·41 ev. He found similar values of activation energy for self-conductivity in cyclohexane and cyclohexadienes (∼ 0·42 ev).

The value for cyclohexane was 0·16 ev and for 1:3:5-trimethylbenzene—0·19 ev. Forster associated the values of activation energy with those of excitation energy and considered that conductivity in the liquids under investigation was connected with the existence of double bonds and π electrons in the molecules of the liquid.

Guizonnier [173–8], however, is of a different opinion. He conducted extensive research on natural conductivity in different hydrocarbons, in carbon tetrachloride, paraffin oil, silicone oil and other liquids. The values of activation energy obtained, irrespective of the type of liquid used, were about 0·41 ev. Guizonnier assumed that conductivity in those liquids under investigation depended on the admixtures and not on the structure of the liquid molecules. It should, however, be noted that Guizonnier did not use sufficiently purified liquids for his research and the liquids had a high conductivity.

A large series of experiments on the dependence of conductivity in liquids on temperature was mainly conducted in our Institute by Jachym [190–3, 239, 269, 271].† The dependence of both natural and ionization conductivity (by means of x and γ-radiation) was simultaneously measured in liquid cyclohexane [193] (fig. 11.36). These measurements

† See also Jachym A. and Jachym B. [345].

showed that the increase in the ionization current was less dependent than the natural current on the increase in temperature. That is, the activation energy for natural conductivity was almost four times greater than that for ionization conductivity. Figure 11.37 shows the dependence of the

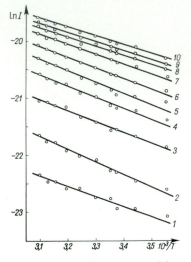

Fig. 11.37. The relationship $I = f(10^3/T)$ on a semi-logarithmic scale for cyclo-hexane at a number of electric field strengths E (in kv cm^{-1}): 1, 0·212; 2, 0·420; 3, 0·825; 4, 1·345; 5, 1·856; 6, 2·488; 7, 3·375; 8, 4·0; 9, 4·4; 10, 5·04. Ref.: Jachym [193].

current on the reciprocal of temperature. The slope of the straight lines (which is less pronounced for the ionization current) is a measure of activation energy. Attention should be drawn to the drop in activation energy both for natural and ionization conductivity. This drop corresponded to an increase in the electric field strength. When the strengths of electric fields were higher in the region of the linear regions of the current–voltage characteristics, the activation energy for ionization conductivity was not only smaller than that for self-conductivity but also smaller than that for viscosity ($W_\eta = 0.13$ ev).

Measurements taken in saturated hydrocarbons (from hexane to hexadecane) of natural and ionization currents (γ radiation ^{137}Cs) in a wider range of temperatures showed another important difference in the dependence of natural and ionization conductivity on temperature. Ionization conductivity is characterized by straight lines, which are almost parallel to each other when $\ln I$ is plotted against $1/T$. In the case of self-conductivity, however, a violent change takes place at a certain temperature, above which a further increase in the natural current is connected with an increase in activation energy. Jachym and Jachym [271] explain the above phenomenon in terms of changes in the quasi-crystalline internal structure of the liquid in this region. It is also possible to interpret such changes as the result of a change in the number of possible molecular vibrations. This change influences the change in the coefficient of viscosity.

In recent years investigations have been made by Kao *et al.* [287, 288] on the dependence of the ionization current on time, on the distribution of the potential within the ionization chamber, on the degree of the impurities in the liquid and on temperature.† They concentrated on experiments with hexane containing admixtures of decane.

A separate series of experiments, between 1963 and 1965 by Bässler *et al.* [214–17, 220] were concerned with natural conduction in liquid hydrocarbons. They used paraffin hydrocarbons (including hexane), a group of aromatic hydrocarbons (benzene, naphthalene, diphenyl and anthracene), aliphatic hydrocarbons (cyclohexane and decalin) and also other liquids such as carbon tetrachloride and chloroform. The liquids were purified both by chemical methods and by means of zone melting. Some of the liquids were degassed. Despite very careful preparation, the natural conductivity in some liquids was relatively high (in hexane, for example, it was $10^{-17}\ \Omega^{-1}\,\mathrm{cm}^{-1}$) and thus showed that they still retained some impurities. Because of the lack of an earthed guard ring at the collector electrode, the system of electrodes used was suspect as other research workers have indicated that this omission might adversely influence both the distribution of the electric field in the liquid-filled chamber and also the value of the current measured.

A number of experiments on the electrical conductivity as a function of temperature in various liquids were conducted by Bässler and his colleagues. They determined the activation energy for those processes and established the following general relationship for natural conductivity:

$$\sigma = \sigma_1 + \sigma_2 = \sigma_{01} \exp\left(-W_1/KT\right) + \sigma_{02} \exp\left(-W_2/KT\right) \qquad (11.25)$$

having found that

$$\sigma_{01} = (2 \pm 1)\,10^{-2}\ \Omega^{-1}\,\mathrm{cm}^{-1} \qquad (11.26)$$

and that $W_1 = 0.87$ ev and that W_1 is independent of the chemical structure of the molecules. σ_{02} is characteristic for each substance.

When discussing σ_1, σ_2, W_1 and W_2 as a function of the type and structure of the liquid, the authors concluded that σ_1 prevails over σ_2 in very pure liquids and that σ_2 is a result of complexes formed of its own and alien molecules.

During experiments the authors found the following relationship between the number of π electrons in an atom N (see table 11.4) and the value of activation energy W_1:

$$W_1 = -0.64\,[\pi] + 1.52\ [\mathrm{ev}], \qquad (11.27)$$

where $W_1([\pi] = 1) = 0.88$ ev and $W_1([\pi] = 2) = 0.24$ ev. A plot of the relationships established by those authors is shown in fig. 11.38.

† See also the new papers from 1968: Felici [364], Sharbaugh and Barker [356], Adamczewski and Fachym [360].

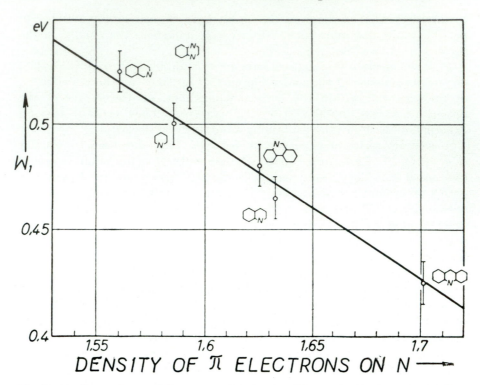

Fig. 11.38. Dependence of the energy of activation (W_1) on the density of π electrons in a molecule. Ref.: Bässler and Riehl [215].

Table 11.4.

Substance	Structure	Number of π electrons per N atom
Pyridin		1·585
Chinolin		1·633
Isochinolin		1·560
Acridin		1·705
Phenanthridin		1·625
Chinoxalin		1·593

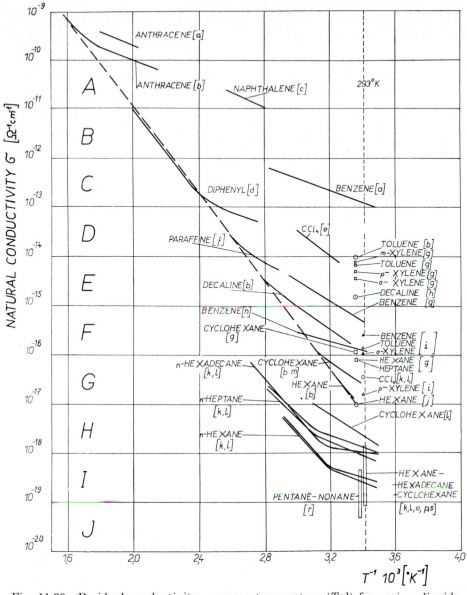

Fig. 11.39. Residual conductivity σ versus temperature (T^{-1}) for various liquids (data of many authors). Regions A, B, C paraffin hydrocarbons of technical purity or pure aromatic hydrocarbons (with π electrons). Regions E, F, some aromatic hydrocarbons, highly purified and dried. Regions H, I, paraffin hydrocarbons, highly purified, dried and degassed. Region J, some samples of standard paraffin hydrocarbons of highest purity under thick lead screen (influence of cosmic rays). Various values of σ_0 in the equation $\ln \sigma = \ln \sigma_0 - (W/kT)$ are probably due to unequal amounts of water in the sample. The ionization of liquids may change the value of the liquid conductivity by several orders of magnitude (from H, I regions to A, B). [a] Rabinowitsch, [b] Bässler and co-workers, [c] Bornmann, [d] Schlecht, [e] Guizonnier and co-workers, [f] Riehl, [g] Förster, [h] Meyer, [i] Zając, [j] Le Blanc, [k] A. Jachym, [l] B. Jachym, [m] Kalinowski, [n] Sułocki, [o] Gzowski, [p] Terlecki, [r] Adamczewski, [s] Blanc and co-workers.

In the case of carbonyl bonds, W_1 increases linearly and there is also an increase in the wave number $\bar{\nu}$ of the infra-red oscillations in this group:

$$W_1 = 8{\cdot}7 \times 10^{-4}\,\bar{\nu} - 0{\cdot}9 \text{ ev.} \tag{11.28}$$

Bässler and his colleagues discovered that in several liquids there were changes of slope in the curves. This indicated that at a certain temperature, a change occurs in the activation energy of the conduction process. That is, the second term in the eqn. (11.25) begins to prevail.

Nowak [122] also investigated natural and induced electrical conduction in benzene. In addition, Nowak measured the ion mobilities in benzene at various temperatures. References will be made later to some of the results obtained.

Variations in the viscosity and density of the liquid, corresponding to changes of temperature, should be taken into account when considering an accurate solution of this problem. The following should also be taken into consideration: variations in the mobility, recombination and diffusion coefficients of the ions and the possible increase in both the energy of the photoelectrons and their influence on the value of the current. Systematic research on this subject has only begun recently.

At the end of this chapter there is a comparison of results by various research workers for the conductivity of different dielectric liquids as a function of temperature, T^{-1} (fig. 11.39) [266].

Chapter 12

The Mobility of Ions

12.1. *Introduction.* The coefficients of ion mobility u and of diffusion D are important parameters in the following equation which pertains to the electric current in gases and dielectric liquids:

$$i = e(n_1 u_1 + n_2 u_2)E - D_1 \operatorname{grad} n_1 + D_2 \operatorname{grad} n_2, \qquad (12.1)$$

where 1 refers to the positive ions and 2 to the negative ions. The coefficient of ion recombination α is also of considerable importance in this phenomenon. These three coefficients characterize the substance under investigation and determine the mechanism of the transport of ions within the substance, thus determining the mechanism of electrical conduction at low fields. A great deal of experimental and theoretical research has been devoted to the investigation of these coefficients in gases and their dependence on molecular structure, temperature, pressure, electric field strength, etc. In 1903 Langevin [70] developed a number of experimental methods for the purpose of measuring the coefficients. He also theoretically formulated both the dependence of the coefficients on the other physical parameters characterizing the gas and the relationships between the coefficients themselves (on the basis of Einstein's earlier theories).

Extensive research is now being done on this subject because of its relevance to plasma physics (cf. Brown [67]). The theory of ions with reference to electrolytes was also treated in depth many years ago. This was possible owing to numerous, large-scale experiments which are of great practical importance in electrochemistry. Such experiments, however, were limited in the case of dielectric liquids because of the problems involved in obtaining pure dielectric liquids and the disturbing influence of both impurities and admixtures. It should be stressed that a thorough investigation of the effects of ionization in liquids can only be performed in dielectric liquids because the phenomenon of molecular dissociation into ions, caused by electrochemical effects, does not occur in these liquids. This investigation, therefore, supplements research on ionized gases subjected to high pressure and, to some extent, also the research on solid semiconductors.

12.2. *Methods of Measurement.* Early work on ion mobilities in dielectric liquids was done indirectly by means of analysing current–voltage characteristics in low fields. In such a case, as is already known, the intensity of the current can be expressed by the equation:

$$i = eE(n_1 u_1 + n_2 u_2). \qquad (12.2)$$

When the electric charges of both signs are of equal density, the equation is as follows:

$$i = eEn(u_1 + u_2). \tag{12.3}$$

It is possible to determine the sum of ion mobilities $(u_1 + u_2)$ from eqn. (12.3) when n has been derived from the saturation current and the recombination of ions has been taken into consideration. Such a method could only serve to give an estimation of the sum of ion mobilities. The values obtained oscillated between 0.92×10^{-4} and 11.7×10^{-4} cm^2 v^{-1} sec^{-1}. These values corresponded in order of magnitude to the values of ion mobilities in electrolytes and were about 1000 times smaller than those for gas ions.

Jaffe [14] and Bijl [53] were the first to use direct methods when experimenting on the mobility of ions. Their experimental methods were similar to those applied in research on gases. The essence of such methods was the measurement of the transit time of the edge of the ion layer which was produced in a dielectric liquid by an ionizing agent. The transit time was measured as a function of the separation of the electrodes in a parallel-plate capacitor and the electric field strength. These methods are discussed later when the author's works are described. Many early scientific works in this field suffered from the failure to establish exact experimental conditions which could influence the results obtained (for example, the purity grade of the liquid, its viscosity, temperature, etc.). These investigations of the mobility of ions in different liquids, rarely had any methodical connection with each other.

12.3. *The Dependence of the Mobility of Ions on the Viscosity of the Liquid.* The first systematic investigations on the mechanism of the motion of ions in dielectric liquids were carried out by Adamczewski [1, 18, 41, 42] in the period 1932–34. He measured the mobility and recombination of ions in five pure compounds belonging to the saturated hydrocarbon group of the type C_nH_{2n+2}, ranging from pentane $(n = 5)$ to nonane $(n = 9)$. He accurately determined the physical and chemical properties of these liquids (i.e. their boiling point, density, viscosity and electrical conductivity). In view of the significance of the viscosity coefficient in relation to the mobility of ions, a precise determination of this coefficient was of great importance.

Although the methods used by the author for measuring the mobility of ions were similar to those applied in the investigation of gases, it was, however, necessary to modify them because of the considerably smaller values of ion mobilities in liquids (about 1000 times smaller). The apparatus used by the author to measure the mobility of ions was based on that described in §11.4 in fig. 11.7. The measurements themselves were taken in the ionization chamber K_1, the detailed design of which is given in fig. 10.4. It was possible to change continuously the separation of the electrodes from 0 to 20 mm by means of a micrometer screw. An auxiliary capacitor K_2 (filled with solid pure paraffin) was used to collect the charge of ions while the electric field was active. The voltage was taken from a battery source (up to 2000 v). In most cases the following electrometers of an older type were in use: the quadrant, Hoffmann and quartz-fibre electrometers. These, however, had high voltage and charge sensitivities. All the insulators connected with the electrometer were made of amber and the whole measuring device was placed in an electro-static screen. An x-ray apparatus (cf. §11.4) was employed as the source of ionization. A specially designed automatic device was used for a number of interrelated functions such as switching the voltage and unearthing the electrometer. This device was capable of repeating these functions in a time of 0·1 to 1 sec, having an accuracy of 0·001 sec.

The ionization of the whole volume of the liquid inside the parallel plate capacitor was the fundamental principle on which the measurement was based. After a steady state had been reached (in about 15 sec) the irradiation was shut off and the voltage was applied to the capacitor plates. The ions produced in the liquid began to move the electrode of the opposite sign where they deposited their charge. One of the electrodes was connected to an additional capacitor and by means of an amber-insulated switch it was possible to connect it to the electrometer. This was necessary because large charges were induced on the electrode while the voltage was switched on and off. Although the charges could damage the electrometer during the first moments after switching, compensation took place at a later stage.

The author used three methods in the course of his work: (a) the collection of the charge of ions of the same sign; (b) the collection of the remainder of the charge which had still not moved to earth during the time t of the field action (the so-called Bijl method [53]) and (c) the Langevin method, i.e. the changing, during the measurement, of the polarity of the applied field so that some ions of both signs were collected on the collector electrode. The disturbing influence of diffusion is counter-acted by this method. The mobility of the ions in all three methods is given in diagrammatic form in fig. 12.1. The shaded parts represent the ion layers collected on the electrode B during the measurements. The lower half of fig. 12.1 illustrates the collected ion charge as a function of the time t of field action. The formulae for these relationships are as follows:

Method (a)

$$
\begin{aligned}
&(1) \quad Q(t) = neu\,Et \quad &&\text{for} \quad t < \tau, \\
&(2) \quad Q = ned \quad &&\text{for} \quad t \geqslant \tau;
\end{aligned}
$$

Method (b)

$$
\begin{aligned}
&(1) \quad Q(t) = ned - neu\,Et \quad &&\text{for} \quad t < \tau, \\
&(2) \quad Q = 0 \quad &&\text{for} \quad t \geqslant \tau;
\end{aligned}
$$

$$(12.4)$$

Langevin method (c)

$$
\left.
\begin{aligned}
&(1) \quad Q(t) = ne[u_1 Et - (d - u_2 Et)] && \text{for} \quad t \leqslant \tau_2, \\
&(2) \quad Q(t) = neu_1 Et && \text{for} \quad \tau_2 < t < \tau_1, \\
&(3) \quad Q \;\;\; = ned && \text{for} \quad t \geqslant \tau_1.
\end{aligned}
\right\} \qquad (12.5)
$$

u_1 and u_2 denote the mobilities of ions, E—the electric field strength ($E = U/d$), d—electrode separation and τ—arrival time of the edge of the ion layer of a given type ($\tau = d/uE$). In order to carry out the series of measurements necessary to obtain the whole relationship $Q(t) = f(t)$, a substantial number of experimental points is required for all three

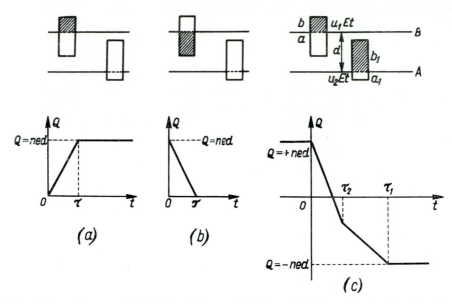

Fig. 12.1. Above: pictorial description of ionic motion in an ionization chamber after cutting-off the source of ionization and applying a voltage. Below: curves of the charge collected as a function of field duration by three methods: (*a*) charge collection method, (*b*) Bijl's method, (*c*) Langevine's method. Ref.: Adamczewski [40].

methods. Each of the points must be repeated at least two or three times under fully reproducible conditions of ionization. An automatic measuring device strictly controlled (by means of a chronograph) to switch off the radiation and to switch the voltage of the chamber on and off, etc., must be used at the same time. The advantage of the modern methods described below is that they can record the whole measuring cycle on one oscillogram. Examples of a few curves corresponding to the collected ion charge, as a function of the time of application of the field, are given in figs. 12.2, 12.3 and 12.4 (cf. Adamczewski [40, 41]). Figure 12.2 shows a series of curves obtained in heptane, by means of the Langevin method, for the inter-electrode separation $d = 0.6$ cm and the field strength

$E = 1510 \text{ v cm}^{-1}$. It can be seen that the curves are characterized by three distinct bends which correspond to the arrival at the electrode of the edge of one of the three layers of various ions. This is in accordance with the theoretical diagram shown in fig. 12.1. From the duration of time corresponding to each of the bends, it is possible to determine the

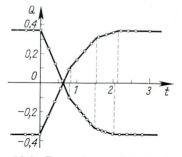

Fig. 12.2. Dependence of ionic charge collected Q on time t for heptane; $d = 0.6$ cm, $E = 1510 \text{ v cm}^{-1}$, Q is in e.s.u. cm^{-3} and t is in seconds. Ref.: Adamczewski [41].

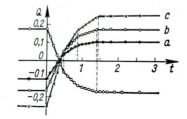

Fig. 12.3. Dependence of Q on t for pentane at various ionization beam intensities; $d = 0.4$ cm, $U = 307$ v, Q is in e.s.u. cm^{-3} and t is in seconds. Ref.: Adamczewski [41].

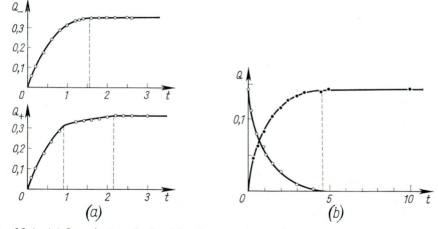

Fig. 12.4. (a) Q against t relationships for negative and positive ions at $d = 0.6$ cm, $U = 918$ v; Q is in e.s.u. cm^{-3} and t is in seconds. (b) Q against t for hexane using two different experimental methods; Q is in e.s.u. cm^{-3} and t in seconds. Ref.: Adamczewski [41].

mobilities of the three kinds of ions. Figure 12.3 shows a similar system of curves for pentane. In one series of measurements three different x-ray beams were used: (a) 25 kv, 5 mA; (b) 30 kv, 5 mA and (c) 40 kv, 5 mA. It can be seen that in such cases the number of ions changes while their mobilities remain the same. Figure 12.4 (a) shows the curves reflecting the collection of the charge of unipolar ions as a function of time (in accordance with the theoretical diagram given in fig. 12.1). Since it was impossible to determine the sign of each of the three different types of ions on the basis of the Langevin curves, the author was forced to rely on his own measurements. Two bends were clearly visible in the case of positive ions, but there was only one bend for negative ions because the effect of ion diffusion greatly distorted the image of the curve.

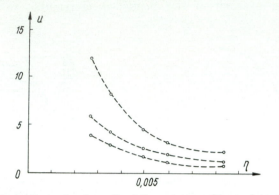

Fig. 12.5. Mobility u (in $cm^2\, v^{-1}\, sec^{-1}$) of three types of ions plotted against liquid viscosity η (in poise) for normal saturated hydrocarbons. Ref.: Adamczewski [41].

Figure 12.4 (*b*) illustrates the application of both Bijl's method and the charge collection method. In both methods the influence of ion diffusion was so great that it was impossible to see the bends of the curves without difficulty and thus to determine precisely the arrival time of the ions. From the several hundred diagrams obtained by the author for five compounds of the saturated hydrocarbon group of the type C_nH_{2n+2} (from $n = 5$ to $n = 9$ at room temperature), the following conclusions could be drawn:

(1) Two kinds of positive ions were found in all liquids but only one kind of negative ion which had an intermediate mobility.

(2) The values of mobility for ions of one kind systematically varied with the changing viscosity of the liquid (the values decreased with increasing viscosity).

(3) The following experimental relationship was found between the ion mobility u and the viscosity coefficient of the liquid η:

$$u = A\eta^{-3/2}. \tag{12.6}$$

For all three kinds of ions A is $A_1{}^+ = 1 \cdot 5 \times 10^{-7}$, $A_2{}^- = 0 \cdot 85 \times 10^{-7}$ and $A_3{}^+ = 0 \cdot 53 \times 10^{-7}$, u being expressed in $cm^2\, v^{-1}\, sec^{-1}$ and η in poise.

The numerical results (taken from the author's work) of ion mobility for five compounds of the C_nH_{2n+2} group are listed in table 12.1. Figure 12.5 shows a diagram of these relationships for the three different kinds of ions. As emphasized above, this was the first systematic research concerning the mechanism of the mobility of ions in dielectric liquids. This research was carried out in one group of liquids in which the physical and chemical properties, and particularly the viscosity which is very important in this case, systematically changed with the increase in the length of the molecular chain by one CH_2 group.

Table 12.1

Liquid	Symbol	M	ρ	η	t_w (°C)	u_1	u_2	u_3	α	$\alpha/\sum u$
						($10^{-4}\ \mathrm{cm^2\ sec^{-1}\ v^{-1}}$)			($\mathrm{cm^3\ sec^{-1}\ ion^{-1}}$)	
Pentane	C_5H_{12}	72·14	0·630	0·00260	35–37·5	12·1	6·8	3·8	2·13	3·28
Hexane	C_6H_{14}	86·17	0·663	0·00334	68·5–69·6	8·4	4·5	2·8	1·5	3·24
Heptane	C_7H_{16}	100·19	0·709	0·00480	96–98	4·7	2·8	1·8	0·88	3·23
Octane	C_8H_{18}	114·22	0·725	0·00595	122–123·5	3·4	1·8	1·2	0·69	3·67
Nonane	C_9H_{20}	128·25	0·750	0·00753	150–151	2·2	1·2	0·8	0·49	3·71

M—molecular mass, ρ—density, η—coefficient of viscosity, t_w—boiling temperature, u—mobility of ions, α—coefficient of ion recombination.

The experimental formula (12.6) established by the author determined the change in ion mobility with the change in viscosity for this group of liquids. This formula differs from the well-known Walden law $u\eta = \text{const.}$ and from the Stokes law $u = e/6\pi\eta r$. The author, however, was aware of the fact that the experimental relationship he found may, to some extent, depend on the variation of the radius of ions in different liquids, defined by the variation of r in the form $r \sim \sqrt{\eta}$. If the ions were to be regarded as spheres, the calculated radii would oscillate within the limits of 2·86–14 Å. This would correspond approximately, in terms of the order of magnitude, to the molecular dimensions but it would not be satisfactory for the picture of ion mobility.

In order to decide whether the relationship (12.6) is also valid for one single liquid whose viscosity varies with changes in temperature, the author [42] carried out experiments in paraffin oil which has a high viscosity. It was impossible for these measurements to be as precise as those taken previously because the high viscosity of the liquid ($\eta \sim 1$–5 poises) caused difficulties. In this case the results revealed that the behaviour of ion mobility is consistent with the Stokes–Walden law $u = A\eta^{-1}$. The relationship $u = \text{const.}/\eta$ satisfies the equation of a straight line passing through the origin. The activation energy for viscosity and ion motion was equal to 12·8 kcal/mol. and the radius of an ion, estimated on the basis of Stokes law, was about 3·8 Å. The results of this investigation as well as the method of measurement are described further on in § 12.7 where viscous liquids are discussed.

Two different methods were used to investigate the mechanism of the mobility of ions in dielectric liquids. The first aimed at changing the viscosity of the liquid by means of varying the molecular structure within one group of liquids. The second method modified the temperature of a liquid. As can be seen from the data, this research required further analysis in order to explain both the applicability of the formula (12.6) and also the significance of the ion dimensions in the Stokes equation. The author developed this subject [45] and Gzowski [31] systematically investigated this type of dependence on the viscosity, structure and temperature of liquids in the saturated hydrocarbon group. This work is described in detail below. In the last few years several works have been published on ion mobilities in specific dielectric liquids. Only the few using original measuring methods will be described in this book.

12.4. *Other Work on Ion Mobility.*†   In 1937 Reiss [22] elaborated a new method of measuring the mobility of ions produced in a solution of anthracene in hexane, irradiated by ultra-violet rays. The use of anthracene was necessary because ultra-violet radiation in pure hexane does not produce a photoelectric effect.

† See also the new papers: Essex and Secker [357], Buttle and Brignell [341], Miller, Howe and Spear [352].

The principle of Reiss's method is given in fig. 12.6 and a diagram of his apparatus in fig. 12.7. The electrodes consist of two concentric hollow metal cylinders having the radii r_1 and r_2 and placed in a brass vessel A. When a voltage U is applied to the electrodes the electric field is given by the formula:

$$E_r = \frac{U}{r \ln (r_2/r_1)}.$$
(12.7)

In this volume a synchronous motor M forces the liquid layer to rotate with angular velocity ω. The ultra-violet light from a quartz lamp Q produces a layer of ions at a certain fixed point on the surface of the liquid. While the liquid rotates this layer moves spirally in the electric field towards the outer earthed electrode. A probe connected with the electrometer is placed near the electrode.

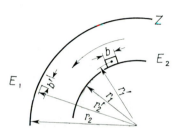

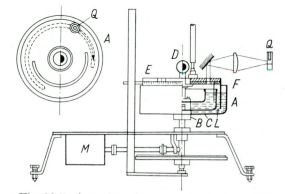

Fig. 12.6. The principle of the Reiss method for measuring ionic mobility. Ref.: Reiss [22].

Fig. 12.7. Apparatus for measuring ionic mobilities: A, brass vessel; B, amber insulator; C, high-voltage electrode; D, voltage input; E, insulated cover; F, probe; L, paraffin oil bath; M, synchronous motor; Q, light source. Ref.: Reiss [22].

The mobility of the ion is derived from the formula:

$$u = \frac{r_2'^2 - r_1'^2}{\varphi_2} \frac{\omega}{U} \ln \frac{r_2}{r_1}.$$
(12.8)

r_1' and r_2' are illustrated in fig. 12.6 and φ_2 denotes the angle between the zero and maximum positions of the ion layer. Figure 12.8 shows diagrams of the dependence of the current i on the angle φ which defines the position of the probe. The maxima on the curves correspond to the centres of the ion layers.

The following values of the mobilities of ions were obtained by Reiss in a solution of 10^{-4} mol of anthracene in 1 litre of hexane:

$$u_+ = 9 \cdot 9 \times 10^{-4} \quad \text{and} \quad u_- = 13 \times 10^{-4} \quad (\text{cm}^2 \, \text{v}^{-1} \, \text{sec}^{-1}).$$

Contrary to the results obtained by Jaffe [15], Bijl [53] and Adamczewski [40, 41], Reiss stressed in the interpretation of his results that negative ions had a higher mobility than positive ions. According to Reiss, the

13

differences can be attributed either to the influence of admixtures (which had been confirmed experimentally) or to the difference in the energy of ionizing radiation (x, γ and ultra-violet). It should be stressed, however, that Jaffé, Bijl and Adamczewski, in particular, worked with uniform liquids of a very high purity grade. In fact, these liquids were several degrees purer than the solutions of anthracene in hexane investigated by

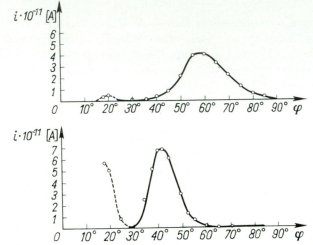

Fig. 12.8.   Dependence of current i on the probe setting angle φ. Current peaks indicate the times of ion arrivals. The upper curve is for positive ions, the lower for negative ions. Ref.: Reiss [22].

Reiss. In addition, Reiss's complicated method of measuring the mobilities, which required a motion of the liquid, could cause further discrepancies in the results. It is necessary to emphasize, however, that later research (LeBlanc [57] and Terlecki [34]) corroborated Reiss's results for ultra-violet and x-radiation.

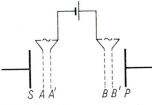

Fig. 12.9.   The Tyndall–Powell method for measuring ionic mobilities. AA' and BB' are wire grids, S and P are electrodes. Ref.: Meyer and Reif [54].

A method devised by Tyndall and Powell [55] for use in gases was applied in 1958 by Meyer and Reif [54] for the measurement of ion mobilities in dielectric liquids. In accordance with this method, the ions produced at one of the electrodes (S in fig. 12.9) by α-particles from a ^{210}Po source are subjected to the action of a constant electric field which shifts them to the electrode C. The ions are also subjected to the action of an alternating field applied across the grids AA' and BB'. The electrometer connected to the electrode C indicates the current flowing when the transit time of the ions between the grids AB is equal to half of the period of field oscillation between the grids AA' and BB'. The plot showing the dependence of the current on the oscillation frequency of the alternating electric field reveals a number of maxima (fig. 12.10) which correspond to the various transit times of ions.

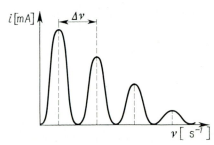

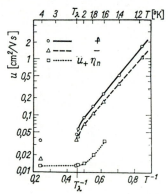

Fig. 12.10. Curve of frequency dependence of current i in ionic mobility measurements. Ref.: Meyer and Reif [54].

Fig. 12.11. Temperature dependence of positive and negative ion mobility in liquid helium. Ref.: Meyer and Reif [54].

If the distance AB is d and the velocity of the ion is v ($v = uE = uU/d$) then for the nth maximum $d/v = (2n+1)/2\nu_1$, where ν_1 denotes the frequency which is necessary for the first maximum to appear. Ion mobility is calculated from the formula:

$$u = \frac{d^2 \Delta \nu}{U}. \tag{12.9}$$

The following numerical results (in $cm^2\,v^{-1}\,sec^{-1}$) were obtained by Meyer and Reif for liquefied helium and argon: $u_- = 0 \cdot 07$, $u_+ = 0 \cdot 09$ at temperature $T = 1 \cdot 18°K$ for helium and $u_+ = 6 \times 10^{-4}$ at temperature $T = 90°K$ and under the pressure $p = 29 \cdot 2$ atm. for argon. Figure 12.11 shows both the plot of the dependence of the mobilities of negative and positive ions on temperature in ^4He and also the change in the product u caused by variations in temperature.

A number of measurements of the current–voltage characteristics and the mobilities of ions in transformer oils were taken in 1956 by Sato *et al.* [56]. They used the apparatus shown in fig. 12.12. The charge carriers were produced by the electrons emitted from a heated cathode placed above the liquid. A low vacuum (of the order 10^{-3} mm Hg) was produced between the cathode and oil. Under the influence of the electric field, the electrons emitted from the metal of the cathode passed into the liquid where they travelled to the anode in the form of heavy negative ions. The plot showing the dependence of the charge collected by the anode Q on the time t of collection is given in fig. 12.13 for two different electrode separations. The value u, of the order of 10^{-4} $cm^2\,sec^{-1}\,v^{-1}$ for the mobility of ions in oil, was thought to be consistent with Adamczewski's work. Sato *et al.*, however, did not take into consideration the great difference in the viscosities of both liquids. This means that in the measurements the ion mobilities should be much lower. On the other hand, the principle of the method itself is very interesting. It would be

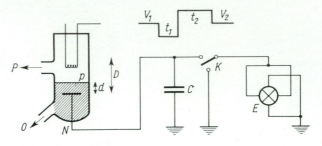

Fig. 12.12. Diagram of the apparatus for measuring ionic mobility in oil. N, denotes vessel filled with liquid; O, to the oil reservoir; P, to the pump; p, liquid level; D, electrode gap; d, depth of liquid layer; C, capacitor (230 pF); E, electrometer; K, switch; V_1, V_2, voltage pulses. Ref.: Sato *et al.* [56].

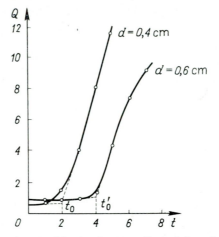

Fig. 12.13. Time dependence of ionic charge collected Q; t_0 is the time of the ion's arrival. $V_1 = 850$ v, $t_2 = 10$ sec, $p = 2 \cdot 2 \times 10^{-3}$ mm Hg. Ref.: Sato *et al.* [56].

useful to apply this method with greater precision under various experimental conditions, determining the distribution of the field, the viscosity of the liquid, variations of the collected charge as a function of the separation of electrodes and introducing a supplementary grid electrode on the surface of the liquid, etc.

In 1958 Gzowski and Terlecki [12] designed a method to determine the mobility of ions on the basis of one picture which is registered on the oscillograph during the transit time of a narrow layer of ions produced in the ionization chamber by x-rays. A diagram showing this method is given in fig. 12.14. The ion layer, having a thickness a and produced at the distance b from the upper electrode, is shifted towards the electrodes by the uniform field whose strength is $E = U/d$ (U denotes the voltage applied to the capacitor and d—the separation of the electrodes $d = a+b$). The electric field E in the capacitor is responsible for moving the ion at the expense of the energy of the voltage source:

$$eE\,ds\cos\alpha = Ui\,dt, \tag{12.10}$$

where α denotes the angle between the direction of the ion velocity v and

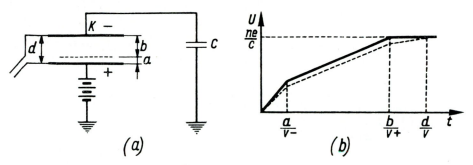

Fig. 12.14. The layer method for ionic mobility measurements. a, Ion layer distribution in the ionization chamber; b, curve of $U = f(t)$ dependence for collected ions. Ref.: Gzowski and Terlecki [12].

the electric field strength E. The current $i = eEv/U = ev/d$ induced in the circuit produces, in the time $dt < a/v,\ b/v_+$, an increase U of the potential on the capacitor C:

$$\Delta U = \frac{\Delta Q}{C} = \frac{ne}{dC}\,(v_- + v_+)\,\Delta t. \tag{12.11}$$

The velocities of the positive and negative ions are represented by v_+ and v_-. The plot of the relationship $U = f(t)$ shows a line consisting of some straight sections. Bends on the line appear when one type of the ions has reached the electrode, i.e. in the times:

$$\tau_1 = \frac{a}{v_-}; \quad \tau_2 = \frac{b}{v_+}; \quad \tau_3 = \frac{d}{v}. \tag{12.12}$$

When there are only ions of one kind in the whole volume of the ionization chamber, the current in the chamber has the following form:

$$i = \frac{\sigma v}{d}\int_0^{d-v\tau} dx = \frac{\sigma v}{d}\,(d - v\tau), \tag{12.13}$$

where σ denotes the linear density of the charge. The voltage pulse on the capacitor C for one kind of ions is:

$$U_c = \frac{\sigma v}{Cd}\int_0^\tau (d - v\tau)\,d\tau = \frac{\sigma v}{Cd}\,(d\tau - \tfrac{1}{2}v\tau^2). \tag{12.14}$$

When the ionization chamber contains two types of ions in a layer of the thickness a and when that layer is placed in the centre of the capacitor:

$$U = U_+' + U_-' + U_+'' + U_-'' \qquad (12.15)$$

where $U_+' = \sigma a v_+ t / Cd$; $U_-' = \sigma a v_- t / Cd$ for times from 0 to $(d-a)/2v_+$ and from 0 to $(d-a)/2v_-$, and

$$\left.\begin{aligned}
U_+'' &= \frac{\sigma v_+}{Cd}\left(a\tau_+ - \tfrac{1}{2}v_+\tau_+^2\right), \\[2mm]
U_-'' &= \frac{\sigma v_-}{Cd}\left(a\tau_- - \tfrac{1}{2}v_-\tau_-^2\right)
\end{aligned}\right\} \qquad (12.16)$$

for times from $(d-a)/2v_+$ to $(d+a)/2v_+$ and from $(d-a)/2v_-$ to $(d+a)/2v_-$. After all the ions have been collected on the capacitor C, the potential will be $U = \sigma a / C$.

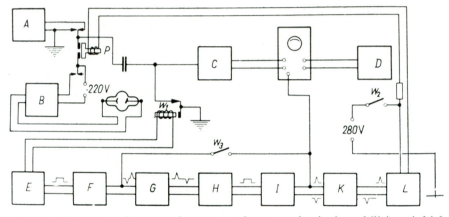

Fig. 12.15. Schematic diagram of apparatus for measuring ionic mobilities. A, high-voltage supply; B, x-ray source; C, d.c. amplifier; D, time marker; $E, F, G_1, H_1, I, J, K, L$, electronic control system; W_1, W_2, W_3, switches. Ref.: Gzowski and Terlecki [12].

A plot of a relationship of this kind is shown in fig. 12.14 (*b*). The measurements were taken in such a way that the bends in the curve $U = f(t)$ were determined for two different cases in both of which the ion layer a was moved from its original position by a distance Δs. This displacement was achieved by moving the lead diaphragm which transmitted x-rays. The mobility of ions u was found from the formula:

$$u = \frac{\Delta s\,d}{\Delta t\,U}. \qquad (12.17)$$

A diagram of the apparatus used by Gzowski and Terlecki is given in fig. 12.15. An example of the oscillograph records they obtained is shown in fig. 12.16. The ion layer was produced at the lower electrode which was connected with the negative pole of the battery. The negative ions therefore moved to the collector electrode from the bottom upwards.

Figure 12.17 shows the oscillograph record for the motion of the positive ions—i.e. in the case when the lower electrode was connected to the positive pole of the battery. The time interval between the dots on the oscillograph record is about 0·04 sec and the times of arrival measured were approximately 0·5 sec. Using this apparatus, Gzowski carried out a large series of measurements of the mobilities of ions as a function of temperature in various compounds of the saturated hydrocarbon group. He was thus able to determine the activation energy of the mobility of ions of both signs in various liquids.

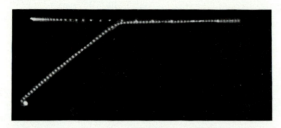

Fig. 12.16. Oscillogram showing the motion of negative ions. Ref.: Gzowski and Terlecki [12].

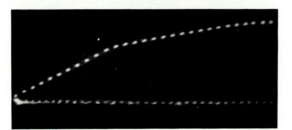

Fig. 12.17. Oscillogram showing the motion of positive ions. Ref.: Gzowski and Terlecki [12].

12.5. *The Dependence of the Mobility of Negative Charge Carriers (Electrons or Ions) on the Electric Field Strength.*† LeBlanc [57], Chong and Inuishi [112] and Terlecki [33, 34] introduced new methods of measuring the mobilities of charge carriers (electrons or negative ions) in liquids. In order to liberate the free electrons from the cathode they made use of the photoelectric effect and also examined the mobility of these charge carriers in liquids. They irradiated the cathode with radiation which gives no photoelectric effect in saturated hydrocarbons. The photoelectrons therefore only existed in a very thin layer close to the cathode. LeBlanc irradiated the cathode from the side whereas Terlecki who applied very high fields, irradiated the cathode through the semi-transparent anode.

The principle of LeBlanc's apparatus is shown in fig. 12.18 (*a*) and an outline diagram of the apparatus in fig. 12.18 (*b*). Figure 12.19 (*b*) shows the shape of the light pulses he applied for 6 msec. Figure 12.19 (*b*) gives

† See also Essex and Secker [357], Miller, Howe and Spear [352].

an example of the oscillograph records from which LeBlanc determined ion or electron mobilities. At a temperature of 17°c the value of mobility for hexane was found to be 13×10^{-4} while for heptane, at 27°c, it was 13×10^{-4} (the values are given in $cm^2 v^{-1} sec^{-1}$). Even when the difference in temperature during the measurements has been taken into consideration, LeBlanc still fails to explain why the difference in the mobilities of ions in hexane and heptane is so small.

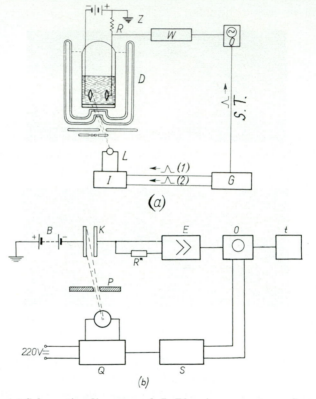

Fig. 12.18. (a) Schematic diagram of LeBlanc's apparatus. L, quartz lamp; G, control system; W, electrometer; O, oscillograph; R, resistor; D, thermostat; I, lamp current source; WPC, sweep trigger. (b) Electrical equipment for LeBlanc's apparatus. B, battery; K, ionization chamber; P, screen; R, resistor (time scale in msec). Ref.: LeBlanc [57].

A block diagram of Terlecki's apparatus is given in fig. 12.20. Examples of oscillograph records obtained by Terlecki are shown in fig. 12.21. The first refers to the investigation of the time of arrival of the ion layer front while the second concerns the investigation of the arrival time of the tail. From the bends in the curves Terlecki calculated the ion mobilities u_- for hexane $(9 \cdot 8 \times 10^{-4} \, cm^2 \, v^{-1} \, sec^{-1})$, for octane $(7 \times 10^{-4} \, cm^2 \, v^{-1} \, sec^{-1})$ and for decane $(3 \times 10^{-4} \, cm^2 \, v^{-1} \, sec^{-1})$. Terlecki was mainly concerned with examining the dependence of ion mobility on the electric field strength.

Research on both high field electrical conduction and electrical breakdown in liquids (cf. Chapter 19) suggests that this particular dependence is of great importance in formulating the theory of the mechanism of these phenomena. Terlecki's results for this relationship in decane are listed in

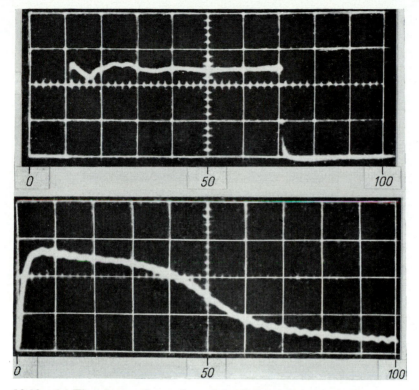

Fig. 12.19. (*a*) The shape of an ultra-violet light pulse. (*b*) Oscillogram of an electric current pulse in liquid. Ref.: Le Blanc [57].

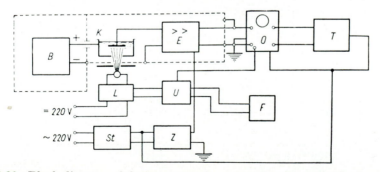

Fig. 12.20. Block diagram of the apparatus for measuring the mobility of negative-charge carriers in high electric fields. K, ionization chamber; L, quartz lamp; B, battery; E, electrometer; O, oscillograph; T, time marker; U, control system; Z, stabilized supply; F, photographic camera; St, stabilizer. Ref.: Terlecki [33, 34].

table 12.2. It can be seen, within the limits of experimental error, that there is no variation of the mobility in the field region from 5 kv/cm to 324 kv/cm. Neither has he found any distinct change in ion mobility in hexane within the field limits

$$9\text{–}100 \;\; \text{kv/cm} \;\; (u = 9\!\cdot\!8 \times 10^{-4} \; \text{cm}^2\,\text{v}^{-1}\,\text{sec}^{-1} \pm 5\%)$$

nor in octane within the field limits

$$8\text{–}192 \; \text{kv/cm} \;\; (u = 7\!\cdot\!0 \times 10^{-4} \; \text{cm}^2\,\text{v}^{-1}\,\text{sec}^{-1} \pm 5\%).$$

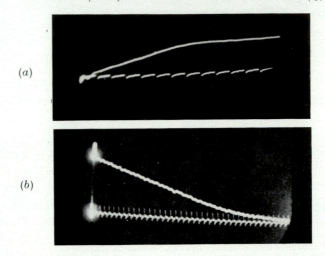

(a)

(b)

Fig. 12.21. (a) Oscillogram showing the motion of negative charge carriers in hexane; the arrival time of the front of the charge layer can be measured. (b) Oscillogram of the motion of negative charge carriers in hexane. The arrival time of the end of the charge layer can be measured. Ref.: Terlecki [33].

Table 12.2. The dependence of the mobility of negative ions in decane on the electric field strength

E (kv/cm)	$u \times 10^4$ (cm² v⁻¹ sec⁻¹)		E (kv/cm)	$u \times 10^4$ (cm² v⁻¹ sec⁻¹)	
5·3	—	2·9	170	—	2·83
12	3	—	200	3·0	—
32	—	3·0	230	3·03	3·25
68	3·1	—	261	3·2	2·7
105·3	—	3·05	287	2·7	2·9
138·6	3·16	—	324	2·85	—

After Terlecki [33, 34].

Plots of these relationships are given in fig. 12.22 (a). Circles denote the values obtained with an increase in field strength (cols. 2 and 5 of table 12.2) and crosses correspond to the values obtained with decreasing field strength (cols. 3 and 6 of table 12.2). It can be seen from fig. 12.22 (a) that these relationships are straight lines parallel to the abscissa. When discussing these results, Terlecki stressed that with increasing field strength a number of new experimental difficulties are apparent, although they are not visible when measurements are taken in low fields. These difficulties

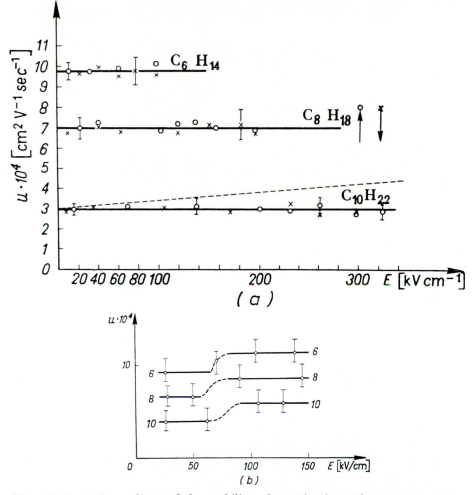

Fig. 12.22. (*a*) Dependence of the mobility of negative ions (electrons) u on the electric field strength E; the broken line is the predicted curve. Ref.: Terlecki [33, 34]. (*b*) Dependence of the mobility of negative ions (slow) on the field in hexane, octane and decane with a very noticeable change in mobility in the region 40–60 kv cm^{-1}. The figures 6, 8 and 10 denote the number of carbon atoms in the molecule. Ref.: Terlecki [33].

include the very strong influence of both small dust particles and traces of liquid impurities or gas admixtures, the necessity of a precise parallel adjustment of the electrodes, the difficulty of switching the voltage on and off, etc. These experimental problems may sometimes lead to discrepancies in the results obtained. Terlecki, for example, mentions occasional experimental results obtained in some liquid samples (shown in fig. 12.22 (*b*)), in which enhanced ion mobility was observed in the region from 40 kv/cm to 70 kv/cm when the values of ion mobility, both below and above this region, were constant. It has not yet been established whether this effect is to be attributed to accidental impurities in the liquid or chamber, or whether it conceals a physical effect characteristic of liquids.

Terlecki also compared in two different ways the mobilities of the negative carriers produced under the same experimental conditions. Firstly by means of ultra-violet radiation which gives a photoelectric effect at the cathode and secondly by ionizing a layer of the liquid with a narrow beam of x-rays. In both cases the same values were obtained for the mobility of the negative carriers, i.e. $(1 \cdot 0 \pm 0 \cdot 1) \times 10^{-3}$ cm^2 v^{-1} sec^{-1}. Terlecki also made an interesting attempt to explain the conditions under which both fast and slow negative ions can appear in the same liquid. He discovered that in *n*-nonane in which ion mobility was established as $3 \cdot 8 \times 10^{-4}$ cm^2 v^{-1} sec^{-1}, it is possible to induce a decrease in mobility by the introduction of a small quantity of nonene (1% in molar proportion); the number of slow negative ions then increases in proportion to the concentration of nonene. The mobilities of positive ions ($u = 1 \cdot 9 \times 10^{-4}$ cm^2 v^{-1} sec^{-1}) did not change under these conditions. It would therefore seem that in order to explain the phenomenon of the appearance and disappearance of various kinds of negative ions in dielectric liquids, it is necessary to devote a more careful examination to the influence of the traces of impurities left by foreign substances, both liquids and gases (such as oxygen, air, vapours of alcohol and ether, which are most frequently used to rinse the ionization chambers, etc.). Wizevich and Fröhlich [58], Davidson and Larsh [59], Morant [49], LeBlanc [57], Terlecki [33], Gzowski [31] and other scientists have discussed the influence of gases, and particularly oxygen. Although this influence is well known owing to various experiments on the mobility of ions in gases, it has not received any systematic treatment.

In 1960 the Japanese scientists Chong, Sugimoto and Inuishi [157] obtained similar results for hexane. The results of their investigation are shown in the fig. 12.23.

In order to establish whether a change in the mobility of ions occurs during the breakdown process in the liquid, an investigation of the $u = f(E)$ relationship in higher fields, up to 10^6 v cm^{-1}, would be very useful. This investigation is of special importance for fields above 300–400 kv cm^{-1} where a new mechanism appears for slowing down the electrons which results in the fluorescence of the liquid (cf. § 18.3). Fields of 1000–1200 kv cm^{-1} constitute another important limit for this phenomenon. It is only in these fields that there occurs an avalanche ionization induced by the application of voltage pulses (cf. § 18.3). These measurements will, of course, encounter many new problems during experiments and it is possible that different results will be obtained. At this stage, only an approximate estimation of the values of the mobility of ions can be expected at these fields. High-voltage generators, with a range of several million volts as well as wide separations between the electrodes, should be used to conduct these investigations in order both to reduce the disturbing influence of surface effects on the results of the experiments and to increase the volume effect occurring in the bulk of the liquid.

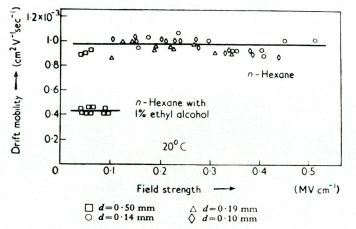

□ $d = 0 \cdot 50$ mm   △ $d = 0 \cdot 19$ mm
○ $d = 0 \cdot 14$ mm   ◇ $d = 0 \cdot 10$ mm

Fig. 12.23. The dependence of drift mobility upon field strength. Ref.: Chong and Inuishi [112].

12.6. *The Dependence of the Mobility of Ions on Temperature. Activation Energy.* Using the apparatus described above, Gzowski [31, 32] in 1961 carried out a large series of experiments on the dependence of the mobilities

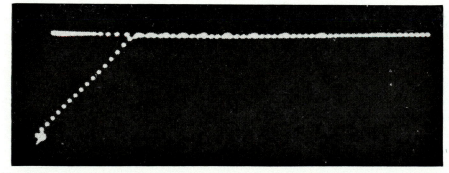

Fig. 12.24. Oscillogram for negative ions in heptane: temperature $t = 20°$c, voltage applied to the chamber $U = 1020$ v, electrode separation $d = 1$ cm, the gap between the collecting electrode and the front of ionic layer $S = 9 \cdot 3$ mm, $\tau = 0 \cdot 8$ sec, mobility $u_- = 6 \cdot 6 \times 10^{-4}$ cm^2 v^{-1} sec^{-1}. Ref.: Gzowski [31].

of ions on temperature in various compounds of the saturated hydrocarbon group. This enabled him to determine the activation energies for ions of both signs in different liquids. Figure 12.24 shows an oscillograph

record for heptane at a temperature of 20°c (Gzowski [31]). An ion layer
was formed at the lower (negative) electrode so that the distance between
the layer front and the collector electrode *s* was 9·3 mm and the separation between the electrodes was 1 cm. The chamber was filled with
hexane and the voltage applied was $U = 1040$ v. It was possible to find
the mobility of the negative ions after the number of time marks (each
one being 0·04 sec) had been counted up to the point where the parabolic
course began. At a temperature of 9°c the front of the negative ion layer
arrived at the electrode in a time of 1·12 sec. The mobility was
$u = 7·9 \times 10^{-4}$ cm²/v sec. At a temperature of 48°c the values were
0·72 sec and $1·2 \times 10^{-3}$ cm²/v sec, respectively. Figure 12.25 shows an

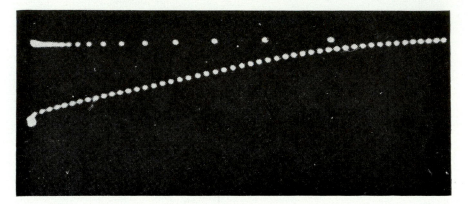

Fig. 12.25. Oscillogram for positive ions in heptane: temperature $t = 20°$c, mobility
$u_+ = 4·3 \times 10^{-4}$ cm² v⁻¹ sec⁻¹, electrode separation $d = 1$ cm, voltage applied
to the chamber $U = 1045$ v, $\tau = 52 \times 0·04$ sec $= 2·08$ sec. Ref. Gzowski [31].

oscillograph record of the positive ions in heptane taken at a temperature
of 20°c. The positive ion mobilities are recorded on the picture in the
form of a broadened spectrum and this is typical of all the liquids under
investigation. At a temperature of 20°c the mobility of the fastest positive
ions was $4·3 \times 10^{-4}$ cm²/v sec and at 40°c—$6·7 \times 10^{-4}$ cm²/v sec. The
distance between the electrodes was 1 cm and the distance from the front
of the ion layer to the upper electrode was 9·3 mm. The voltage
$U = 1045$ v. At a temperature of 20°c the arrival time of the front of the
positive ion layer was 2·08 sec and at 40°c—1·36 sec.

Gzowski conducted a series of experiments, at a temperature ranging
from 8°c to 51°c, using such normal saturated hydrocarbons as hexane,
heptane, octane, nonane and decane as well as mixtures of nonane and
hexane and octane and hexane. The results of Gzowski's research are
listed in table 12.3. Figures 12.26 and 12.27 show on semi-logarithmic
plots the dependence of the mobility of ions on temperature $u = f(10^3/T)$.
It is possible to determine the activation energies of particular types of
ions on the basis of these plots.

Table 12.4 lists the values of activation energy for the viscosity and mobility of ions in several liquids. These values are determined from the following two relationships:

$$\eta = A \exp(W_1/RT) \quad \text{and} \quad u = B \exp(-W_2/RT).$$

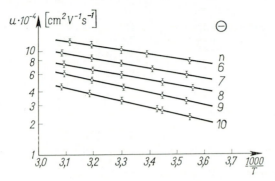

Fig. 12.26. Dependence of the mobility of negative ions on temperature (u as a function of $10^3/T$) for a number of saturated hydrocarbons. Ref.: Gzowski [31].

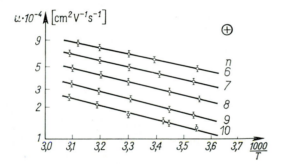

Fig. 12.27. Dependence of the mobility of positive ions on temperature (u as a function of $10^3/T$) for a number of saturated hydrocarbons. Ref.: Gzowski [31].

In order to examine the dependence of ion mobility on the viscosity of the liquid, Gzowski drew two plots. The first one corresponded to the Stokes–Walden law $u = f(\eta^{-1})$ (fig. 12.28), while the second was consistent with Adamczewski's formula $u = f(\eta^{-3/2})$ (fig. 12.29). As can be seen from the plots, the negative ion mobilities follow Walden's law and the mobilities of the positive ions obey Adamczewski's formula (12.6). It is a characteristic feature of Gzowski's work that the viscosities of the liquid were changed in two different ways: (*a*) by changing the structure of the liquid (when it passed from one liquid to another at a constant temperature) and (*b*) by varying the temperature in one liquid.

When discussing the results Gzowski concluded that the positive ions in all the liquids may be expected to satisfy Adamczewski's [45] relationship

Table 12.3. The mobilities u_+ and u_- of positive and negative ions in a group of saturated hydrocarbons at different temperatures

Type of liquid	t (°C)	$10^4 u_-$ (cm² v⁻¹ sec⁻¹)	$10^4 u_+$	η (cP)
n-Hexane Boiling temperature: 68–69°C Density at 20°C: 0·658 g/cm³	9	7·9	4·7	0·357
	22	9·2	5·8	0·312
	30·3	10·0	6·6	0·289
	40	11·3	7·6	0·268
	48	12·0	8·4	0·253
n-Heptane Boiling temperature: 97·5–98·5°C Density at 20°C: 0·648 g/cm³	10	5·9	3·6	0·458
	20	6·6	4·2	0·409
	30	7·5	4·9	0·367
	40	8·4	5·7	0·333
	51	9·5	6·7	0·302
n-Octane Boiling temperature: 125–126°C Density at 20°C: 0·702 g/cm³	8	4·4	2·4	0·625
	17	5·1	2·8	0·560
	30	5·8	3·5	0·479
	39	6·5	4·1	0·432
	50	7·4	4·8	0·386
n-Nonane Boiling temperature: 150–151°C Density at 20°C: 0·746 g/cm³	10	3·3	1·7	0·830
	17	3·7	1·9	0·740
	29·2	4·5	2·4	0·635
	40	5·2	2·8	0·555
	50	6·0	3·3	0·491
n-Decane Boiling temperature: 171–172°C Density at 20°C: 0·729 g/cm³	9	2·3	1·3	1·200
	17	2·6	1·4	1·03
	19	2·7	1·5	1·01
	30	3·3	1·7	0·85
	40·8	3·9	2·1	0·74
	51	4·5	2·5	0·64
Mixture Hexane–octane, 1 mol : 1 mol Density at 20°C: 0·686 g/cm³	17	5·8	—	0·479
	19	6·0	3·5	0·468
	30	6·8	4·2	0·422
	39	7·6	4·8	0·384
	50·5	8·5	5·5	0·347
Mixture Hexane–nonane, 1 mol : 1 mol Density at 20°C: 0·710 g/cm³	9·8	4·8	3·0	0·545
	18	5·4	3·4	0·499
	30	6·1	4·1	0·444
	39·5	6·9	4·6	0·410
	50	7·9	5·5	0·387
Mixture Hexane–nonane, ¼ mol : 1 mol Density at 20°C: 0·736 g/cm³	11·2	3·9	2·4	0·705
	16·5	4·2	2·7	0·660
	20	4·4	—	0·640
	29·2	4·9	3·3	0·568
	39	—	3·7	0·510
	50	6·1	4·3	0·455

After Gzowski [31].

in the form:

$$u = 3 \cdot 85 \times 10^{-4} \exp(0 \cdot 174n)\,(kT)^{-3/2} \exp[-0 \cdot 0139(n + 5 \cdot 05)/kT], \quad (12.18)$$

where n denotes the number of carbon atoms in a molecule and k— Boltzmann's constant in ev deg^{-1}. The values of mobility found in Gzowski's experiments are given in fig. 12.30 where they are plotted on the

Table 12.4

Type of liquid	Activation energy of viscosity (kcal/mol)	Activation energy of positive ion mobility (kcal/mol)	Activation energy of negative ion mobility (kcal/mol)
n-Hexane	$1 \cdot 5 \pm 0 \cdot 1$	$2 \cdot 5 \pm 0 \cdot 25$	$1 \cdot 8 \pm 0 \cdot 09$
n-Heptane	$1 \cdot 8 \pm 0 \cdot 11$	$2 \cdot 7 \pm 0 \cdot 27$	$2 \cdot 0 \pm 0 \cdot 1$
n-Octane	$2 \cdot 1 \pm 0 \cdot 12$	$2 \cdot 9 \pm 0 \cdot 29$	$2 \cdot 1 \pm 0 \cdot 1$
n-Nonane	$2 \cdot 4 \pm 0 \cdot 14$	$3 \cdot 1 \pm 0 \cdot 31$	$2 \cdot 7 \pm 0 \cdot 13$
n-Decane	$2 \cdot 7 \pm 0 \cdot 16$	$3 \cdot 2 \pm 0 \cdot 32$	$2 \cdot 9 \pm 0 \cdot 14$
Mixture of hexane and nonane, $\frac{1}{4}$ mol : 1 mol	$2 \cdot 0 \pm 0 \cdot 12$	$2 \cdot 7 \pm 0 \cdot 27$	$2 \cdot 1 \pm 0 \cdot 1$
Mixture of hexane and nonane, 1 mol : 1 mol	$1 \cdot 7 \pm 0 \cdot 11$	$2 \cdot 5 \pm 0 \cdot 25$	$2 \cdot 0 \pm 0 \cdot 1$
Mixture of hexane and octane, 1 mol : 1 mol	$1 \cdot 7 \pm 0 \cdot 4$	$2 \cdot 5 \pm 0 \cdot 25$	$2 \cdot 0 \pm 0 \cdot 1$

After Gzowski [31].

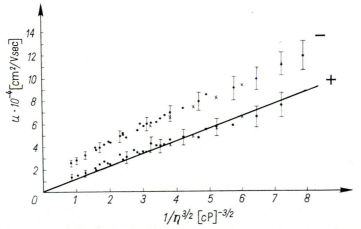

Fig. 12.28. The dependence of mobility u on $1/\eta^{3/2}$ for saturated hydrocarbons. Ref.: Gzowski [31].

straight lines obtained from the formula (12.18). It can be seen in the case of the four liquids that the experimental results are in agreement with theoretical data. Hexane, however, required the introduction of a certain constant numerical factor. Positive ions and neutral molecules therefore

14

have different activation energies. This is at least true for the form in which they appear in the measurements of layer viscosity. Gzowski also discussed in detail the possibility of interpreting the types of ions in liquids. He made use of the works by Frenkel [101], Swan [51], Terlecki [33], Stacey [38], von Hippel [60] and Dewhurst [61] which are concerned with the appearance of positive ions of different kinds in liquids.

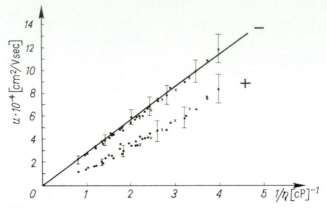

Fig. 12.29. The dependence of mobility u on $1/\eta$ for saturated hydrocarbons; positive ions obey the relationship $u_+ = A\eta^{-3/2}$ and negative ions the relationship, $u_- = B\eta^{-1}$. Ref.: Gzowski [31].

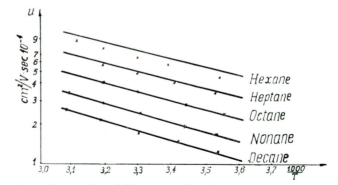

Fig. 12.30. Dependence of mobility u on T with n as a parameter according to eqn. (12.18). The crosses are experimental points. Ref.: Gzowski [31].

So far as aromatic hydrocarbons are concerned, experiments have been largely limited to benzene C_6H_6. In 1960 Chong and Inuishi [112] measured the mobilities of negative charge carriers in hexane and benzene, as a function of temperature. By means of short flashes of ultra-violet light they produced an electronic layer at the cathode and they measured the time taken for this layer to cross the ionization chamber. They obtained relationships of the type:

$$u = u_0 \exp\left(-W/kT\right). \tag{12.19}$$

W in hexane was $0\cdot16$ ev ($3\cdot7$ kcal mol^{-1}) while for the viscosity the activation energy was 0.08 ev (2.03 kcal mole^{-1}) in hexane and 0.11 ev (2.55 kcal mol^{-1}) in benzene. The numerical values of ion mobility at room temperature were 10^{-4} cm^2 v^{-1} sec^{-1} in hexane and 4.5×10^{-4} cm^2 v^{-1} sec^{-1} in benzene. It should be emphasized that on the strength of the experimental relationships obtained by Chong and Inuishi, the law $u \times \eta^{-n} = $ const. is valid; $n = 2$ for hexane and $n = 1.5$ for benzene. This confirms to a large extent the validity of Adamczewski's formula (12.6), in contrast to the results consistent with the Stokes–Walden law for which $n = 1$.

Between 1962 and 1966 Nowak [122] conducted extensive research on electrical conduction in benzene ionized by x-rays. He analysed the dependence of both positive and negative ion mobilities on temperature, concentrating on the coefficient of ion recombination and also determining the activation energies of these processes. Nowak's results are listed in table 12.5. Nowak's results show that the values of ion mobilities in benzene at room temperature are $u_+ = 2.36 \times 10^{-4}$ cm^2 v^{-1} sec^{-1} and $u_- = 6.48 \times 10^{-4}$ cm^2 v^{-1} sec^{-1}. He found that the values of n in the equation $u\eta^{-n} = $ const. are slightly more than 1 for the positive ion and almost 1 for the negative ion.

Table 12.5. The mobilities of positive ions u_+, and negative ions u_-, the viscosity coefficient η and ion recombination coefficient α in benzene at different temperatures T†

T (°K)	$u_- \times 10^4$ (cm^2 v^{-1} sec^{-1})	$u_+ \times 10^4$ (cm^2 v^{-1} sec^{-1})	η (cP)	α (cm^3 sec^{-1})
285.5	5.77	2.06	0.845	—
289.0	6.28	2.36	0.800	1.51
293.5	6.48	2.36	0.750	1.45
300.0	6.95	2.41	0.685	1.47
303.0	7.31	2.53	0.660	1.68
307.0	7.66	2.55	0.630	1.72
311.0	7.87	2.85	0.600	—
319.0	8.40	3.13	0.550	1.87
324.5	8.58	3.26	0.520	1.87

† Benzene $\sigma_w \sim 10^{-14}$ Ω^{-1} cm^{-1}—Nowak [112].

It should be emphasized that experiments on electrical conduction in benzene involve many additional difficulties. The reason for this is that natural conductivity in the purest benzene is greater by a factor 10^4 than that in pure saturated hydrocarbons (due to the presence of π electrons in benzene molecules). For this reason Nowak used a double ionization chamber in which the effects of natural conduction were compensated. It is necessary to use quartz insulators in the chamber because some insulators (e.g. ebonite and amber) are attacked chemically by benzene.

Zając [213] continued Nowak's work and used a modified version of the apparatus. He conducted a similar series of measurements for benzene, toluene, ortho- and para-xylenes in addition to naphthalene solutions in benzene. The following more important conclusions can be drawn from Zając's work:

(1) Current–voltage characteristics are similar in form to that given in the formula $I = G_0 d(1 + \gamma dE)$. In the limits of experimental error γ is a constant and equals $1 \cdot 81 \times 10^{-4}$ v^{-1}. G_0 increases in proportion to the molar volume within the limits from $6 \cdot 57 \times 10^{-12}$ A cm^{-3} for benzene to $9 \cdot 26 \times 10^{-12}$ A cm^{-3} for para-xylene.

(2) In these liquids the mobilities of negative ions satisfy Walden's law $u\eta = C$ ($C = 4 \cdot 47 \times 10^{-4}$ cP cm^2 v^{-1} s^{-1}).

(3) On the other hand, the mobilities of positive ions deviate slightly from Walden's law ($u\eta^\kappa = C_1$; $\kappa = 1 \cdot 07$). The relationships $\ln u = f(T^{-1})$ are straight lines and their angles of inclination vary for different liquids. In Zając's opinion this variation is accounted for by the structure of molecules. Some examples of the most important results obtained by Zając concerning the group of aromatic hydrocarbons are given in table 12.6.

Table 12.6. Molar volume of the liquid [$M\rho^{-1}$ (20°c)], the number of ions produced n_0; G_0 and γ from the equation $I = f(E)$

Liquid	$M\rho^{-1}$ (20°c) (cm^3 mol^{-1})	$10^{12} \times G_0$ (A cm^{-3})	$10^{-7} \times n_0$ (cm^{-3} s^{-1})	$10^4 \times$ (v^{-1})
Benzene	88·94	6·57	4·11	1·77
Toluene	106·35	7·78	4·86	1·68
Ortho-xylene	120·63	8·95	5·59	1·80
Para-xylene	123·20	9·26	5·79	1·86
Naphthalene (0·05 mol) + benzene (1·00 mol)	90·62	8·75	5·47	1·88
Naphthalene (0·10 mol) + benzene (1·00 mol)	91·70	9·45	5·91	1·85

After Zając [213].

In 1965–66 Hummel *et al.* [204, 205] used Gzowski's and Terlecki's [12] method to take a series of measurements of the mobility of ions in various dielectric liquids. X-ray flashes (50 kv) were used to irradiate a layer of liquid 1 mm thick in a time of 0·5 sec. The space between the electrodes was about 1·5 cm. The ionization chamber used differed from others in that the earthed ring was placed midway between the electrodes. This arrangement was designed to ensure a uniform distribution of the electric field within the chamber.

The following values were obtained for the mobilities of ions in hexane at a temperature of 24°c:

$$u_- = (1 \cdot 27 - 1 \cdot 32) \times 10^{-3} \text{ cm}^2 \text{ v s}^{-1}, \left. \right\}$$
$$u_+ = (0 \cdot 66 - 0 \cdot 71) \times 10^{-3} \text{ cm}^2 \text{ v s}^{-1}. \left. \right\} \quad (12.20)$$

These values approximately correspond to those obtained by Gzowski and as far as the positive ions are concerned, they are in agreement with Adamczewski's data.

Hummel *et al.* also investigated the dependence of ion mobilities on temperature and established their activation energy. The plot of these dependencies is given in fig. 12.31. It can be seen from the diagram that

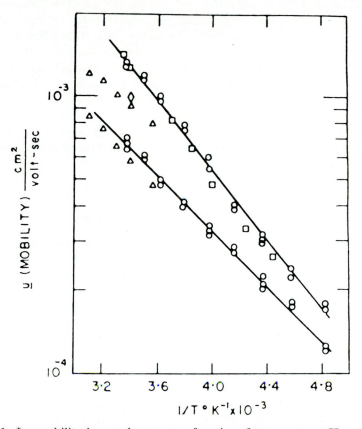

Fig. 12.31. Ion mobility in pure hexane as a function of temperature. Upper points, negative ion; lower points, positive ion; ○, Hummel work; □, LeBlanc (photo-electrons), ◇, Chong and Inuishi (photoelectrons); △, Gzowski. Ref.: Hummel *et al.* [204].

the positive ions are in complete agreement with the formula given by Adamczewski ($u\eta^{3/2}$ = const.) and verified by Gzowski. The way in which the values of the mobilities of negative ions change, following variations in temperature, differs somewhat from Gzowski's conclusions but values tend to satisfy the Adamczewski's formula.

The most important difference between the measurements made by Adamczewski and by Hummel *et al.* is not in the numerical values of the ion mobilities but in their general dependence on temperature, and more particularly in the change of activation energy (W_-) and (W_+) in relation to the activation energy of viscosity (W_η). The following values found by Hummel *et al.*, $W_\eta = 0.073$ ev, $W_- = 0.125$ ev and $W_+ = 0.097$ ev, comply with the general formula:

$$u_\pm = A\eta^{-3/2}. \tag{12.21}$$

Adamczewski arrived at this formula on the basis of his experiments in 1936 but it was later presented in greater detail (45). According to Adamczewski $W_+ = \frac{3}{2}W_\eta$; this value corresponds to 0.11 ev which is approximately the same as both the values given by Hummel *et al*. Special attention should be given to the relationship for negative ions $u_-\eta^m = $ const. $(m \neq 1)$. This relationship, however, has been questioned by some scientists whose results were more in agreement with Walden's law $u\eta = $ const. For both positive and negative ions Adamczewski suggested the general formula $u\eta^m = $ const. $(m = 1.5)$.

Hummel *et al*. also measured the mobility of ions both in carbon tetrachloride and in 1,4-dioxane. They established the following values at room temperature:

$$\left.\begin{array}{l} u_- = 0.33 \times 10^{-3};\ u_+ = 0.40 \times 10^{-3}\ \mathrm{cm^2\,v^{-1}\,s^{-1}}\text{---for CCl}_4, \\[2mm] u_- = 0.48 \times 10^{-3};\ u_+ = 0.8 \times 10^{-3}\ \mathrm{cm^2\,v^{-1}\,s^{-1}}\text{---for 1,4-dioxane.} \end{array}\right\} \tag{12.22}$$

The fact that impurities influenced (up to 10%) the measured values of ion mobility in certain liquids was confirmed by Hummel *et al*. (this influence had already been discovered by Terlecki, Gzowski and others), but they did not confirm the influence of oxygen dissolved in liquid on the mobility of negative ions. They devoted a separate section of their research to the ionization of hexane by β-rays from ^{39}Ar (with a maximum energy of 0.57 мev). The value obtained for the conductivity was $3.6 \times 10^{-14}\ \Omega^{-1}\,\mathrm{cm^{-1}}$ when the intensity of activation was 2.83×10^{10} ev/cm^3 s. This is the same value as that for x-radiation with an energy of 1.5 мev. Hexane was ionized with an electromagnetic radiation induced by the capture of K, L, M electrons in ^{37}Ar dissolved in liquid. The following information was obtained on the basis of the dependence of conductivity on the activation of dissolved radioactive gas: the radiation output (fig. 12.32), the distribution of energy for a number of electrons of different energies (fig. 12.33) and the range of electrons in various substances (fig. 12.34). Energy loss is calculated from the formula:

$$dP = \frac{n_e\,d\sigma}{-dE_k/dx} = \frac{dW}{2W^2 \ln(2E_k/J)}, \tag{12.23}$$

where E_k is the kinetic energy of the primary electron, W—the energy loss and dP—the number of energy loss events lying between W and $W + dW$, per unit total energy lost by the primary. J is the mean stopping potential of the molecules in the medium, n_e—the number of electrons per unit volume and $d\sigma$—the cross section for electron–electron collisions leading to energy losses between W and $W + dW$.

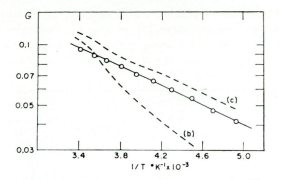
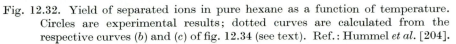

Fig. 12.32. Yield of separated ions in pure hexane as a function of temperature. Circles are experimental results; dotted curves are calculated from the respective curves (b) and (c) of fig. 12.34 (see text). Ref.: Hummel *et al.* [204].

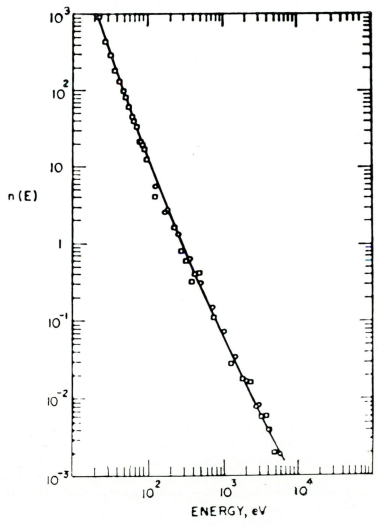

Fig. 12.33. Calculated distribution of energy losses in hexane. □, Magee *et al.* [308]; ○, present work. Ref.: Hummel *et al.* [204].

Watson measured the dependence of conductivity and ion mobility on temperature in two liquids belonging to the group under investigation. This was done by means of injecting electrons into the liquid. It was found that the mobility of ions failed to satisfy Walden's law $u\eta = \text{const.}$,

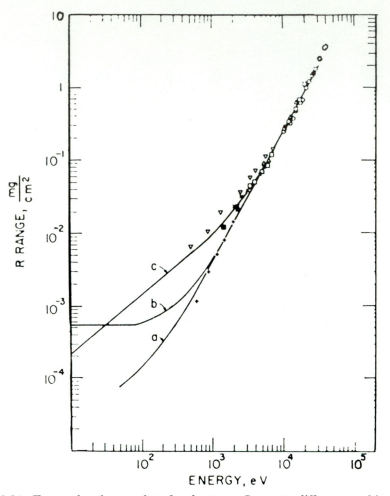

Fig. 12.34. Extrapolated range data for electrons. Lane: □, different combinations of Al and plastic; ■, collodion. Schonland: ●, Al; ○, Al. Young: △, Al. Davis: +, protein. Ref.: Hummel *et al.* [204].

but tended to conform to the relationship $u\eta^m = \text{const.}$ where m oscillates between 1·07 and 1·37 (fig. 12.35). It is evident that Watson's result was closely related to Adamczewski's result obtained as early as 1936. The latter discovered during his experiments that Walden's law did not apply to ions in saturated hydrocarbons. These ions obey the law $u\eta^{3/2} = \text{const.}$ Chong and Inuishi [112] among others confirmed this, as did Gzowski [31, 32] in the case of positive ions.

In 1965 Secker and Lewis [228] conducted a series of measurements of the current–voltage characteristics in hexane. They applied the special technique of injecting a charge into the liquid by means of a beam of x-rays. Their conclusion was that the current of ionization in the liquid

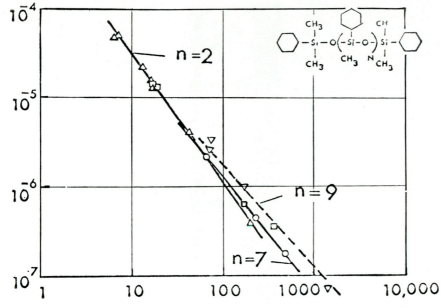

Fig. 12.35. Dependence of charge carrier mobility on viscosity. Ref.: Watson [230].

can be formulated in the following way:

$$I = I_0[1 + CE - \exp(-E/E_0)], \qquad (12.24)$$

where E_0 is a constant. For $E > 1 \text{ kv cm}^{-1}$ the exponential expression can be omitted and in such a case:

$$I = I_0(1 + CE). \qquad (12.25)$$

Secker and Lewis took into consideration the phenomenon of liquid movement in the electric field which has been observed by Gray and Lewis [203].† Figure 12.36 shows a photograph from their experiment in which a drop of dye was injected into a test cell and seen to move under the influence of the electric field. In the opinion of Secker and Lewis the velocity of the movement of the liquid in the electric field, whose strength was E and when the current was I, was:

$$v = C_1 \frac{E}{u} \ln \left(C_2 \frac{I}{E} \right), \qquad (12.26)$$

where C_1 and C_2 are constants.

The velocity of the liquid movement was estimated to be $6 \times 10^{-4} E$ cm sec^{-1} for the movement of positive ions. Such a phenomenon could cause an artificial decrease in the measured transit time of the ions. As a result

† See also Essex and Secker [357], Buttle and Brignell [341], and also Ostroumov [248, 249 from part IV].

of this the calculated mobilities of the ions would be greater and this would be of particular importance in very high fields. In the vicinity of breakdown intensities, Secker and Lewis expected that the increase in the

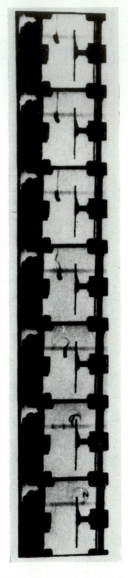

Fig. 12.36. Dye motion in cell. Frame interval is $\frac{2}{3}$ sec. First frame is shown at the top. The field is applied at the second frame and maintained thereafter. Ref.: Gray and Lewis [203].

mobilities of ions would exceed, by two orders of magnitude, the values obtained in low fields. According to the above-mentioned scientists, the velocity of liquid movement depended on the distribution of the potential

(V) inside the ionization chamber. They stated that the velocity was:

$$V = \frac{1}{3} \frac{u\epsilon}{I} \left\{ \left[\left(\frac{v}{u}\right)^2 + \frac{2Id}{u\epsilon} \right]^{3/2} - \left(\frac{v}{u}\right)^3 \right\} - \frac{v}{u} d, \qquad (12.27)$$

where d denoted the separation of the electrodes and ϵ the permittivity of the medium (the meaning of other symbols has already been given).

In the opinion of the authors the non-uniform distribution of the field accounts for the great discrepancy of the results obtained for the mobilities of ions and especially for the fields below 3 kv cm^{-1}. For negative ions in higher fields about 2×10^{-3} cm^2 v^{-1} sec^{-1} was obtained and for positive ions, about 1×10^{-3} cm^2 v^{-1} sec^{-1}. It is the present writer's opinion that these results show the influence of measuring techniques on the values of ion mobilities obtained by various scientists.

It should be emphasized that Langevin's method, which so far has only been applied to liquids by Adamczewski [40, 41], generally eliminated the possible influence of liquid movement in the electric field. In addition the influence of non-uniform potential distribution and that of the difference between the mobilities of negative charge carriers of two different types was also eliminated. This was made possible owing to a change in the direction of the electric field while the charge of the collected ions was being measured.

12.7. *Measurements of the Mobilities of Ions in Highly Viscous Liquids.* Experiments on the mobility of ions in highly viscous liquids have considerable significance for high-voltage techniques in view of the research into the strength of transformer oils. It is also possible to determine on the basis of these experiments whether the relationship (12.6) is also valid for a liquid in which the viscosity varies only with changes in temperature. In 1937 Adamczewski [42] conducted preliminary researches into a highly viscous liquid—paraffin oil. These measurements were not so accurate as the previous ones because of the experimental difficulties resulting from the high viscosity of the liquid ($\eta \sim 1$–5 P). The density of the paraffin oil was $0 \cdot 885$ g cm^{-3} and self-conductivity was of the order of 10^{-18} Ω^{-1} cm^{-1}. The oil was dried with metallic sodium and filtered through a fine Schott filter by means of a vacuum pump.

In collaboration with Świętosławska-Ścisłowska [62] Adamczewski devised a technique for measuring the mobility of ions. A thin layer of liquid was ionized at the lower electrode of a capacitor which was connected to a source of voltage ($E = 2000$ v cm^{-1}). The x-ray beam (40 kv, 5 mA) passed through a narrow slit (1–1·5 mm) in a thick lead diaphragm (3 cm) and the rate at which the charge was collected was observed on an electrometer. The results of these measurements are shown in fig. 12.37 in the form of the relationship $Q = f(t)$ and those of the relationship $i = f(t)$ in fig. 12.38. The theoretical forms for these measurements are given in fig. 12.39. It can be seen that the arrival of the ion layer at the collector electrode is marked by a distinct increase in the electric charge. Unfortunately the tail of the layer is not clearly visible as there is a considerable spreadout due to the variety of ions and diffusion. In this work the numerical values of the mobility of ions in paraffin oil were 1.3×10^{-6}–2.0×10^{-6} cm^2 sec^{-1} v^{-1}.

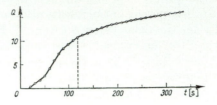

Fig. 12.37. Increase in charge Q with time t. Ref.: Świętosławska-Ścisłowska and Adamczewski [62].

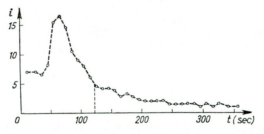

Fig. 12.38. Variation of current i (in A cm$^{-2} \times 10^{-14}$) with time t (in sec) after radiation has been cut-off. Ref.: Świętosławska-Ścisłowska and Adamczewski [62].

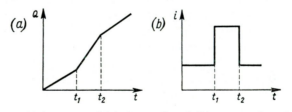

Fig. 12.39. Theoretical curves of (a) charge Q and (b) current i against time t after radiation has been cut-off. Ref.: Świętosławska-Ścisłowska and Adamczewski [62].

Adamczewski [63] later carried out a number of measurements on the dependence of ion mobility in paraffin oil on temperature. The method of measuring used was similar but γ-radiation from radium (12 mg) replaced the x-rays. The whole layer of liquid (without the electric field) was irradiated from above and after a stationary state of ion density had been reached the voltage was applied across the electrodes. The results of the measurements are given in table 12.7.

Figure 12.40 shows a system of curves $i = f(t)$ at various temperatures. As can be seen from the diagram, at room temperature it is possible to give a relatively precise definition of the limits of an ion layer. At lower temperatures, however, it is very difficult to determine the arrival time of the ions at the electrode and the time can only be estimated within a large margin of error.

Table 12.7

T ($^\circ$K)	η (P)	$u \times 10^7$ (cm^2 v^{-1} sec^{-1})	T ($^\circ$K)	η (P)	$u \times 10^7$ (cm^2 v^{-1} sec^{-1})
277·8	5·56	0·95	289·3	2·25	10·9
280·3	4·50	4·87	292·0	1·73	12·0
284·6	3·18	7·14	297·0	1·27	18·0
286·0	2·85	7·4			

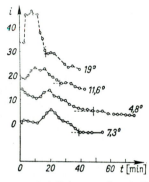

Fig. 12.40.   Dependence of current i (in A cm$^{-2} \times 10^{-14}$) on time t at a number of temperatures in paraffin oil. Ref.: Adamczewski [42].

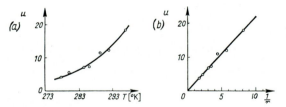

Fig. 12.41.   (a) Dependence of ionic mobility u on temperature for paraffin oil. (b) Dependence of ionic mobility u on $1/\eta$ (u is in units of cm v^{-1} sec$^{-1} \times 10^{-7}$). Ref.: Adamczewski [45].

The results of measuring $u = f(t)$ and $u = f(1/\eta)$ are given in fig. 12.41 (a, b). The mobilities of ions varied from 0.95×10^{-7} cm^2 v^{-1} sec^{-1} for $T = 277.8^\circ$K ($\eta = 5.56$ P) to 18×10^{-7} cm^2 v^{-1} sec^{-1} for $T = 297^\circ$K ($\eta = 1.27$ P). The product $u\eta = 2.2 \times 10^{-6}$ cm^2 v^{-1} sec^{-1} P is constant within the limits of experimental error.

This relationship corresponds to Walden's and Stokes's laws, the ion radius being constant and of the order of $3 \cdot 8 \times 10^{-8}$ cm. It is very difficult to compare the results of this work with those of previous research carried out in highly viscous liquids because in all previous experiments only the total of ion mobilities was estimated on the basis of the current–voltage characteristics and the value of the viscosity coefficient had not been determined with precision.

Table 12.8. Ion mobilities in highly viscous liquids

Liquid	η (P)	T (°K)	u (cm² v⁻¹ sec⁻¹)	Reference
Vaseline oil	—	293	$0 \cdot 97 \times 10^{-8}$	Białobrzeski [17] (estimate)
Vaseline oil	—	293	$2 \cdot 2 \times 10^{-7}$	Szivessy and Schaffer (estimate)
Paraffin oil	$1 \cdot 73$	293	$(1 \cdot 3 - 2 \cdot 0) \times 10^{-6}$	Świętosławska and Adamczewski [62] (measurement)
Paraffin oil	$1 \cdot 27$	297	$1 \cdot 8 \times 10^{-6}$	Adamczewski [42] (measurement)
Paraffin oil	$5 \cdot 56$	$277 \cdot 8$	$0 \cdot 095 \times 10^{-6}$	
Paraffin oil	$1 \cdot 60$	293	$(2 \cdot 21 - 4 \cdot 1) \times 10^{-6}$	Jachym [190, 191]
Paraffin oil	$0 \cdot 30$	323	$(1 \cdot 7 - 1 \cdot 9) \times 10^{-6}$	(measurement)
Paraffin oil	$0 \cdot 138$	323	$(23 \cdot 3 - 46 \cdot 7) \times 10^{-6}$	
Paraffin oil	$4 \cdot 4$	277	$(0 \cdot 65 - 1 \cdot 5) \times 10^{-6}$	,,
Paraffin oil + octane	$0 \cdot 253$	$293 \cdot 5$	$(7 \cdot 73 - 14 \cdot 5) \times 10^{-6}$	,,
	$0 \cdot 09$	$320 \cdot 5$	$(24 \cdot 2 - 38 \cdot 9) \times 10^{-6}$	
Paraffin oil + hexane	$0 \cdot 045$	$292 \cdot 6$	$(21 \cdot 2 - 57 \cdot 1) \times 10^{-6}$	
	$0 \cdot 021$	326	135×10^{-6}	,,
Silicon oil (OE-4018)	$0 \cdot 3$	291	$(13 \cdot 4 - 21 \cdot 0) \times 10^{-6}$	Jachym [192]
	$0 \cdot 154$	323	$(25 \cdot 8 - 37 \cdot 5) \times 10^{-6}$	,,
Silicon oil (DC-703)	$0 \cdot 765$	288	$(1 \cdot 34 - 10 \cdot 0) \times 10^{-6}$	,,
	$0 \cdot 183$	323	$(6 \cdot 35 - 42 \cdot 0) \times 10^{-6}$	,,
Calculated from Adamczewski's formula (12.19)	1	—	$0 \cdot 288 \times 10^{-6}$	Adamczewski [45]
	4	—	$0 \cdot 033 \times 10^{-6}$	,,

In 1962–63 Jachym [163, 190–2] conducted a long series of experiments on the mobility of ions in highly viscous liquids as a function of viscosity and temperature. His measuring technique was based on the apparatus (described above) used by Gzowski and Terlecki [12] and on the principle adopted in the works of Świętosławska-Ścisłowska and Adamczewski [42, 62, 63]. Jachym succeeded in filling the gap in research on the mobility of ions in liquids with a viscosity coefficient ranging from $\eta = 1 \cdot 27$ P to $\eta = 0 \cdot 00775$ P. He established which of the formulae given above for the relationship $u = f(\eta)$ was valid. Table 12.8 lists several values of ion mobilities for highly viscous liquids and includes recent measurements taken by Jachym [191, 192].

Figure 12.42 shows a plot of the relationship $u = f(1/T)$ which is based on the formula (12.18) (taken from Adamczewski's work [45]). In this formula the value of the numerical constant was slightly changed in order to demonstrate the general agreement, as far as temperature was concerned, with the results of LeBlanc [57] which are marked by crosses in the diagram.

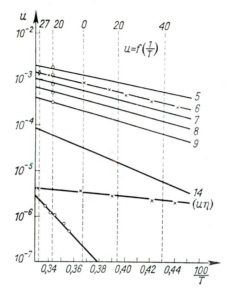

Fig. 12.42. Dependence of the mobility of negative ions u on temperature (u against $100/T$). The crosses are LeBlanc's experimental points; the bottom line shows Adamczewski's results for paraffin. Ref.: Adamczewski [45].

A summary of the results of experiments carried out by various scientists on the mobilities of ions in dielectric liquids according to the formula $u = A\eta^{-1}$ (as a logarithmic plot) is given in fig. 12.43.† Several values of mobility for gases and vapours can be found in the upper right-hand corner of the diagram. The coefficients of viscosity vary within the limits of 10^{-5}–6 P and ion mobilities within the limits of 10–10^{-7} cm² v⁻¹ sec⁻¹. The diagram suggests that the Stokes–Walden law is obeyed over a very wide range of values for the coefficient of viscosity. Adamczewski's law $u = A\eta^{-3/2}$ is valid, however, in specific regions and especially for one group of liquids (when the change in the structure of molecules causes a change in the viscosity of the liquid). It is an interesting fact that the extrapolation of this relationship leads to the region of the mobilities of saturated hydrocarbon vapours. This phenomenon can be explained by the fact that the ion should be a resistant rigid sphere for monoatomic

† See also the new papers from 1967–1968: Jachym A. and Jachym B. [345], Adamczewski and Jachym [360], Adamczewski [335].

liquids (e.g. helium). X-ray photographs and changes in the activation energy of viscosity indicate that since very long molecular chains of various oils are rolled up around the mass centre of the molecule many

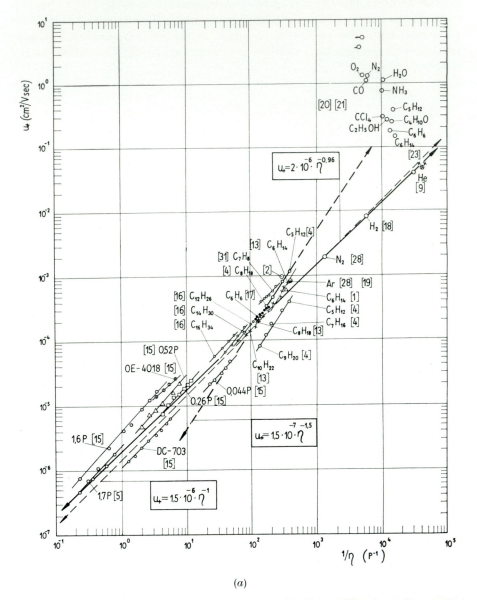

(a)

Fig. 12.43. u plotted again $1/\eta$ on a log–log scale for: (a) positive ions, (b) negative ions, (c) positive and negative ions in very viscous liquids. The experimental values were measured by Adamczewski [41], I.A.; Jachym [190], B.J.; Gzowski [31], O.G.; Nowak [122], W.N.; Terlecki [33], J.T.; LeBlanc [57], L.B.; Bijl [53], H.B.

After Adamczewski and Jachym [163].

times, these chains should also have a spherical ion structure. Only in the case of short chains of lower saturated hydrocarbons can the influence of the chain length of the molecule on the ion radius be expected, together with a dependence on viscosity, to deviate from Stokes's law. This phenomenon would be most likely to occur if an ion appeared in the

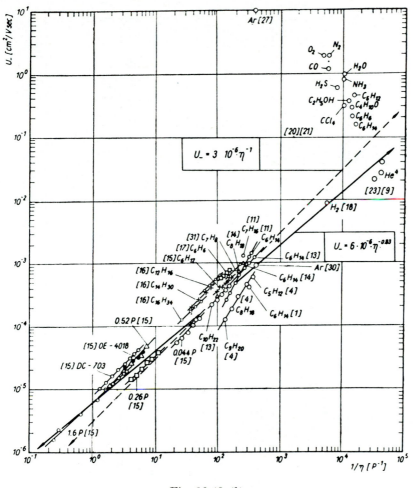

Fig. 12.43. (*b*)

form of a polaron. Further investigation is, however, required, especially in the region of very low viscosities and for specified compounds in one group of liquids.

Figure 12.44 shows a nomogram drawn by Adamczewski in order to facilitate the geometrical determination of the mobility values of both positive and negative ions in a group of saturated hydrocarbons of the C_nH_{2n+2} type. On the left-hand side of the drawing the first straight line denotes the temperature t in accordance with both the Celsius and the

15

$10^3/T$ scale. The middle line denotes the number n of carbon atoms in the molecule. The values n for the determination of the mobility of positive ions are shown on the left-hand side of the middle line while those for negative ions are on the opposite side. The third straight line, on the right-hand side, illustrates the ion mobilities in a logarithmic scale.

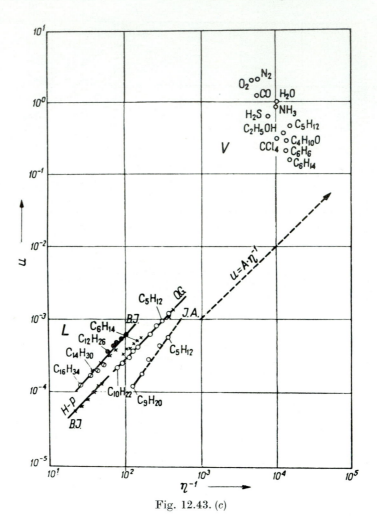

Fig. 12.43. (c)

In order to determine the mobilities of ions for a compound having a number n of carbon atoms in a molecule and a temperature of $t°c$, it is necessary to draw a straight line through the two appropriate points of the first and second straight lines and to extend it until it intersects the line u. A series of straight lines passing through $n = 7$ is shown in the diagram. This makes possible the determination of the positive ion mobilities for heptane at a temperature ranging from $+10°c$ to $+50°c$. A second series of straight lines originating from $t = +20°c$ enables the determination of the positive ion mobilities in the group of saturated hydrocarbons from hexane ($n = 6$) to decane ($n = 10$) at room temperature.

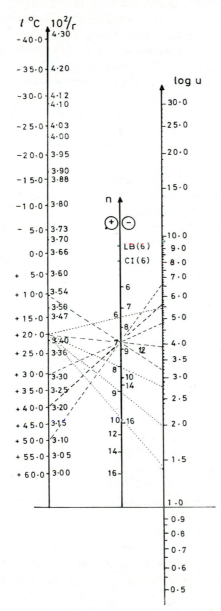

Fig. 12.44. Nomogram for calculating ionic mobilities in saturated hydrocarbons.

The values of the mobility of ions determined by the two series of straight lines correspond with considerable accuracy to the values given by Gzowski [31, 130] in table 12.3. The saturated liquid hydrocarbons investigated by different scientists were not synthetically made but were obtained by the distillation of mineral oil. For this reason they generally contained a certain amount of neighbouring hydrocarbon admixtures. The number n therefore, given by various scientists, did not represent exactly the same chemical compound. This indicates the necessity of moving the number n on the central scale when required. The values of n given in the diagram are based on Gzowski's work [31] which provides the most extensive experimental data. In order to obtain the results of three different series of measurements of ion mobilities (Gzowski [31, 32], LeBlanc [57], Chong and Inuishi [112]), three different values are shown in the diagram for negative ions in hexane ($n = 6$). These measurements were taken within a very wide temperature range varying from $-40°$c to $+50°$c. The results can be obtained from the diagram by selecting an appropriate value for n.

At the bottom of the line n, points marked for $n = 12$, 14 and 16 are further apart than the points in the centre of the line. It is possible to use these points to find the negative ion mobilities which were recently obtained by Jachym for these hydrocarbons [4, 190]. By introducing certain corrections on the scale of n (e.g. subtracting approximately 0·5–0·8 from the values of n) it is possible to obtain the values of ion mobilities measured by Adamczewski in 1935.

Other examples of nomograms, for finding the coefficients η, u, α and D as a function of temperature and η will be presented in Chapter 15.

12.8. *Research on Ionization Conductivity and the Mobility of Ions in Liquefied Gases.*† Considerable attention has been devoted in recent years to research on the mechanism of electron and ion movement in liquefied gases at very low temperatures (down to $0·83°$к). A number of measurements were taken in such noble liquefied gases as argon, krypton and xenon within limits of pressure varying from 1 atm. to about 90 atm. and extensive investigations have also been made with helium. Many new results, typical of these substances and the conditions of temperature have been found and a number of theoretical contributions have been made on the basis of the quantum theory. The latter can be applied to these liquids in view of the simplicity of their molecular structure (they are either mono- or diatomic).

Since the properties of liquefied gases at low temperatures substantially differ from those of standard dielectric liquids, only a few of the most important results of the above-mentioned works will be given here. Detailed information on this subject can be found in the work of Careri *et al.* [247].

Several works published at the end of 1962 were concerned with experimental and theoretical research on ion mobility in liquefied noble gases such as ^{3}He, ^{4}He, Ar, Kr and Xe at very low temperatures. A précis

† See also the new papers from 1967–1968: Halpern, Lekner, Rice and Gomer [361], Lekner [351], and Miller, Howe and Spear [352].

of the values given in tables 12.9 and 12.10 is the result of measurements, taken by Meyer *et al.* [117], of the mobility of ions in liquid helium at temperatures from 1·21°K to 4·3°K. In accordance with the work done by Zinovieva [123], the author has listed in the last column of table 12.10 the values of viscosity of these liquids under normal pressure.

The following table was also given by Meyer *et al.*:

Table 12.9

T (°K)	4·2	3·2	3·0	2·2	1·2	Liquid
ηu_+	1·93	—	1·73	1·54	—	Helium I
ηu_-	0·79	—	0·94	1·08	—	Helium I
ηu_+	—	1·68	—	—	2·25	^{3}He
ηu_-	—	0·6	—	—	1·0	^{3}He

Ref.: Meyer *et al.* [117].

At the end of 1962 Davis *et al.* [118] produced interesting work which included theoretical and experimental investigations on the mobility of positive ions in liquid Ar, Kr and Xe at temperatures ranging from 90°K to 192°K and within pressure limits from 4·7 atm. to about 92 atm. These investigations served to verify the author's formula for the ion mobility u in these substances. Owing to the fact that liquids of this type can be regarded as monoatomic it is relatively easy to establish their theoretical

Table 12.10. The mobilities of positive and negative ions in liquid ^{3}He and ^{4}He at different temperatures

Substance	T (°K)	p (atm.)	$u_+ \times 10^2$ (cm^2 v^{-1} sec^{-1})	$u_- \times 10^2$	ρ (g/cm^3)	$\eta \times 10^5$ (P)
^{3}He	1·21	1·15	7·65	3·64	0·0842	2·14
	1·79	1·63	8·25	4·04	0·0834	~1·92
	~2·20	2·05	8·17	4·06	0·0830	1·83
	2·94	~3·12	8·28	4·32	0·0811	1·68
	~3·12	3·41	8·14	4·04	0·0803	
	2·94	1·34	9·36	3·50	0·0714	1·68
	3·20	1·09	8·83	3·09	0·0629	1·61
^{4}He	2·2	1·0	~5·22	~3·73		2·55
	2·2	~8·1	3·96	3·60		
	2·2	14·2	2·90	2·88		
	2·2	~20·8	2·47	2·62		
	2·2	27·5	1·85	2·06		
	3·0	1·0	4·4	2·60		3·30
	3·0	7·4	3·4	2·75		
	3·0	14·0	3·05	2·51		
	3·0	27·7	2·24	2·06		
	4·2	1·0	5·33	2·2		3·00
	4·2	7·5	4·17	2·8		
	4·2	14·7	3·60	2·6		
	4·2	27·8	2·86	2·37		

Ref.: Meyer *et al.* [117].

formulae. The ion mobilities obtained by Davis *et al.* on the basis of Rice's and Allnatt's theory [120] are given in §16.4. Table 12.11 gives a selection of the values obtained by Davis *et al.* for ion mobilities u at various temperatures T and under differing pressures p. Special attention should be drawn both to the application in these works of a method of measuring ion mobilities in order to investigate the structure of liquefied gases and also to the phenomenon of the superfluidity of helium (i.e. the mechanism of ions acting as vortices). It is evident that both positive and negative ions moving in liquid helium can interact with the already

Table 12.11. Mobilities of positive ions in liquid Ar, Kr and Xe

Substance	T (°K)	p (atm.)	$u \times 10^4$ (cm² v⁻¹ sec⁻¹)
Argon	90·1	4·7	6·61
		29·2	5·98
		68·8	4·22
	111	33·0	12·2
		81·6	10·4
	145	44·9	26·1
		92·1	22·5
	145	34·1	6·87
		61·9	6·82
Krypton	168·5	34·3	12·19
		62·0	11·5
	184·3	32·3	10·6
		68·3	11·0
		91·6	10·5
Xenon	184·2	7·5	2·85
		27·7	2·60
		53·4	2·16
	192·1	8·5	3·29
		27·9	3·02
		55·9	2·84

Ref.: Davis *et al.* [118].

existing vortices and that at great velocities they themselves become a source of 'charged whirl'. Experiments with 'hot' ions—i.e. ions whose velocity in the electric field was approximately the same as that in thermal motion—were carried out in two different ranges of temperatures, both high ($T \approx 10°$K) and low ($T \approx 0.3°$K).

From 1964 to 1966 more attention was paid to the influence of molecular rotations on the viscosity of the liquid. According to Davies and Matheson [261], it is possible to explain in this way the three different dependencies of viscosity on temperature which can be found in various liquids and sometimes also in the same liquid at different temperatures.

The Arrhenius relationship ($\eta = \eta_0 \exp(W/kT)$) is, for example, satisfied for a whole range of temperatures in such liquids as liquid neon, argon, krypton, neopentane, etc. The molecules of these liquids have a spherical structure and are therefore free to rotate in all directions.

In the case of other liquids, such as liquid propane and toluene, Arrhenius's relationship is only satisfied above a certain temperature, below which it is no longer applicable. Another group of liquids is represented by isopropylbenzene which obeys the Arrhenius relationship at high temperatures but has two regions for lower temperatures. In the first region when the drift velocity $v_D < 29$ m/sec it is proportional to E with the exception of small but periodic non-continua which appear with the values representing multiples of $5 \cdot 2$ m/sec for positive ions and of $2 \cdot 42$ m/sec for negative ions. When $v_D = 29$ m/sec there is a sharp drop in the curve $v_D = f(E)$ and the values of v_D decrease by two orders of magnitude. This is connected with the formation of singly quantized rotating rings. When the values of E are higher, v_D reaches a minimum value and thereafter increases slowly.

An investigation on the influence of the flow of a normal and superfluid liquid in an opposite direction to the transverse flow of ions led to the conclusion that when the heat current had a critical value, only the flow of negative ions was strongly affected. It was therefore suggested that negative ions have a large active cross section for interaction with the lines of the vortex. This was confirmed by Careri and others. Since negative ions proved to be so useful in the investigation of the phenomenon of vortices, Careri began to analyse rotating liquid helium. He directed a flow of ions both transversely towards and parallel to the axis of rotation. He discovered that the rotation of liquid helium exerted a great influence on negative ions which exhibited a great anisotropy. When ions flow transversely towards the axis of rotation the current increases together with an increasing angular velocity; when the flow of ions is parallel to the axis of rotation this effect does not occur.

The following recent works devoted to the mechanism of ionization and transportation in dielectric liquids at low temperatures should be mentioned: (1) the work of Bruschi *et al.* [258] on the dependence of the critical velocity of negative ions in liquid helium on temperature; (2) the theoretical research of a group of scientists led by Rice and Jortner [243] on the properties of an excited electron in liquid helium and its influence on the helium atom and (3) Swan's [262, 263] research on the ionization of argon by α-particles (the average energy necessary to produce a pair of ions is equal to 26 ev, as in gas).

Figure 12.45 shows the plots of changes in the velocity of ions in helium II corresponding to an increase in the strength of the electric field E. As can be seen from the diagram, a violent non-continuum occurs in field E ($0 \cdot 75$ kv/cm). This indicates a change in the mechanism of transporting ions. The change of mobility u together with the change of field strength is shown in fig. 12.46.

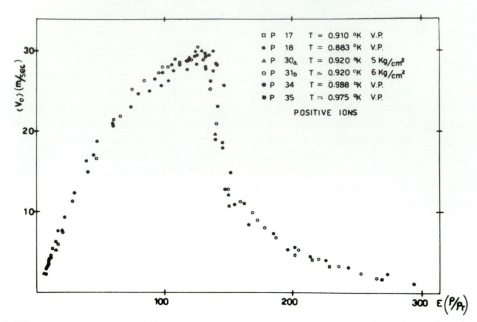

Fig. 12.45. The drift velocity (v_D) of positive ions plotted against the reduced
electric field $E(\rho/\rho_r)$ for different runs at the temperature and pressure
indicated. The term (ρ/ρ_r) is the ratio of the total density ρ of the fluid to
the roton part of normal fluid density ρ_r. Ref.: Careri *et al.* [247].

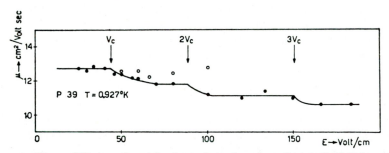

Fig. 12.46. The mobility u of positive ions plotted against electric field E for one run.
Open circles represent the measurements taken during bath disturbances, and
point to the existence of an upper metastable level. The curve which best fits
the stable conditions is indicated and the discontinuities are shown by
arrows. Ref. Careri *et al.* [253].

Chapter 13

Recombination of Ions

13.1. *General Law of Recombination.* The phenomenon of the recombination of ions is of great importance in the electrical conduction in an ionized substance in all three states of aggregation (gaseous, liquid and solid). Recombination is a collision of ions of opposite signs followed by the neutralization of their electric charges; the ions become either atoms or molecules which are electrically neutral.

In the most general and simple case when, as a result of ionization, equal densities n of ions having opposite signs are uniformly produced in the volume, recombination causes the decay of ions in a period of time in accordance with the following equation:

$$\frac{dn}{dt} = -\alpha n^2, \tag{13.1}$$

where α denotes the coefficient of recombination. Its dimension is $cm^3 \, sec^{-1} \, ion^{-1}$. The solution of eqn. (13.1) enables the ion density n to be calculated from the formulae:

$$n = \frac{n_0}{1 + \alpha n_0 t} \quad \text{or} \quad \frac{1}{n} = \frac{1}{n_0} + \alpha t, \tag{13.2}$$

where n_0 denotes the initial ion density (for $t = 0$). The second form of (13.2) is very convenient for the purpose of these calculations because it represents a straight line intersecting the ordinate at the point $1/n_0$. The slope of the straight line determines the coefficient of ion recombination α. This is called the decay method. Two other methods are also sometimes used: the build-up of charge method and the ion equilibrium method. In the first of these methods a constant number q of ions in cm^3/sec is produced in the ionization chamber. The increase in charge as a function of the time of irradiation follows the formula:

$$\frac{dn}{dt} = q - \alpha n^2. \tag{13.3}$$

After eqn. (13.3) has been integrated, the following relationship is obtained:

$$n = k \frac{\exp(2k\alpha t) - 1}{\exp(2k\alpha t) + 1}, \tag{13.4}$$

where $k = \sqrt{(q/\alpha)}$ and q is the number of ion pairs produced in cm^3/sec.

For small values of t the density of ions increases linearly with t and for large values it reaches a certain constant value n (saturation) defined from the formula:

$$\frac{dn}{dt} = q - \alpha n_\infty{}^2 = 0, \qquad (13.5)$$

hence $n_\infty = k = \sqrt{(q/\alpha)}$. This is a stationary state in which the number of ions lost in the process of recombination is equal to the number of ions produced in the same period of time by the source of ionization. It is also possible to determine the coefficient of the recombination of ions (the equilibrium method) from eqn. (13.5).

The law (13.1) for the recombination of ions does not always appear in such a simple form. Several other expressions were suggested in which the exponent is other than 2 (cf. Sutherland [65], Jaffe [15], Bijl [53] and Loeb [134, 135]).

13.2. *Different Types of Recombination.* It is possible to distinguish between several different types of recombination such as electron, ion, volume, initial, preferential, columnar and wall recombination. Electron recombination results from the collision between a free electron and a positive ion. In gases it mainly occurs when ions have very high densities. The coefficient α then oscillates within wide limits depending on the conditions of the experiment (10^{-14}–10^{-6} cm^3 sec^{-1} ion^{-1}). Ion and volume recombination follow the simplest law of recombination described in § 13.1 (in gases α is of the order of 10^{-6} cm^3 sec^{-1} ion^{-1} and in liquids—about 10^{-9} cm^3 sec^{-1} ion^{-1}). Initial recombination can occur when, immediately after ionization has taken place, ions of opposite signs are closer to each other than could be expected on the basis of uniform distribution. Recombination of this type occurs in gases at low pressure and follows irradiation with x, γ and β-rays.

Preferential recombination occurs most frequently when electrons or negative ions are produced in the vicinity of the positive ions and instead of being uniformly distributed are formed into groups in certain specified regions. They may then recombine before diffusion or the electric field has moved them further away. Recombination of this type occurs in gases under high pressure and in liquids, especially in cases when, due to the action of ionizing radiation in the substance, low-energy δ electrons are produced forming small compact groups of ions. In order to explain the difference between initial and preferential recombination it is necessary to introduce both the concept of the average distance r_0 between ions of opposite sign as well as that of the radius d_0 of the sphere of effective interaction between the ions. At this radius their kinetic energy balances the potential energy resulting from the Coulomb forces.

If ionization results in a uniform distribution of groups of ions of both signs (macroscopic isotropy) in the substance, while the density of ions in these groups considerably exceeds the average density of the whole volume of the ionized substance (microscopic anisotropy), initial recombination takes place when $r_0 > d_0$ (e.g. in gases under normal pressure). Preferential recombination takes place when $r_0 < d_0$ (e.g. in gases under a pressure higher than 2 atm. and in liquids). Columnar recombination is a special case of preferential recombination. It occurs most frequently in cases when the substance is ionized by heavy particles (protons, α-particles and fragments) and particularly in gases under high pressure and in liquids. Both micro and macroscopic anisotropy occurs under such circumstances. Jaffé [15] formulated in 1913 a theory of columnar ionization and recombination in gases and liquids and more detail is given in § 16.5.

Wall recombination occurs when the charge carriers are close to and diffuse towards the walls of the vessel and are subject to intensified recombination with the walls. Recombination of all kinds depends to a large extent on the kind of substance, its state of aggregation, temperature, pressure, viscosity and the type of ionizing radiation used. Although the phenomenon of ion recombination in gases has been thoroughly investigated both theoretically and experimentally, there is only a small number of works describing this process in liquids. Jaffé [14, 15] and Bijl [53] were the first to work on this subject.

13.3. *The Dependence of the Coefficient of Ion Recombination on the Viscosity of Liquids.* Adamczewski [1, 40, 43] was the first to investigate systematically the dependence of the coefficient of ion recombination on the viscosity of the liquid in the group of saturated hydrocarbons of the $C_n H_{2n+2}$ type. These investigations supplemented Adamczewski's research on the similar dependence of the mobility of ions in the same group of liquids. The recombination of ions was measured by means of the apparatus described in § 12.3 (fig. 11.7) to which certain changes were introduced. An x-ray lamp with a 5 mA electron beam intensity at 30 kv voltage was used as the source of radiation.

The author's method of measuring the recombination coefficient involved the following measurements: after the liquid had been ionized in order to obtain a steady state the ionizing radiation was cut off and the voltage was applied to the plates of the capacitor after a time t (the time of recombination of the ions). The voltage was selected in such a way that the time of ion collection would be shorter by at least one order of magnitude than the time t. The charge of the ions collected on the capacitor C which was connected to the collector electrode was measured by means of the electrometer E connected by the switch P_2. The longer the time of recombination t the less was the charge Q of collected ions (fig. 13.1). In plotting the relationship $1/Q = f(t)$ a straight line intersecting the axis of ordinates at the point $1/Q_0$ is obtained in accordance with eqn. (13.2). It is possible to calculate the coefficient α from the slope of this line. The results of the author's measurements taken in the group of saturated hydrocarbons are given in figs. 13.1, 13.2 and 13.3. Figure 13.3 shows that the experimental points are arranged very neatly on straight lines and that the slopes of the lines vary systematically from one liquid to another. In short, the slopes decrease with an increase in the molecular chain length thus indicating a reduction in the coefficient of recombination. Table 12.1

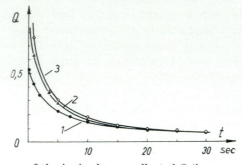

Fig. 13.1. Dependence of the ionic charge collected Q (in e.s.u. cm^{-3}) on the recombination time for ions in nonane; $d = 0.3$ cm; $U = 1780$ v; 1, 40 kv, 5 mA; 2, 35 kv, 5 mA; 3, 30 kv, 5 mA. Ref.: Adamczewski [68].

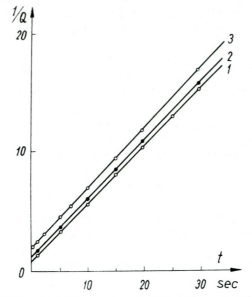

Fig. 13.2. Plot of $1/Q$ against t for nonane. 1, 40 kv, 5 mA; 2, 35 kv, 5 mA; 3, 30 kv, 5 mA. Ref.: Adamczewski [68].

lists the numerical values of the recombination coefficients found by Adamczewski [40, 43] as well as their ratio to ion mobility. It can be seen from the table that this is constant within the limits of experimental error.

Such an experimental method is only suitable for investigating the phenomenon of general ion recombination which occurs when the spatial distribution of ions is uniform. This is determined by the long time necessary to irradiate the liquid (between 10 and 20 sec) before the stationary state is reached. Recombination of other types (electron, initial and preferential) occurs in a considerably shorter length of time (about 10^{-14}–10^{-10} sec).

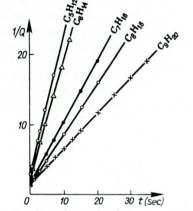

Fig. 13.3. Plot of $1/Q$ against t for five saturated hydrocarbons.
Ref.: Adamczewski [68].

Figure 13.4 shows a diagram of the coefficient of ion recombination as a function of liquid viscosity (curve I) and of its reciprocal (curve II). A more detailed analysis of the results showed that the relationship $\alpha = f(\eta)$ can be expressed as:

$$\alpha = B\eta^{-3/2}, \tag{13.6}$$

where $B = 3 \times 10^{-4}$ c.g.s. units.

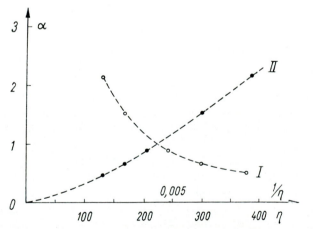

Fig. 13.4. Dependence of recombination coefficient α (in cm^3 sec^{-1} ion^{-1}) on viscosity (in poise). Curve I, α against η; curve II, α against $1/\eta$. Ref.: Adamczewski [68].

It should also be stressed that the plots of the dependence of the collected initial charge of heavy ions Q_0 on the separation of the electrodes in the ionization chamber are straight lines intersecting the origin of the coordinates. This means that the effect of ionization is volumetric and that the diffusion of ions has little bearing on the measurements. In 1939 Adamczewski [68] conducted another series of experiments in order to determine the importance of preferential recombinations in dielectric liquids. During the time of application of the electric field in the capacitor the liquid was ionized by short pulses of ionizing radiation in times of 0·04 sec and 0·001 sec. Adamczewski also discussed the relevance of columnar and preferential recombination (cf. § 16.6).

13.4. *New Research on the Phenomenon of Ion Recombination.*   In 1956 Fowler [104, 129] conducted a series of measurements on the recombination of charge carriers in solid polyethylene compounds (cf. § 16.9).

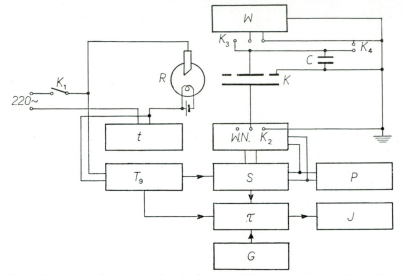

Fig. 13.5. Apparatus for measuring ionic recombination: C, capacitor; K, ionization chamber; k_1, k_2, k_3, k_4, switches; R, x-ray tube; WN, high-tension battery. Ref.: Gazda [69].

Fowler's curves $1/Q = f(t)$ are similar to Adamczewski's despite the fact that the ionized substances were in different phases in both cases and that in Adamczewski's research electron-hole recombination was of major importance.

Gazda [69, 309] has recently undertaken research on the recombination of ions in dielectric liquids. His main purpose was to investigate the dependence of the ion recombination coefficient on temperature and to determine the activation energy of this process. Gazda's apparatus was based on the method described above for collecting the ion charge. He measured the ion recombination coefficient as a function of temperature for various liquids in times ranging from several milliseconds to seconds. The design of his apparatus is shown in fig. 13.5. The experiment proceeded in the following way: as soon as the x-ray apparatus was switched off a pulse from the generator was simultaneously relayed both to the delay unit and to the trigger unit. When the trigger opened the gate, oscillations were passed from the RC generator to the amplifier and to the scaler which registered the pulses of known frequency. After an arbitrary time t (anything from 4 sec to 10 sec) the pulse is sent from the delay unit to an electronic switch. While the electrometer is disconnected, a voltage is applied to chamber K by means of the electromagnetic relay k_2. The pulse produced after the voltage has been applied is sent to the switch thus closing the gate to oscillations from the generator. After 2 sec the voltage is switched off by the switch k_2. As a result, the charge induced by the voltage disappears and only the charge collected during the application of the applied field remains on the capacity C (C represents the total capacitance due to the electrometer, collector electrode and measuring capacitor). The electrometer is manually connected with the circuit and the potential is measured. The electronic scaler measures the time between the disconnection of the x-ray apparatus and the switching on of the field.

The numerical values of the recombination coefficients obtained by Gazda differed slightly from those found by Adamczewski [43] and were more closely related to those published in earlier works. It is possible that discrepancies in the values of this coefficient are caused by the appearance of fast negative charge carriers in several liquid samples. This phenomenon has not yet been satisfactorily explained. The appearance of fast negative ions or electrons could also be the result of the removal of some liquid or gas admixtures (e.g. oxygen) from the liquid. The introduction of some new admixtures (e.g. ether) during the purification process of the liquid or the cleaning of the chamber may be another explanation of this phenomenon. It is clear that there are no basic differences in the results obtained by various scientists as far as the measurements of the mobilities of heavy positive ions are concerned. Certain scientists, however, using different samples of liquid have found that considerable discrepancies exist in the results of the measured mobilities of negative ions.

In the process of recombination the presence of faster negative ions may be responsible for a considerable change in the value of the ion recombination coefficient. This change may cause the discrepancies observed in the results. It is also possible that the discrepancies in the numerical values of the recombination coefficient may be the result of more important factors connected both with the non-uniformity of ionization and the possibility of the presence of various kinds of recombination (cf. § 16.5, 16.6 and 16.9).

This phenomenon can be divided into different types, all of which are characterized by a different coefficient. The frequency of the appearance of these coefficients depends on both the spatial and time distribution of free electrons and positive and negative ions. This fact has received attention in the theory of ionization in gases under high pressure and in liquids as well as in the theory of electronic semiconductors and in plasma physics. Bates and Kingston [102], for example, have shown that recombination caused by electron–electron collisions may be of great importance in gaseous plasma. The coefficient of recombination may then vary depending on the square of the number n of electrons per cm³, i.e. within the limits of $n = 10^5$–10^{13} electrons per cm³, the recombination coefficient is 10^{-13} and for $n = 10^{13}$–10^{19} the coefficient increases almost exponentially up to 10^{-9}.

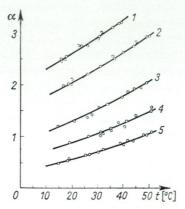

Fig. 13.6. Temperature dependence of ionic recombination coefficients α for saturated hydrocarbons. 1, C_6H_{14}; 2, C_7H_{16}; 3, C_8H_{18}; 4, C_9H_{20}; 5, $C_{10}H_{22}$. Ref.: Gazda [69].

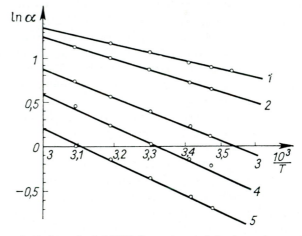

Fig. 13.7. $\ln \alpha$ plotted against $10^3/T$ for saturated hydrocarbons. The gradients of the straight-line denote activation energies. 1, C_6H_{14}; 2, C_7H_{16}; 3, C_8H_{18}; 4, C_9H_{20}; 5, $C_{10}H_{22}$. Ref.: Gazda [69].

In 1955 Sosnowski's work [143] on electronic semiconductors emphasized the great importance of collision recombination which depends on the square of the number of electrons (see also §16.9). The experimental results of the ion recombination coefficient as a function of temperature are given in table 13.1. Gazda [69, 194] compiled the table and estimated the error of the calculated values of the coefficient to be about 10%.

Figure 13.6 gives the plots of the coefficient of ion recombination in the group of saturated hydrocarbons as a function of temperature. Figure 13.7

shows the dependencies of the same values as a function of $1/T$, in order to determine the activation energy of recombination. The plot of the dependence of the coefficient α on $\eta^{-3/2}$ is given in fig. 13.8. It is clear

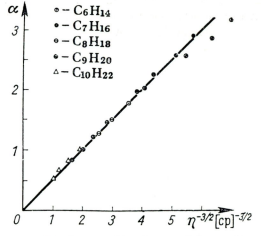

Fig. 13.8. Dependence of ionic recombination coefficient α (in $cm^3\,sec^{-1}\,ion^{-1}$) on $\eta^{-3/2}$ (in centipoise). Ref.: Gazda [69].

from the drawing that almost all the experimental points obtained for various coefficients of viscosity, as a result of changes in liquid or temperature, confirm this relationship. This can be presented in the following form:

$$\alpha = B_1(kT)^{-3/2}\exp\left(-\frac{3}{2}\frac{F_1}{kT}\right), \qquad (13.7)$$

where

$$B_1 = (7\cdot7 \pm 3\cdot8)\times10^{-10} \quad \text{and} \quad \tfrac{3}{2}F_1 = -0\cdot174\,nkT + 0\cdot0139\,(5\cdot05 + n);$$

α in this case is expressed in cm^3 per ion pair sec.

Table 13.1. Coefficients of ion recombination in the group of saturated hydrocarbons at different temperatures

Hexane		Heptane		Octane		Nonane		Decane	
Temperature (°c)	α	Temperature (°c)	α	Temperature (°c)	α	Temperature (°c)	α	Temperature (°c)	α
15·5	2·48	19	2·0	14·6	1·18	18·5	0·86	18·5	0·535
17	2·58	25·8	2·27	17·2	1·17	26·4	0·98	24·2	0·605
20	2·67	30·3	2·34	20·6	1·27	31	1·0	30·2	0·71
23	2·74	36·2	2·52	26	1·35	35	1·14	35·8	0·78
30·2	2·89	40	2·70	33·8	1·61	41·0	1·3	39·8	0·87
36	3·1	45·5	2·90	39·5	1·75	45·1	1·42	41·3	0·92
39·5	3·21	50	3·08	44·3	1·92	51·1	1·7	49·2	0·01
—	—	—	—	48·8	2·03	—	—	50·6	1·1

Ref.: Gazda [69].

16

In 1963 Nowak [122] measured the ion recombination coefficient (cf. §12.6) in order to complete his research on the mobility of ions in benzene. In view of the great value of self-conductivity in benzene he estimated the values of the collected charge of ions on the basis of those oscillograph records reflecting the relationship $l = f(t)$. He found that at room temperature $\alpha = 1 \cdot 45 \text{ cm}^3 \text{sec}^{-1} \text{ion}^{-1}$.

Chapter 14

The Diffusion of Ions

14.1. *The General Law of Diffusion.* The diffusion of ions is the third factor of great importance for electric conduction. It mainly occurs in certain regions, e.g. at the edge of a dense ion layer, where the concentration of ions is not uniform. The fundamental law of diffusion can be expressed by the formula:

$$\frac{\partial n}{\partial t} = D\nabla^2 n = D\left(\frac{\partial^2 n}{\partial x^2} + \frac{\partial^2 n}{\partial y^2} + \frac{\partial^2 n}{\partial z^2}\right),\tag{14.1}$$

where D denotes the coefficient of diffusion, n—the concentration of ions and t—the time of diffusion. According to the kinetic theory of gases the coefficient of diffusion D is proportional both to the mean free path λ and the coefficient of ion mobility u:

$$D = \tfrac{1}{3}\lambda u.\tag{14.2}$$

On the other hand, the coefficients of mobility and diffusion in gases are related to each other by the Nernst–Einstein formula:

$$\frac{u}{D} = \frac{Ne}{RT},\tag{14.3}$$

where N denotes the Avogadro constant and R is the molar gas constant.

14.2. *Different Types of Diffusion.* Apart from the simplest and more common type of diffusion there also exist more complex kinds. These include the diffusion of neutral molecules within their own environment (so-called self-diffusion), diffusion in foreign environments, diffusion of ions in electric or magnetic fields, the diffusion of electrons, ambipolar diffusion, thermal diffusion, etc.

Ambipolar diffusion occurs when there is a high density of ions of both signs and when their mobilities differ considerably. This can happen, for instance, in a heavily ionized gas where there is a large number of electrons and positive ions. Since electrons have higher mobilities than positive ions, the distribution of the spatial charge changes after some time (cf. fig. 14.1). This results in the appearance of new electrostatic forces which slow down the electrons but accelerate the positive ions. There is an identical flow of charges of both signs:

$$\mathbf{\Gamma}_+ = \mathbf{\Gamma}_- = n_+ v_+ = n_- v_-.\tag{14.4}$$

Hence the following equations:

$$\mathbf{v}_+ = \frac{-D_+}{n_+}\nabla\mathbf{n}_+ + u_+\,\mathbf{E}_s, \quad \mathbf{v}_- = -\frac{D_-}{n_-}\nabla\mathbf{n}_- - u_-\,\mathbf{E}_s, \tag{14.5}$$

and when $n_+ = n_- = n$ and $v_- = v_+ = v$ the following is obtained by eliminating E_s:

$$v = -\frac{D_+u_- + D_-u_+}{u_+ + u_-}\frac{\nabla\mathbf{n}}{n}. \tag{14.5a}$$

The expression:

$$D_a = \frac{D_+u_- + D_-u_+}{u_+ + u_-}. \tag{14.6}$$

defines the coefficient of ambipolar diffusion. This is a diffusion of two kinds of molecules interacting in such a way that they diffuse with the

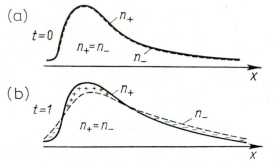

Fig. 14.1. Charged particle density distributions when (a) $t = 0$, and (b) $t = 1$ as a function of x at high ion densities. Ref.: Brown [67].

same velocity. Except for a different coefficient, the equation of ambipolar diffusion can be formulated in a similar way to that of free diffusion:

$$\frac{\partial n}{\partial t} = D_a \nabla^2 n. \tag{14.7}$$

In 1952 Brown and Rose [88] examined the coefficient of ambipolar diffusion D_a in gases by means of the ultra-high frequency technique.

Ion diffusion in dielectric liquids has only recently been investigated by using direct methods. For this reason, the process of self-diffusion (for selected neutral molecules) to which attention has been paid in the past few years is described here in detail.

14.3. *Methods of Measuring the Coefficient of Self-diffusion.* The following techniques have been applied in measuring the coefficient of self-diffusion in liquids:

(1) The capillary method based on the application of molecules traced with radioisotopes.

(2) The method of spin echo.

(3) The method of porous partition.

In 1955 Fishman [71] used the first method to determine the coefficients of diffusion for pentane D_5 and heptane D_7. He obtained the values $D_5 = 4 \cdot 14 \times 10^{-5}$ and $D_7 = 2 \cdot 08 \times 10^{-5}$ (in $cm^2\,sec^{-1}$ at $0°c$) and also the temperature dependencies in the form:

$$\left. \begin{aligned} D_5 &= 8 \cdot 11 \times 10^{-4} \exp\left(-\frac{1597}{RT}\right), \\ D_7 &= 1 \cdot 35 \times 10^{-3} \exp\left(-\frac{2246}{RT}\right). \end{aligned} \right\} \qquad (14.8)$$

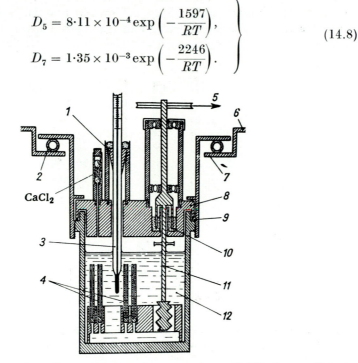

Fig. 14.2. Cross section of an apparatus for measuring self-diffusion coefficients. 1, Ground-glass joint filled with mercury; 2, cushioning piece suspending the chamber in the thermostat; 3, Beckmann thermometer; 4, capillary tube; 5, driving belt; 6, thermostat cover; 7, thermostat cavity; 8, chamber lid; 9, rubber washer; 10, mercury seal for stirrers; 11, stirrer; 12, liquid. Ref.: Umiński and Dera [97].

Naghizadeh and Rice [165] used the same method in their experiments to determine the coefficients of self-diffusion in the liquefied gases Ar, Kr, Xe and CH_4. They arrived at the following formula:

$$\log \bar{D} = 0 \cdot 05 + 0 \cdot 07\bar{p}\,\frac{1}{T}\,(1 + 0 \cdot 1\bar{p}), \qquad (14.8a)$$

where $\bar{D}$ denotes the reduced value of the coefficient of diffusion, $\bar{p}$—the pressure (in atm.) and $\bar{T}$—the absolute temperature.

Umiński and Dera [72, 97] have recently used this method in our Institute. A liquid traced with a radioactive element (e.g. ^{14}C) was introduced into one or a number of capillaries (figs. 14.2 and 14.3) whose lengths and diameters were defined. The capillaries were subsequently immersed in the same liquid which was free of radioactivity. Diffusion occurs in two directions at the mouth of the capillary. Molecules from the normal liquid travel to the capillary and replace those molecules which have passed from the traced liquid into the normal liquid. The amount of traced liquid in the capillary decreases as a result of this process. Having removed the liquid from the capillary after a time interval t and having examined the liquid's radioactivity, Umiński and Dera found that the concentration of active molecules had diminished from the initial value n_0 to n_1.

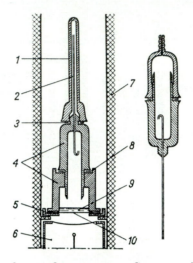

Fig. 14.3. Cross section of part of an apparatus for measuring the activity of liquid samples from diffusion measurements. 1, Micropipette cover; 2, micropipette; 3, rubber washer; 4, metal body; 5, selenium cement; 6, Geiger–Müller counter; 7, counter screen; 8, rubber washer; 9, liquid being investigated; 10, mica window. Ref.: Umiński and Dera [97].

Once the equation for diffusion (14.8) has been solved it is clear that in linear cases:

$$\frac{\partial n}{\partial t} = D\,\frac{\partial^2 n}{\partial x^2},\tag{14.9}$$

$$\frac{n_1}{n_0} = \frac{8}{\pi^2}\sum_{m=0}^{m=\infty}\frac{1}{(2m+1)^2}\exp\left(-2m+1\right)^2\pi^2\frac{Dt}{4l^2},\tag{14.10}$$

where l denotes the length of the capillary, t—duration of the process of diffusion and D—the coefficient of diffusion.

Umiński and Dera [72, 97] used 10–12 parallel capillaries which were connected with each other (fig. 14.2). Each capillary had a length of about 50 mm and a diameter of 0·7–0·9 mm. and their temperature was accurately controlled. Depending on the temperature and the kind of liquid used diffusion lasted anything from 40 to 60 hours. The dependence of the diffusion coefficient D on temperature T was determined and from the plots $D = f(1/T)$ the activation energies of self-diffusion were established. The specific activity of the liquid was about $2\ \mu\text{ci/ml}$. This activity was measured by means of a G.M. counter with a window $1\cdot4\ \text{mg/cm}^2$ thick.

The method of spin echo is more modern and accurate but requires more expensive equipment. Since this method demands considerable knowledge of the theory of nuclear resonance it will not be described in detail. More extensive data will be found in the work by Hahn [74] and others. The basic principle of the spin echo method can be summarized as follows: a sample of the liquid under investigation is placed inside a wire coil whose axis is perpendicular to the magnetic field of a permanent magnet. In a state of equilibrium the directions of the nuclear moments in the liquid sample are almost parallel to the constant field. Under the influence of a short radio-frequency current pulse applied to the coil, the vectors of magnetic nuclear moments begin to precess around the direction which is perpendicular to the direction of the constant intensity of the magnetic field. It is possible to assume in the beginning that all the nuclear moments are in the same phase of motion. Owing to unavoidable non-uniformities of the field, however, the frequency of the precessions of the nuclear moments begins to differ. After some time has elapsed they are all in different phases and this results in the signal in the measuring coil being reduced to zero. A subsequent pulse (180° pulse) is applied to the coil after the time τ. This pulse turns the directions of nuclear moments by a further 90°. As a result, those magnetic moments which had the longest delay in precessive motion precede the remaining moments. Since the frequencies of precession remain constant all the moments meet in their phases after the time 2τ and the induced pulse reaches its maximum. Under such circumstances a so-called 'spin echo' appears. While measuring diffusion in the area occupied by the sample, a supplementary gradient of the magnetic field is produced by means of an auxiliary winding. This gradient combined with the process of diffusion results in the displacement of those nucleons having magnetic moments in time intervals from 0 to τ and from τ to 2τ. Therefore changes in phase in the first and second time intervals are different. The number of nucleons which had the same phase in the absence of the supplementary gradient at the moment 2τ is now reduced. According to Carr and Purcell [124] the amplitude of that pulse which corresponds to spin echo is determined by the formula:

$$U_{\text{max}} = A(2\tau)\exp\left(-\frac{2\nu^2 G^2 D\tau^3}{3}\right), \qquad (14.11)$$

where D denotes the coefficient of diffusion, A—a constant, G—the gradient of the field, τ—the time interval between the 90° and 180° pulses, $\nu = \mu/hI$ where h is Planck's constant, μ—magnetic moment of the nucleus and I—its moment of inertia. The coefficient of diffusion D is then obtained from the graph $\ln\left[U_{\text{max}} A(2\tau)\right]$ as a function of τ^3 for $G = 0$.

14.4. *The Relationship between the Coefficient of Self-diffusion on both the Viscosity and Temperature of the Liquid.* Research determining the viscosity of the liquid and involving measurements at various temperatures is of particular importance for the investigation of the mechanism of particle motion in liquids. Such research has been conducted by Umiński and Dera [97] in our Institute. Their results are presented in

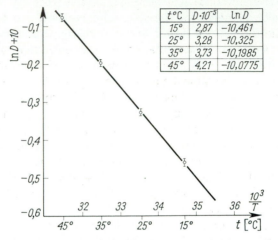

$t\,°C$	$D\cdot10^{-5}$	$\ln D$
15°	2,87	−10,461
25°	3,28	−10,325
35°	3,73	−10,1985
45°	4,21	−10,0775

Fig. 14.4. Dependence of molecular self-diffusion coefficient D on temperature for methyl acetate. Ref.: Umiński and Dera [97].

Table 14.1. Coefficients of self-diffusion for methyl acetate and ethyl acetate

t (°C)	Methyl acetate		Ethyl acetate	
	η (cP)	D ($10^{-5}\,\mathrm{cm^2\,sec^{-1}}$)	η (cP)	D ($10^{-5}\,\mathrm{cm^2\,sec^{-1}}$)
15	0·4024	2·87	0·473	2·38
25	0·3617	3·28	0·426	2·77
35	0·3270	3·73	0·379	3·22
45	0·2970	4·21	0·341	3·62
55	—	—	0·309	4·11

After Umiński and Dera [97].

table 14.1 and are given in a semi-logarithmic plot in fig. 14.4. It is easy to calculate from the values listed in tables 14.1 and 14.2 that $D_0\eta = \mathrm{const.} \times T$ within the limits of experimental error.

Douglass and McCall [73] used the method of nuclear spin in a series of measurements of the coefficient of self-diffusion in saturated hydrocarbons at various temperatures. The values they obtained are given in table 14.2.

Table 14.2. Coefficients of self-diffusion D and of the viscosity and activation energies of diffusion W_D and of viscosity W_η in saturated hydrocarbons

	D $(10^{-5}\ \mathrm{cm^2\ sec^{-1}})$	W_D (kcal/mol)	η $(10^{-3}\ \mathrm{P})$	W_η (kcal/mol)
Pentane	5·45	1·54	2·2	1·8
Hexane	4·21	2·07	2·9	1·78
Heptane	3·12	2·19	3·8	1·98
Octane	2·00	2·42	5·1	2·09
Nonane	1·70	3·08	6·7	2·40
Decane	1·31	3·56	8·5	2·51
Octadecane (50°c)	0·46	3·94	23·4	3·85
Dicetyl (100°c)	0·30	5·64	53·6	4·79

After Douglass and McCall [73].

Douglass and McCall determined the activation energies for the process of diffusion in these liquids from the relationship $D = f(1/T)$ using the equation:

$$W_D = -R \frac{d \ln D}{d(1/T)}. \tag{14.12}$$

They compared these energies of activation with those for viscosity which were determined from the formula:

$$W_\eta = \frac{R \, d \ln \eta}{d(1/T)}. \tag{14.13}$$

The values calculated in this way are also given in table 14.2.

On the basis of the results obtained by Douglass and McCall in the course of their experiments, it is possible to express the relationship between the coefficient of self-diffusion D_0 and the viscosity of the liquid η at room temperature in the following way:

$$D_0 = 1\cdot15 \times 10^{-7} \eta^{-1} \quad (\mathrm{cm^2\ sec^{-1}}). \tag{14.14}$$

For any temperature T this relationship is:

$$D_0 = 3\cdot93 \times 10^{-10} T \eta^{-1} \quad (\mathrm{cm^2\ sec^{-1}}). \tag{14.14a}$$

The substitution of $\eta = f(n, T)$ (cf. formula (2.9)) gives:

$$D_0 = 8\cdot38 \times 10^{-8} k^{-1} \exp\left(-\frac{F_1}{kT}\right), \tag{14.15}$$

where $F_1 = -0\cdot116 \, nk \, T + 0\cdot0093 \, (n + 5\cdot05)$.

The theoretical interpretation of the experimental relationships given above is based on Einstein's well-known formula:

$$D = \frac{kT}{Z}, \tag{14.16}$$

where Z for spherical molecules is defined by Stokes's law

$$Z = 6\pi\eta r,$$

provided that they are rigid and large in comparison to the molecules of the solution. A combination of these two formulae leads to the Einstein–Stokes equation:

$$\frac{D\eta}{T} = \frac{k}{6\pi r}, \tag{14.17}$$

where k is Boltzmann's constant.

Basset [144] introduced the following correction into this formula for small spherical molecules:

$$Z = 6\pi\eta r \frac{\beta r + 2\eta}{\beta r + 3\eta}, \tag{14.18}$$

where β denotes an experimentally defined factor whose values may be anything from 0 to ∞.

On the basis of the above formula Rossi and Bianchi [145] calculated in 1961 the value of r for a number of liquids. They discovered that r for saturated hydrocarbons changed from 2·5 Å (for n-decane) to 5·16 Å (for dotriacontane).

14.5. *The Measurement of the Coefficient of Ion Diffusion.* In 1963 Gazda [125] used a direct method to measure the coefficients of ion diffusion in dielectric liquids and measured these coefficients in hexane. The fundamental principle of this method is the production in a liquid of a layer (whose density is not too high) of ions of one sign. It is necessary to observe the decay of the charge of ions caused by the phenomenon of ion diffusion which takes away a part of the ions to the earthed electrodes. Low density is necessary because at a high density, apart from the effects of diffusion, Coulomb repulsion may occur and increase the velocity of separating the ions. This process can be described by the equation:

$$\frac{1}{\rho_2} - \frac{1}{\rho_1} = 4\pi u (t_2 - t_1), \tag{14.19}$$

where ρ_i denotes the density of ions in the time t_i and u—the mobility of ions.

Gazda produced a layer of ions of one sign by means of a razor-edge beam of x-rays. A thin layer of liquid, close to one of the electrodes, was ionized and the voltage was applied across the ionization chamber in order to remove from the chamber all the ions of one sign and to extend those of the opposite sign to the whole volume of the chamber. The voltage was then switched off and it was necessary to wait for the time t to elapse before diffusion could occur. The voltage was again applied after the time t in order to collect all the ions remaining in the chamber. The coefficient of diffusion was calculated on the basis of the relationship between $\ln Q$ and t. Gazda's apparatus used for measuring the coefficient of ion diffusion was largely based on his apparatus for measuring the coefficient of ion recombination which was described in §13.1. He calculated the density of ions between the plates of the capacitor from eqn. (14.1). After the introduction of several simplifications, the expression:

$$Q = \frac{8}{\pi^2} Q_0 \exp\left(-\pi^2 Dt/l^2\right) \tag{14.20}$$

was found for the charge Q of these ions. In a case when Coulomb repulsion plays a minor rôle, i.e. when the density of the ions is not too high,

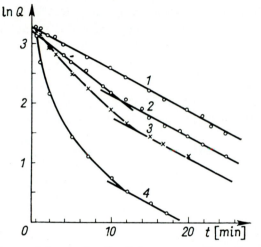

Fig. 14.5. Dependence of charge density Q (in e.s.u. cm^{-3}) on time t (in minutes) for hexane. Ref.: Gazda [125].

the coefficient of ion diffusion D can be determined on the basis of the graph (14.20) given as a semi-logarithmic plot. Only in these circumstances can the plot of this relationship be linear.

Figure 14.5 shows a system of the plots $\ln Q = f(t)$ for various ion densities from $8 \cdot 4 \times 10^6$ ions/cm^3 (curve 1) to 168×10^6 ions/cm^3 (curve 4). It can be seen from the graph that the coefficient of diffusion can be established for the densities of ions not exceeding $8 \cdot 4 \times 10^6$ ions/cm^3. For higher densities of ions diffusion occurs faster and the relationship $\ln Q = f(t)$ ceases to be linear; the determination of the coefficient of ion diffusion from this formula is then impossible.

In measurements taken at $25°$c and with an experimental error not exceeding 10%, Gazda found:

$$D_+ = 1 \cdot 45 \times 10^{-5} \quad (\text{cm}^2\,\text{sec}^{-1})$$

for positive ions and

$$D_- = 2 \cdot 75 \times 10^{-5} \quad (\text{cm}^2\,\text{sec}^{-1})$$

for negative ions. It is interesting to note that in the works previously described the coefficient of diffusion for neutral molecules in hexane was higher $(D = 4 \cdot 2 \times 10^{-5}\ \text{cm}^2\,\text{sec}^{-1})$. Gazda compared his experimental results with Nernst's and Einstein's theoretical conclusions, according to which the ratio u/D should be equal to e/kT. Using the results of Gzowski's research [31], Gazda accepted the following values for the mobility of ions:

$$u_+ = 6 \cdot 1 \times 10^{-4} \, \text{cm}^2 \, \text{v}^{-1} \, \text{sec}^{-1}; \quad u_- = 9 \cdot 5 \times 10^{-4} \, \text{cm}^2 \, \text{v}^{-1} \, \text{sec}^{-1}$$

and obtained:

$$\frac{u_+}{D_+} = 42 \cdot 5 \, \text{v}^{-1}; \quad \frac{u_-}{D_-} = 34 \cdot 5 \, \text{v}^{-1},$$

while at $25°\text{c}\, e/kT = 39 \cdot 4 \, \text{v}^{-1}$. This means that the results are generally consistent.

This research is being extended to include the dependence of the coefficient of ion diffusion on temperature and it seems from preliminary results (Korpalska [161]) that this dependence can be written in the form:

$$D_i = AT\eta^{-3/2} \quad (\text{cm}^2 \, \text{sec}^{-1}), \tag{14.21}$$

where $A = 1 \cdot 03 \times 10^{-5}$.

This combined with the formula (2.9) for the viscosity of liquids in the saturated hydrocarbon group gives the following:

$$D_i = 4 \cdot 84 \times 10^{-8} \, T(kT)^{-3/2} \exp\left(-\frac{3}{2}\frac{F_1}{kT}\right). \tag{14.22}$$

The value of F_1 has already been determined (14.15).

It is thus evident that there exist two different laws for the dependence of the coefficient of diffusion on both viscosity and temperature. For neutral molecules, i.e. for the process of self-diffusion, $D_0 = A_1 T\eta^{-1}$ and for the diffusion process of heavy ions $D_i = A_2 T\eta^{-3/2}$. The temperature dependencies are therefore different and the activation energy for ion diffusion is $1 \cdot 5$ times greater than that for self-diffusion.

Following a different course of research, Davis *et al.* [118] reached similar conclusions. They calculated the coefficients of diffusion for positive ions in liquid Ar, Kr and Xe, on the basis of the measurements of ion mobilities and Einstein's formula $u = eD/kT$. These coefficients were compared with the coefficients of self-diffusion which had been calculated by Naghizadeh and Rice [165] for the same liquids on the basis of the kinetic theory of liquids. The values of these coefficients are given in table 14.3. They obey the relationship $D_i = AT \exp(-B/T)$. The values of B are also given in the table.

Table 14.3. The coefficients D_0 and D_i for liquid Ar, Kr and Xe
$(p = 15 \text{ atm.})$

Liquefied substance	T (°K)	D_0 (10^{-5} cm² sec⁻¹)	D_i (10^{-5} cm² sec⁻¹)	D_0/D_i	B_i (°K)	B_0 (°K)
Argon	90·1	2·35	0·474	5·0	468	358
Krypton	145·0	2·78	0·875	3·2	553	408
Xenon	184·3	2·48	0·442	5·6	960	610

After Davis, Rice and Meyer [118].

As can be seen from the table, the ratio B_i/B_0 for liquid Ar, Kr and Xe is 1·31, 1·36 and 1·57, respectively. Therefore, the activation energy for both the diffusion and mobility of ions is from 1·3 to 1·57 times greater than that for viscosity. This result is consistent with the formula $u = A\eta^{-3/2}$ which was established by Adamczewski [40, 41] in 1935.

Chapter 15

A Summary of the Results of Experiments on the Mobility, Recombination and Diffusion of Ions. The Activation Energy of these Processes

15.1. *Results of Experiments.* This chapter contains both a general summary of the results of experiments on the mobility, recombination and diffusion of ions and on electrically neutral molecules in a group of dielectric liquids and especially in saturated hydrocarbon of the type C_nH_{2n+2}. The most important conclusions concerning this research will also be summarized. Some of these results were published in Adamczewski's articles [162, 182], and in the works of Mathieu *et al.* [209] and Mathieu *et al.* [210].

In the case of saturated hydrocarbons of the type C_nH_{2n+2} whose molecules have n carbon atoms and whose chain length $h = 4 + 1 \cdot 22(n-1)$ (Å) it is possible to establish within the limits of $5 \leqslant n \leqslant 14$ the following general relationships for both positive and negative ions:

$$u_+ = A_1 \eta^{-3/2}, \quad u_- = A_2 \eta^{-3/2}$$

or

$$u_- = A_3 \eta^{-1}, \quad \alpha = B_1 \eta^{-3/2}, \quad D_i = C_i T \eta^{-3/2}. \tag{15.1}$$

The following relationships of the Walden law pattern can be established for very viscous liquids (oils) and for electrically neutral molecules (in self-diffusion):

$$u = A\eta^{-1}, \quad D_0 = C_2 T \eta^{-1}. \tag{15.2}$$

On the basis of the formula (2.9) in which the coefficient of viscosity:

$$\eta = 4 \cdot 7 \times 10^{-3} \, kT \exp\left(\frac{F_1}{kT}\right), \tag{15.3}$$

where

$$F_1 = -0 \cdot 116 nkT + 0 \cdot 0093(n + 5 \cdot 05), \tag{15.4}$$

the above relationships can be written in the form:

$$u_+ = 3 \cdot 12 \times 10^3 \, A(kT)^{-3/2} \exp\left(-\frac{3}{2} \frac{F_1}{kT}\right), \tag{15.5}$$

$$u_- = A(kT)^{-1} \exp\left(-\frac{F_1}{kT}\right), \tag{15.6}$$

$$\alpha = B(kT)^{-3/2} \exp\left(-\frac{3}{2} \frac{F_1}{kT}\right), \tag{15.7}$$

$$D_i = 4 \cdot 84 \times 10^{-8} \, T(kT)^{-3/2} \exp\left(-\frac{3}{2} \frac{F_1}{kT}\right), \tag{15.8}$$

and

$$D_0 = 8{\cdot}35 \times 10^{-8}\,k^{-1}\exp\left(-\frac{F_1}{kT}\right), \tag{15.9}$$

$$\tfrac{3}{2}F_1 = -0{\cdot}174nkT + 0{\cdot}0139(n + 5{\cdot}05).$$

On figs. 15.1, 15.2, 15.3 and 15.4 the nomograms are shown for the determination of the coefficients η, u, α and D as a function of temperature, the number of carbon atoms in a molecule and on the viscosity of the liquid.

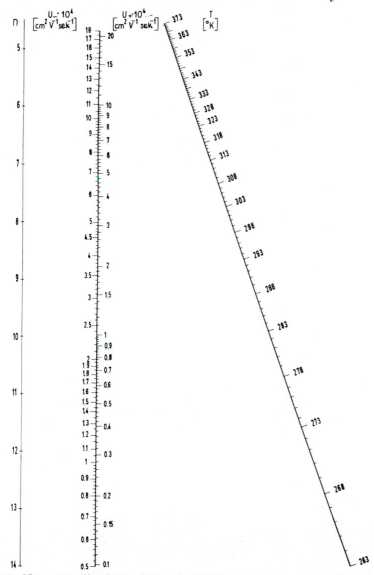

Fig. 15.1. Nomogram for determining the mobility of negative (u_-) and positive (u_+) ions in C_nH_{2n+2} saturated hydrocarbons for various temperatures according to Adamczewski's equation. Ref.: Adamczewski and Kozłowski [230].

In order to interpret correctly experimental results the fine structure of ions whose radii change with η ($r \approx \sqrt{\eta}$) has to be taken into consideration. On the other hand, the ion radius remains constant for neutral

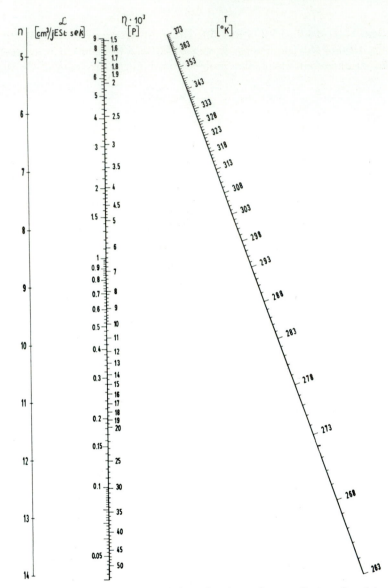

Fig. 15.2. Nomogram for determining the dependence of viscosity (η) and the recombination coefficient (α) on the number (n) of carbon atoms in a molecule and on temperature. Ref.: Adamczewski and Kozłowski [230].

molecules, negative ions and very long molecular chains (which are probably rolled up into balls). It is clear that the energy of activation plays a prominent part in this phenomenon. The general form of this energy can be determined from the relationship:

$$P = A \exp\left(-\frac{W_1}{kT}\right) \quad \text{or} \quad P = A \exp\left(-\frac{W_2}{RT}\right), \qquad (15.10)$$

where k denotes Boltzmann's constant ($k = 8·617 \times 10^{-5}$ ev deg^{-1}). W_1 in this case is measured in ev. The energy of activation W_2 is often expressed in kcal/mole. The constant $R = 1·98 \times 10^{-3}$ kcal/mole (1 ev = 23·05 kcal/mole) is then substituted for k in formula (15.10).

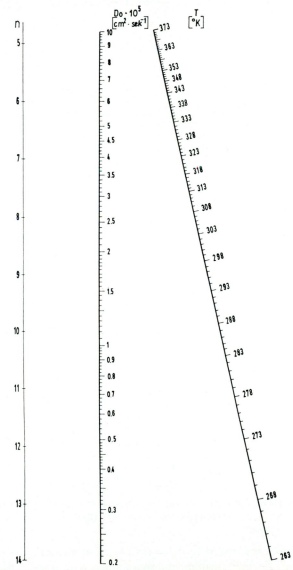

Fig. 15.3. Nomogram for determining the dependences of the self-diffusion coefficient (D_0) on temperature (T) and on the number of carbon atoms (n) in the molecule. Ref.: Adamczewski and Kozłowski [230].

17

In addition to the values obtained theoretically from formula (2.9), table 15.1 gives a list of the values of the activation energy W_η both in ev and kcal/mole for the phenomenon of viscosity and the mobility of ions

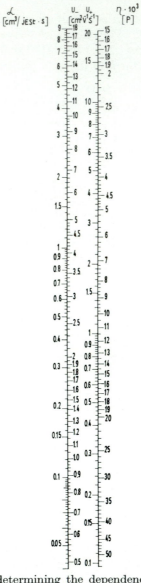

Fig. 15.4. Nomogram for determining the dependence of recombination (α) and mobility (u_-, u_+) coefficients on viscosity (η). Ref.: Adamczewski and Kozłowski [230].

W_u in the group of saturated hydrocarbons. The values of n which define the number of carbon atoms in molecules are given in column 1. Formula (1.2) was used to determine the length of molecular chains of a given compound and the results are to be found in column 2.

Table 15.1. Activation energies W_η and W_{u+} in the group of saturated hydrocarbons of the type C_nH_{2n+2}

n	h (Å)	W_η (theoretical)		W_η (experimental) W_η (theoretical) -0.58	W_u (kcal mole^{-1})
		(ev)	(kcal mole^{-1})		
5	8·85	0·0933	2·16	1·5	3·24
6	10·02	0·1028	2·37	1·8	3·55
7	11·30	0·112	2·56	2·1	3·84
8	12·50	0·121	2·79	2·4	4·18
9	13·70	0·1305	3·05	2·6	4·56
10	15·00	0·140	3·23		4·85
14	19·85	0·177	4·10		6·15
18	24·70	0·215	4·95		7·45
17		0·159	3·67		
28		0·177	4·03		
36		0·189	4·35		
64		0·2315	5·34		

The graphs of the dependencies of the energies of activation W_u, and W_η on n are shown in fig. 15.5. Both the table and the graph indicate that the energy of activation increases systematically with an increase in n. The initial increase is very rapid (up to $n = 12$) but the rate of increase then slows down (for $12 \leqslant n \leqslant 64$). There is a difference between experimental and theoretical values in the region $5 \leqslant n \leqslant 14$. In accordance with formulae (2.9) and (2.10) this discrepancy disappears when $n > 14$.

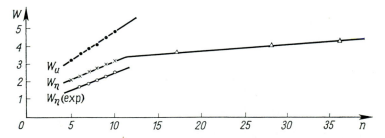

Fig. 15.5. The dependence of the activation energy W (in kcal mol^{-1}) on the number of carbon atoms in the molecule. W_u, activation energy for ionic mobility; W_η, theoretical activation energy for liquid viscosity; $W_{\eta(\exp)}$, experimental data. Ref.: Adamczewski [128].

Since the length of the molecular chain grows systematically with the number (n) of carbon atoms in the molecule, one would suppose that the activation energy increases linearly with the molecular chain length, at least within the limits from $n = 5$ to $n = 10$. For larger values of n the increase in the activation energy is significantly slower, probably because the molecular chains may wind into balls. In figs. 15.6, 15.7 and 15.8 the graphs are shown (after Bässler and coworkers) of the activation energy for the natural conductivity in various dielectric liquids as a function of chemical structure.

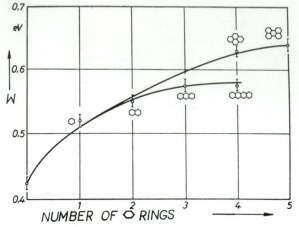

Fig. 15.6. Activation energy (W) as a function of the number of rings in aromatic hydrocarbons. Ref.: Bässler and Riehl [215].

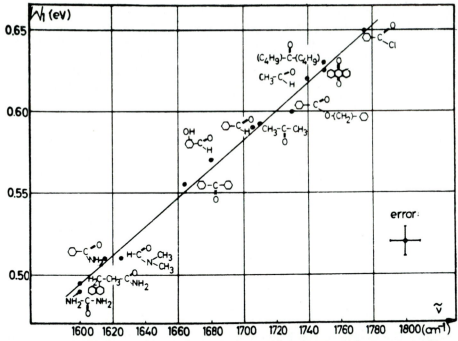

Fig. 15.7. Activation energy (W) as a function of molecular excitation energy in aromatic hydrocarbons. Ref.: Bässler and Riehl [215].

The values of the activation energy W for the ionization current in cyclohexane as a function of the electric field strength E are given in table 15.2 (Jachym [190, 193]).

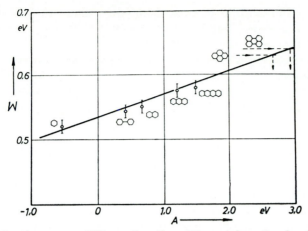

Fig. 15.8. Activation energy (W) as a function of the number of molecular excitation energies in aromatic hydrocarbons. Ref.: Bässler and Riehl [215].

Table 15.2. The activation energy for the current of ionization as a function of the electric field strength

E (kv cm^{-1})	0·212	0·420	0·825	1·343	1·856	2·488	3·375	4·000	4·400	5·05
W (kcal mole^{-1})	3·38	4·27	3·71	3·83	3·87	3·66	3·38	3·18	3·15	3·02

Ref.: Jachym [190, 193].

It can be seen from the table that the energy of activation varies in accordance with the field strength. This may be of great importance for research on the mechanism of electrical conduction and is discussed below in this section. It should be emphasized that a similar experimental effect was observed in 1963 by Forster [147] and Freeman [196] in their measurements of self-conduction currents in benzene, and ionization currents in cyclohexane and paraffin oil.

15.2. *A Theoretical Analysis of the Energy of Activation.* For a long time chemists and physicists have been concerned with the problem of activation energy because of its importance for many physical and chemical processes. Formula (15.10) was most frequently used to determine the activation energy W during the course of experiments. It was generally assumed that A in this formula was independent of temperature and that the value of W was constant for the whole process.

The number of molecular collisions per second in the process of transference under investigation is denoted by A and the effectiveness of the collisions by W. The probability of the molecule being transferred to an adjacent position as a result of collision is expressed by $\exp(-W/RT)$.

The energy of activation is most frequently determined during experiments from the expression $W = -Rd(\ln P)/d(T^{-1})$, i.e. from the slope of the straight line $\log P = \log A - W/RT$.

The numerical values of $\exp(-W/RT)$ at room temperature ($T = 293°\text{к}$) are given below.

W (kcal/mole)	0	10	20	40
$\exp(-W/RT)$	1	10^{-7}	10^{-15}	10^{-29}

The following values are obtained at different temperatures for the activation energy $W = 3$ kcal/mole which approximates to the values most frequently found in the group of saturated hydrocarbons:

T (°к)	250	293	333
$\exp(-3/RT)$	$2\cdot3 \times 10^{-3}$	$6\cdot7 \times 10^{-3}$	1×10^{-2}

If the energy of activation is high, A (in eqn. (15.10)) has little importance in the determination of that energy. On the other hand, when the values of activation energy are low (several kcal/mole) the variation of the value of A in accordance with changing temperatures might result in a change in the energy of activation ranging from 10 to 20%. Other physical and chemical magnitudes, such as the heat of reaction Q, enthalpy H, entropy S and free energy F are also connected with the energy of activation in a specific process.

The laws of thermodynamics and statistical mechanics control the relations between these magnitudes. In accordance with the first principle of thermodynamics, a certain amount of heat dQ, applied to a system under constant pressure, results in a change both in the internal energy energy U and in the volume V:

$$dU = -pdV + dQ \tag{15.11}$$

or

$$d(U + pV) = dQ.$$

The expression $H = U + pV$ is called enthalpy and therefore:

$$dH = dQ. \tag{15.12}$$

According to Eyring's theory [81] the energy of activation W in the expression relating to P denotes the increase of enthalpy ΔH. In the equation:

$$F = H - TS \tag{15.13}$$

enthalpy is connected both with the free energy of the system F and with entropy S. For finite small variations:

$$\Delta F = \Delta H - T\Delta S,$$

hence:

$$W = \Delta H = \Delta F + T\Delta S.$$

Eyring established the following theoretical expression for the number of effective collisions:

$$P = \frac{RT}{Nh} \exp\left(\frac{\Delta S}{R} - \frac{\Delta H}{RT}\right) = \frac{RT}{Nh} \exp\left(-\frac{\Delta F}{RT}\right) \qquad (15.14)$$

or

$$P = \frac{RT}{Nh} \exp\left(-\frac{\Delta H - T\Delta S}{RT}\right). \qquad (15.14a)$$

In this expression N denotes Avogadro's number ($N = 6 \cdot 024 \times 10^{23}$) and h—Plancks' constant ($h = 6 \cdot 624 \times 10^{-27}$ erg-sec). The product of the constants $Nh = 4 \times 10^{-3}$ erg-sec $= 0 \cdot 955 \times 10^{-10}$ cal-sec. Therefore, at room temperature ($T = 293°$к), $RT/Nh = 580/0 \cdot 955 \times 10^{-10} = 6 \cdot 08 \times 10^{12}$ sec^{-1}. The number of collisions per second corresponds therefore to the frequency of atomic and molecular oscillations in the infra-red region.

It can be seen from Adamczewski's semi-empirical formulae for the coefficients of viscosity and ion mobility that there is an obvious similarity between the exponential factor both in Adamczewski's formulae and in theoretical formulae. These formulae can be used to determine the following variations for the group of saturated hydrocarbons (from $n = 5$ to $n = 14$):

(a) Changes in free energy:

$$\Delta F = -0 \cdot 174 nkT + 0 \cdot 0139(n + 5 \cdot 05) \quad \text{(ev)},$$

or

$$\Delta F = -0 \cdot 174 nRT + 0 \cdot 322(n + 5 \cdot 05) \quad \text{(kcal/mole)}$$

$$(R = 1 \cdot 98 \times 10^{-3} \text{ kcal/deg}).$$

(b) Changes in enthalpy (intrinsic energy of activation):

$$\Delta H = 0 \cdot 0139(n + 5 \cdot 05) \quad \text{(ev)}$$

or

$$\Delta H = 0 \cdot 322(n + 5 \cdot 05) \quad \text{(kcal/mole)}. \qquad (15.15)$$

(c) Changes in entropy:

$$\Delta S = 0 \cdot 174 nR.$$

For hexane ($n = 6$) the change in enthalpy is equal to $3 \cdot 54$ kcal/mole and the change in entropy is $1 \cdot 04R$, for $T = 293°$к $T\Delta S = 0 \cdot 601$ kcal/mole and free energy $\Delta F = 2 \cdot 94$ kcal/mole. These results correspond to Gzowski's value of $2 \cdot 5$ [31] obtained during experiments (cf. table 12.4).

15.3. *The Activation Energy of Complex (Competing) Processes.* Each of the processes which take place simultaneously in certain physical and chemical phenomena may have a different activation energy. Although only one value can be determined for the energy of activation, it is possible to separate these processes and to determine the values of A and W for each process by means of a more precise mathematical analysis of the experimental data.

Ruetschi [153], Cremer [152], Nicholas [75] and others studied the problem of the superimposition of various competing mechanisms in the phenomenon of activation energy in other fields. They discussed different aspects of catalysis, the various types of diffusion in solids, displacements of foreign substances in a crystalline lattice, movements of deformations in the lattice, etc. It is worth while discussing how the activation energy W_e, measured in the course of experiments, changes when two competing processes are superimposed in the mechanism of ion motion.

On the basis of works by Cremer [152] and Nicholas [75] it can be assumed that $N(W)$ is the distribution function for a given mechanism. Within the limits of the activation energies W and $W + dW$, the number of processes which take place is proportional to $N(W) dW$. For P the following relationship was established:

$$P = \int_0^\infty N(W) \exp\left(-\frac{W}{RT}\right) dW, \tag{15.16}$$

while W_e and A_e can be expressed as follows:

$$W_e = -R \frac{d(\log P)}{d(T^{-1})} = \frac{\int_0^\infty W N(W) \exp\left(-W/RT\right) dW}{\int_0^\infty N(W) \exp\left(-W/RT\right) dW}, \tag{15.17}$$

$$A_e = P \exp \frac{W_e}{RT} = \int_0^\infty N(W) \exp\left(\frac{W_e - W}{RT}\right) dW. \tag{15.18}$$

The following relationships apply to the simplest case in which only two mechanisms are at work, each having a different but constant value for the energy of activation (W_1 and W_2):

$$\left.\begin{aligned} W_e &= W_1 + \frac{r\varepsilon RT}{1+r}, \\[2mm] A_e &= A_1(1+r) \exp\left(\frac{r\varepsilon}{1+r}\right), \end{aligned}\right\} \tag{15.19}$$

where

$$\varepsilon = \frac{W_2 - W_1}{RT} \quad \text{and} \quad r = \frac{A_2}{A_1} e^{-\varepsilon}, \tag{15.20}$$

r denoting the ratio of the effectiveness of the second mechanism in relation to that of the first mechanism at a temperature T.

It can be seen from the above formulae that even when the values of W_1 and W_2 remain constant, the value of the activation energy W_e (measured under experimental conditions) can change under the influence of various physical parameters which can alter the distribution function $N(W)$. Several examples of the relationship $N(W) = f(W)$ taken from Nicholas's article [75] are given in fig. 15.9. W_e denotes the mean of the 'weighed distribution function' $N(W)\exp(-W/RT)$.

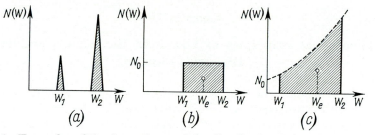

Fig. 15.9. Examples of the dependence of the distribution function $N(W)$ on W for three cases. Ref.: Nicholas [75].

For $r < 10^{-2}$, $W_e = W_1$ and $A_e = A_1$ and for $r > 10^2$, $W_e = W_2$ and $A_e = A_2$. The formula (15.19) is valid for $10^{-2} < r < 10^2$.

It seems probable that the above considerations could be used to explain Jachym's [193] observations on the changes in the activation energy of both ionization and natural currents in dielectric liquids, following variations in the strength of the electric field. Similar results were found by Freeman [196] for several liquids and by Forster [147] for self-conduction currents in benzene. The discrepancies in the results of measurements taken by various scientists for the mobilities of negative ions and for the coefficient of ion recombination can also be explained in the same way. In both cases two kinds of negative charge carriers appear in the process—free electrons and heavy negative ions. These probably have different energies of activation. The distribution function $N(W)$ normally varies with changes in the electric field strength but can also vary as a result of certain experimental conditions. The explanation of this phenomenon should be treated as one of the basic problems in any investigation of the mechanism of electrical conduction in dielectric liquids.

Chapter 16

The Theoretical Principles of Low-field Electrical Conduction in Dielectric Liquids

16.1. *Introduction*. It can be seen from the experimental data presented in previous chapters that ionization conduction in dielectric liquids is closely related to that in ionized gases and particularly in gases under high pressure. The difference between these two phenomena lies in the further increase in current in the so-called 'region of saturation'. There is also a number of other interesting and unexplained phenomena including, in certain cases, the appearance and disappearance of fast and slow negative charge carriers, the different mechanisms of their motion, the number of free electrons produced in the liquid by ionizing agents in relation to ions of various types, excited molecules, radicals, etc. It is possible, of course, that these phenomena may be connected with the increase of the current in the saturation region. However, most theories of electrical conduction in liquids have largely attempted to explain these processes on the basis of the theory of gaseous ions. The theory of semiconductors is seldom taken into consideration.†

16.2. *A General System of Equations for Ionization Conduction*. A complete system of equations comprising all the processes which occur in ionized gases can be expressed in the following general form:

$$\left. \begin{aligned} q - \alpha n_1 n_2 &= -\operatorname{div}(n_1 u_1 E) - D_1 \operatorname{div}\operatorname{grad} n_1, \\ q - \alpha n_1 n_2 &= \operatorname{div}(n_2 u_2 E) - D_2 \operatorname{div}\operatorname{grad} n_2, \\ 4\pi e(n_2 - n_1) &= \operatorname{div} E, \\ i = e[E(u_1 n_1 + u_2 n_2) &- D_1 \operatorname{grad} n_1 + D_2 \operatorname{grad} n_2], \end{aligned} \right\} \quad (16.1)$$

where q denotes the number of ion pairs produced per cm^3 of gas per second, n_1 and n_2 denote the densities of ions of both signs at a given moment, u_1 and u_2—the mobilities of ions, D_1 and D_2—the coefficients of diffusion, α denotes the recombination coefficient, e—the ion charge, E—the electric field strength and i—the density of the current.

The first and second equations determine the state of equilibrium between the number of ions produced by the ionizing agent and the number of ions lost in the processes of recombination and diffusion, as well as those ions removed by the electric field. The third equation is Poisson's. The fourth equation shows the density of the current which is

† See also the new basic monographic paper Rice and Jortner [243], and Morrel Cohen and Lekner, 1967, *Phys. Rev.*, **158**, 305.

determined by the flow of ion charges of both signs. This flow is caused by both the electric field and diffusion.

In the simplest and most frequent case, i.e. when the chamber is a parallel plate capacitor, the system of eqn. (16.1) is as follows:

$$\left.\begin{aligned}
q - \alpha n_1 n_2 &= -u_1 \frac{d}{dx}(n_1 E) - D_1 \frac{d^2 n_1}{dx^2}, \\
q - \alpha n_1 n_2 &= u_2 \frac{d}{dx}(n_2 E) - D_2 \frac{d^2 n_2}{dx^2}, \\
\frac{dE}{dx} &= 4\pi e(n_2 - n_1), \\
i &= e\left[E(u_1 n_1 + u_2 n_2) - D_1 \frac{dn_1}{dx} + D_2 \frac{dn_2}{dx}\right].
\end{aligned}\right\} \quad (16.2)$$

Great mathematical difficulties are encountered in any attempt to find an exact solution for the above system of equations. For this reason an approximate method is normally used for specific cases which can be tested in the course of experiments. If ion diffusion is not taken into consideration, the following relationships can be derived from the system of eqns. (16.2):

$$\left.\begin{aligned}
q - \alpha n_1 n_2 &= -u_1 \frac{d}{dx}(n_1 E) = u_2 \frac{d}{dx}(n_2 E), \\
i &= (u_1 n_1 + u_2 n_2)\, eE, \\
4\pi e(n_2 - n_1) &= \frac{dE}{dx}.
\end{aligned}\right\} \quad (16.3)$$

The following differential equation is obtained after the elimination of n_1 and n_2:

$$\frac{u_1 u_2}{u_1 + u_2} \frac{1}{e} \frac{d}{dx}\left(E \frac{dE}{dx}\right) = q - \frac{\alpha}{e^2 E^2 (u_1 + u_2)^2}\left(i + \frac{u_1 E}{4\pi}\frac{dE}{dx}\right)\left(i - \frac{u_2 E}{4\pi}\frac{dE}{dx}\right). \quad (16.4)$$

If the region is far from being saturated, an approximate solution of eqn. (16.4) is:

$$U = \int_0^\infty E\, dx = IR\left(1 + C\frac{i}{i_s}\right), \quad (16.5)$$

where

$$R = \left(\frac{dU}{di}\right)_{U \to 0}.$$

It is assumed that $C = 1 \cdot 08$ for air. Two regions can be distinguished: for the first $i/i_s < 0 \cdot 6$ and formula (16.5) is valid; for the second $i/i_s > 0 \cdot 7$ and the following formula is valid:

$$U = aR \times 0 \cdot 828 \sqrt{\left(\frac{i \times i_s}{1 - i/i_s}\right)}, \quad (16.6)$$

where a is a constant for a given gas.

It follows from the above equations that there are several regions in the current–voltage characteristics of ionized gases. At the beginning of the characteristic curve the equation for the current can be formulated as a linear dependence on the strength of the field:

$$i = \left[(u_+ + u_-)\,e\left(\frac{dn/dt}{\alpha}\right)^{1/2}\right] E. \tag{16.7}$$

This equation defines the straight line OA in fig. 16.1.

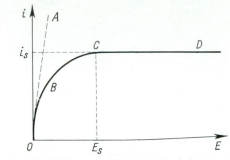

Fig. 16.1. Current–electric field strength characteristic $i = f(E)$ for ionized gases. OA, straight line from eqn. (16.7); OBC, curve from eqn. (16.8); CD, straight line from eqn. (16.9) for the saturation current i_s.

For somewhat higher fields the dependence is parabolic in shape and can be written in the form:

$$i = ey\left[\left(1 + \frac{2d}{y}\right)^{1/2} - 1\right], \tag{16.8}$$

where

$$y = \frac{(u_+ - u_-)^2\,E^2}{2\alpha d}.$$

This equation describes the curve OBC in fig. 16.1.

In sufficiently high fields $(E > E_s)$ when the recombination of ions ceases to be of importance, the density of the current reaches a constant value of saturation and takes the simple form:

$$i_s = qed \tag{16.9}$$

(the straight line CD in fig. 16.1).

In the case of high ion densities it is necessary to take into account the possibility of both the potential and field distribution between the plates of the parallel plate capacitor being distorted.

Figure 16.2 shows a diagram of the influence of the movement of two ion layers of opposite signs on both the field and potential distribution. It is clear that the distribution of both these values ceases to be linear and that distortions are apparent especially near the electrodes. Such distortions are likely to occur when currents are larger than about 10^{-8} A cm^{-2}.

16.3. *The Application of the Similarity Principle.* Further details concerning current–voltage characteristics in ionized gases can be found in the works of Aglintsev [7], Loeb [134, 135], Price [2], Boag and Wilson [132, 133], Mache [76] and others. Similar problems in the case of dielectric liquids have been discussed by Adamczewski [1]. The application of the so-called similarity principle to the system of equations described above yields interesting conclusions when the various experimental results are compared. In 1934 Mache [76] applied this theory to gases and

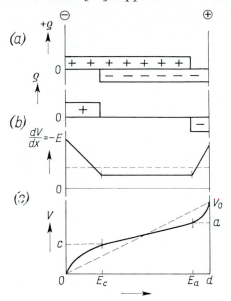

Fig. 16.2. The influence of ion layer shift in a parallel-plate capacitor on: (*a*) charge density, (*b*) field distribution, and (*c*) potential distribution; *d* is the electrode separation. Ref.: von Engel [98].

Adamczewski [1] to liquids. This method can be summarized as follows: a linear change in experimental conditions (i.e. a change in the initial density of ions from q to $q' = \zeta q$, the separation of the electrodes from d to $d' = \lambda d$, electric field strength from E to $E' = \psi E$ and a change in the voltage from U to $U' = \varphi U$) results in a change in the density of ions from n to $n' = \mu n$ and in the current intensity from i to $i' = \gamma i$.

The following simple relationships can be obtained without solving the system of equations even when the terms relating to diffusion are omitted in eqn. (16.1):

$$\zeta = \mu^2; \quad \mu = \frac{\psi}{\lambda} = \frac{\varphi}{\lambda^2}; \quad \gamma = \mu\psi = \frac{\mu\varphi}{\lambda} \tag{16.10}$$

and

$$\frac{i'}{i} = \gamma = \lambda\zeta = \frac{i_s'}{i_s},$$

i.e.

$$\gamma = \lambda\zeta, \quad \varphi = \lambda^2\sqrt{\zeta}. \tag{16.11}$$

On the basis of these simple relationships it is possible to verify the above assumptions for various experimental cases. For example, several curves $i = f(E)$ taken from Adamczewski's work [1] are shown in fig. 16.3. These curves relate to various cases in which $\lambda\zeta = $ const., but the separation of the electrodes and the intensities of ionizing radiation $\neq$ const.

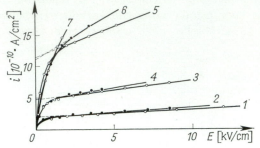

Fig. 16.3.   Current–electric field strength characteristics $i = f(E)$ for various $\lambda\zeta = $ const.

1, 35 kv, 5 mA,	$d = 0{\cdot}1$ cm,	$\zeta = 1{\cdot}94$,	$\lambda\zeta = $	$1{\cdot}94$;
2, 30 kv, 5 mA,	$d = 0{\cdot}2$ cm,	$\zeta = 1$,	$\lambda\zeta = $	$2{\cdot}0$;
3, 35 kv, 5 mA,	$d = 0{\cdot}2$ cm,	$\zeta = 1{\cdot}94$,	$\lambda\zeta = $	$3{\cdot}88$;
4, 30 kv, 5 mA,	$d = 0{\cdot}4$ cm,	$\zeta = 1$,	$\lambda\zeta = $	$4{\cdot}0$;
5, 40 kv, 5 mA,	$d = 0{\cdot}3$ cm,	$\zeta = 3{\cdot}61$,	$\lambda\zeta = $	$10{\cdot}83$;
6, 35 kv, 5 mA,	$d = 0{\cdot}5$ cm,	$\zeta = 1{\cdot}94$,	$\lambda\zeta = $	$9{\cdot}7$;
7, 30 kv, 5 mA,	$d = 1{\cdot}0$ cm,	$\zeta = 1$,	$\lambda\zeta = $	$10{\cdot}0$.

Ref.: Adamczewski [1].

If all the assumptions are valid the corresponding characteristics 1–2, 3–4 and 5–6–7 should be common. It can be seen from the diagram that this is the case, within the limits of error, in low field regions. In the region above saturation, however, distinct deviations are noticeable. In accordance with Adamczewski's formula $i_s = i_0(1 + \gamma Ed)$ (cf. fig. 16.4) the characteristics in the plot $i = f(E)$ are also common in this region. Figure 16.5 illustrates the system of $i = f(E)$ curves and also a parabola (derived from the similarity principle) which indicates the beginning of the saturation of the current. It is clear that there is a satisfactory agreement between theoretical and experimental results. Adamczewski has shown in this way that there is agreement between several other theoretical and experimental cases (e.g. for $\gamma = \lambda$, $\gamma = \zeta$, etc.). He proved that the mechanism of conduction in liquids was, generally speaking, similar to that in gases and was consistent with eqns. (16.2). The distinct deviation which occurs in the region above saturation point is in accordance with Adamczewski's formula [16.67].

16.4. *Theories of Ion Mobility.*† The analysis presented above shows that an important rôle is played in the electrical conduction of ionized dielectric liquids by the mechanism of the transference of charge carriers, i.e. the mechanism of electron or ion mobility. No detailed theory of the

† See also Rice and Jortner [243].

mobility of ions in dielectric liquids has so far been put forward. Attempts have been made to apply the theory of gaseous ions to some simple cases (particularly for liquefied monoatomic gases). This theory, however, is not satisfactory for normal gases, although it sometimes produced results comparable with the experimental data.

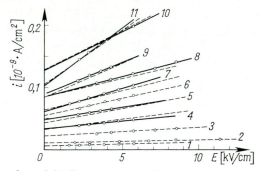

Fig. 16.4. Group of straight lines $i_s = i_0 + cE$ for various experimental conditions (eqn. (11.6)). The open circles are experimental points. Ref.: Adamczewski [1].

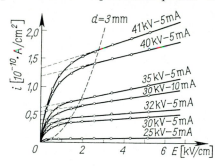

Fig. 16.5. Current–voltage characteristics in hexane for a number of x-ray beam strengths. The broken curve (parabola) indicates the points at which saturation begins. Ref.: Adamczewski [1].

The basic principles of the theory of gaseous ions were first formulated by Langevin [70] and then developed by a number of other scientists including Brown [67]. These principles will be discussed in this section.

If an ion is in an electric field whose strength is E, that ion will then be subjected to the force $F = Ee$ which produces the acceleration $a = Ee/m$ (m denotes the ion mass). In the electric field the motion of the ion superimposes itself on its thermal motion and thus changes the velocity which is produced by a transference of the ion in the direction of the electric field. If the average velocity of ions in their thermal motion is $\bar{v}$, the ion travels the following distance S between the two subsequent collisions:

$$S = \tfrac{1}{2}at^2 = \frac{1}{2}\frac{eE}{m}t^2,\qquad (16.12)$$

where $t = \lambda/\bar{v}$ and λ is the mean free path.

The average drift velocity in the direction of the electric field is:

$$v = \frac{S}{t} = \frac{e\lambda}{2m\bar{v}}E = uE. \tag{16.13}$$

The value of $u = e\lambda/2m\bar{v}$ is generally constant for a given gas at a given temperature and is known as the mobility of the ion. This equation is only an approximation of mobility for a case in which the following conditions have been satisfied: the ion mass must be of the same order of magnitude as that of the molecules of the gas, the collisions must be perfectly elastic, there must be no exchange of the electric charge and no polarizing influence by the ion, etc.

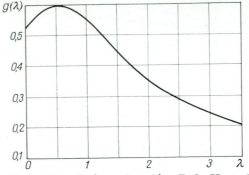

Fig. 16.6. Plot of $g(\lambda)$ against $f(\lambda)$. Ref.: Hasse [99].

Owing to a further development in the theory of gaseous ions it was possible to obtain more precise expressions for the mobility of ions than those obtained by Langevin [70] in the form:

$$u = 0.75 \frac{e\lambda}{M\bar{v}} \sqrt{\left(\frac{m+M}{m}\right)}, \tag{16.14}$$

where both Maxwell's distribution of velocity and the difference between the mass m of the ion and the mass M of the molecule of the gas have been taken into consideration.

Langevin [70] found an even more general expression for a case in which he took into account the effect of polarization produced by an ion in surrounding molecules.

The mobility of ions is then formulated as follows:

$$u = \frac{g(\lambda)}{[\rho(\varepsilon-1)]^{1/2}} \left(\frac{M+m}{m}\right)^{1/2}, \tag{16.15}$$

where

$$\lambda = \left[\frac{8\pi R^4 \rho k T}{(\varepsilon-1)e^2 M}\right]^{1/2},$$

ρ is the density of the gas, ε—dielectric constant, R—the sum total of radii of the gaseous molecule and of the ion and k is Boltzmann's constant. The expression $g(\lambda)$ is a complicated function of λ which is usually determined from the diagram given in fig. 16.6.

The effect of the polarization of gas molecules by an ion is dependent on pressure. For small values of E/p the drift velocity of an ion is proportional to this value which determines the energy acquired by the ion at a distance equal to the length of the free path. The mobility of the ion is constant for a given gas. For high values of E/p the drift velocity is greater than that of thermal motion and the energy acquired by the ion between collisions is proportional to its kinetic energy. Its drift velocity is then proportional to $\sqrt{(E/p)}$ and the ion behaves like an elastic ball.

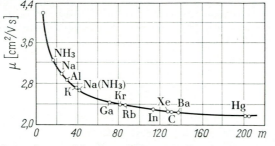

Fig. 16.7. The dependence of ionic mobility (u) of gases on the molecular weight. Ref.: Hornbeck and Wannier [77].

In certain cases experiments have corroborated these simple theoretical hypotheses. Formula (16.15) (cf. fig. 16.7) is satisfied in many gases in respect of the dependence of the mobility of the ion u on its mass m.

Hornbeck and Wannier [77] clearly indicated the influence of the polarization effect of surrounding molecules on the mobility of the ion. Frost [78] obtained the following expression for the dependence of ion mobility on the value of E/p:

$$u = u_0\left(1 + a\,\frac{E}{p}\right)^{1/2}. \tag{16.16}$$

The numerical values for this equation are listed below.

Ion	Gas	u_0	a
He^+	He	9200	0·040
Ne^+	Ne	3500	0·040
Ar^+	Ar	1460	0·0264

In formula (16.14) for ion mobility the mean velocity of the thermal motion is included in the denominator. One would therefore expect an increase in the mobility of the ion in gases to be proportional to the reciprocal of the square root of the absolute temperature. In view of the great importance of ion mobility in gases for plasma physics a great deal of research has been done on this subject. The experimental results, however, have failed to give a uniform or simple picture of the mechanism of ion motion. Moreover, certain reservations must be made before these

18

results are compared with those results obtained in the investigations of ion motion in dielectric liquids because of the different mechanisms of viscosity in both gases and dielectric liquids.

Several scientists have recently studied in depth the theoretical problems of ion mobility in dielectric liquids.

In 1956 Crowe [79] put forward the hypothesis of the 'hopping' electron. This was later discussed by LeBlanc [57] and Terlecki [33] on the basis of their experimental data. According to this hypothesis, the electron in the liquid travels as a free electron the distance λ and is then captured by a molecule (trap). It remains in the trap in a bound state for the time τ and then leaves the trap as a free electron to travel the distance λ until the next capture. Thus the drift mobility of the ion in the electric field can be formulated as follows:

$$u = \frac{u_0 \lambda}{c(\tau + \lambda/c)},$$ (16.17)

where u_0 denotes the ion mobility in the absence of traps and c—the mean velocity of the electron in its thermal motion.

At room temperature $c = 10^7$ cm/sec. The fraction λ/c is so small in comparison with the time τ that it can be omitted in the denominator of (16.17).

If the influence of temperature on the time τ (the length of time in which the electron remains in the trap ($\tau = \tau_0 \exp(W/kT)$) is taken into consideration, eqn. (16.17) can be written as

$$u = \frac{u_0 \lambda}{c\tau_0} \exp\left(-\frac{W}{kT}\right),$$ (16.18)

where W denotes the mean energy with which the electron is captured.

LeBlanc compared his results with eqn. (16.18) and came to the conclusion that $W = 0{\cdot}14 \pm 0{\cdot}02$ ev. LeBlanc offered the following formula for the mobility of the electron in the presence of an electric field with a strength E:

$$u = \frac{u_0 \lambda}{c\tau_0} \exp\left(-\frac{W - Ee\delta}{kT}\right),$$ (16.19)

where δ denotes the length of one edge of the square trap. On the basis of qualitative analysis, LeBlanc obtained the value $\delta = 3$ Å and concluded that this value was generally consistent with the experimental data discussed by Crowe. The latter estimated the mobility of the electron in a field of $1{\cdot}5 \times 10^6$ v cm^{-1} to be about 9×10^{-3} cm^2 v^{-1} sec^{-1}. LeBlanc's measurements suggested that when $E = 0$ the mobility of the electron was $1{\cdot}5 \times 10^{-3}$ cm^2 v^{-1} sec^{-1}. If it is assumed that $\delta = 3$ Å these values would be in agreement with the formula (16.18).

In 1959 Stacey [38] published a report on the problem of ion mobility in liquid argon. His experiments had been conducted in collaboration with Williams [39]. Stacey assumed that the probability of the electron being captured by a neutral argon atom in the time dt was equal to $NQv\,dt$.

N denoted the number of atoms per cm³ of liquid, Q—an active cross section of the process of capture and v—the average velocity of the electron between the two subsequent captures. The probability that the electron will leave the trap in the same time (dt) is $C \exp(-W/kT)\, dt$. W is the energy of capture, k—Boltzmann's constant, T—the absolute temperature and C—a frequency factor of the order of kT/h.

The ratio of the lengths of time in which the electron remains either in a free or in a bound state is $(C/NQv)\exp(-W/kT)$ and the observed velocity is:

$$v' = v \frac{C \exp(-W/kT)}{NQv + C \exp(-W/kT)}, \tag{16.20}$$

when the motion of the negative ions while in existence is neglected.

The mean velocity v of the electron is a function of the electric field strength E. If it is assumed that between two subsequent collisions the electron starts from the beginning and is accelerated by the field, then:

$$v = \left(\frac{e}{m} \frac{E}{2QN} \right)^{1/2}. \tag{16.21}$$

If, as experiments indicate, v' is independent of E and if the electron remains most of the time in the traps, then:

$$NQv \gg C \exp(-W/kT) \tag{16.22}$$

and eqn. (16.20) is reduced to:

$$v' = \frac{C}{NQ} \exp\left(-\frac{W}{kT}\right). \tag{16.23}$$

Unfortunately, Stacey was unable to find the numerical values of Q and W for argon. Both Stacey and Williams conducted their research with temperatures too low for an investigation which required a wide range of temperature. For this reason Stacey reached certain theoretical conclusions based on the following assumptions:

(1) A negative ion has a long lifetime in relation to that of a free electron.

(2) The mobilities of heavy negative ions are of the same order as those of positive ions.

(3) Since the maximum energy of a free electron does not exceed 7 ev its mean energy cannot be greater than 2 ev.

(4) In fields of the order of $E = 10^5$ v cm⁻¹ the ratio of the mobility u_e of the free electron to that of the positive ion u_+ is, according to Williams and Stacey [39], $u_e/u_+ \sim 4 \times 10^3$. Therefore,

$$\frac{C}{NQv} \exp\left(-\frac{W}{kT}\right) \gg \frac{u_+}{u_e}. \tag{16.24}$$

$$\tfrac{1}{2}mv^2 = \frac{eE}{4QN} < 2 \text{ ev}.$$

$$\frac{NQv}{C} \exp\frac{W}{kT} = f, \tag{16.25}$$

where $1 \ll f \ll 4 \times 10^3$.

$$v = fv' \quad \text{and} \quad Q = \frac{e}{m} \frac{E}{2N} \left(\frac{1}{v'f}\right)^2$$

and from eqn. (16.24):

$$W = kT \ln \frac{C}{NQv'}. \tag{16.26}$$

Stacey substituted known numerical values ($e/m = 5 \cdot 3 \times 10^{17}$ electro-static units/g, $m = 9 \cdot 1 \times 10^{-28}$ g, $N = 2 \cdot 1 \times 10^{22}$ cm^{-3}, $v' = 10^6$ cm sec^{-1}, $k = 8 \cdot 6 \times 10^{-5}$ ev/°K, $T = 90°$K and $h = 6 \cdot 6 \times 10^{-27}$ erg-sec) and obtained the following values for Q at $300°$K:

$$Q = 1 \cdot 7 \times 10^{-18} \text{ cm}^2 \quad \text{and} \quad W = 3 \cdot 7 \, kT = 2 \cdot 9 \times 10^{-2} \text{ ev.}$$

According to Stacey, Q can vary within wide limits (by a factor of ten).

He estimated the lifetime of heavy negative ions in argon to be:

$$\frac{1}{C} \exp\left(\frac{W}{kT}\right) \approx 3 \times 10^{-11} \text{ sec} \tag{16.27}$$

(at $90°$K).

Stacey further concluded that Eyring's theory [81] was not applicable to the motion of electrons in a liquid. The reason given for this was that the trap potentials resembled 'holes' rather than normal electric potentials. Such traps could only be formed by argon atoms and not by atoms of impurities. Eyring's theory could not be satisfactorily applied even in the case of positive ions in argon. The behaviour of such ions was in accordance with Stokes's law.

A summary of Stacey's views indicates that the measured electron velocity in liquid argon is determined by the formation of heavy negative ions with a lifetime of 3×10^{-11} sec. These ions are produced as a result of the attachment of the free electrons to neutral argon atoms. An effective cross section of the attachment is 10^{-18} cm^2 and the energy of the attachment is about $0 \cdot 03$ ev. This phenomenon is typical for ionization produced by α-particles and indicates, at least in this region of electric field, that electron multiplication does not occur in liquids.

Davis *et al.* [118] published in 1962 the results of their research on the mobilities of positive ions in liquid Ar, Kr and Xe. Their work was based on Rice and Allnatt's general theory of liquids [120].

This theory involves difficult and complicated mathematical problems and for this reason only the final formula for ion mobility is quoted here:

$$u = \frac{e}{\frac{8}{3}N\sigma^2 g(\sigma) \left[2\pi m m_1 kT/(m+m_1)\right]^{1/2} + \zeta}, \tag{16.28}$$

where e denotes the charge of the electron, N—the number of atoms per cm^3, m—the mass of the ion, m_1—the mass of the atom, σ—the diameter of the ion–atom system [$\sigma = \frac{1}{2}(\sigma_i + \sigma_a)$], $g(\sigma)$ defines the function binding an ion to an atom and $\zeta = kT/D$ is the coefficient of friction defined on the assumption that the action of short-range forces is determined by the Lennard–Jones potential (12–6).

The coefficient of friction ζ is defined by the equation:

$$\zeta = (4\pi mc^3 \rho_m)^{-1} \left[\tfrac{1}{3}\rho_m \int \nabla^2 V(R)\, g(R)\, dR \right]^2, \qquad (16.29)$$

in which ρ_m denotes the density of the liquid, $V(R)$—the intermolecular potential, $g(R)$—the distribution function along the radius R, c—the velocity of sound propagation in the liquid and m—the molecular mass.

Having applied their theory of ion transfer to classic fluids (supplemented and corrected on the basis of wave mechanics) Davis *et al.* calculated the mobilities of both the negative and positive carriers of the electric charge for liquid ^{3}He and ^{4}He. Electrons were considered to be negative-charge carriers and the electric field of the electron both polarizes and combines with the surrounding atoms and changes the density of the surrounding matter. The dielectric constant changes with a change in density and the range of polarization is about 10 Å. The binding of the electronic charge with the medium is dependent on the relative velocity of the electron.

Local fluctuations of density constitute a polarized oscillator of angular frequency ω. The electron is affected by the difference between the local dielectric constant ε and the dielectric constant for high frequency at long distances, ε_∞. This interpretation corresponds to Feynman's polaron model. It is thus possible to determine the effective mass m of the electron from the formula:

$$\frac{m^*}{m_e} = 1 + 0\cdot02\alpha^4. \qquad (16.30)$$

The bond constant α is determined by the formula:

$$\alpha = \tfrac{1}{2}(\varepsilon_\infty^{-1} - \varepsilon^{-1}) \frac{q^2}{\hbar} \left(\frac{2m_e}{\hbar\omega} \right)^{1/2}. \qquad (16.31)$$

Davis *et al.* later corrected their work and found that $m^* = 2m_e$. It was assumed that $\omega = k\theta/\hbar$ and $\theta = T(464/C_v)^{1/3}$. The dielectric constant ε was calculated on the basis of the Clausius–Mossotti equation for a density determined by the first maximum of $g(R)$. The density $g(R)$ at a distance R from an ion is expressed by the equation:

$$g(R) = g^0(R) + \frac{\rho(R) - \rho_0}{\rho_0}, \qquad (16.32)$$

where $g^0(R)$ denotes the pair coupling function in a pure liquid in the absence of an electric charge.

The following expression was found for electron mobility:

$$u_e = \frac{e\hbar^4 \beta^{3/2} a^2}{24\sqrt{(2)}\,\pi^{3/2}\, m^{*3/2}\, N^2\, K_L\, C^2}, \qquad (16.33)$$

where $C = \tfrac{1}{2}\alpha[(\varepsilon+2)/3\varepsilon]^2 q^2$ and K_2 denotes the isothermic compressibility of the liquid.

At a temperature of $2 \cdot 2°$K and at a pressure of 1 atm. the value $u_e = 3 \cdot 72 \times 10^{-2}$ cm^2 v^{-1} sec^{-1} was found for the mobility of the electron in ^{4}He and $u_+ = 5 \cdot 2 \times 10^{-2}$ cm^2 v^{-1} sec^{-1} for the mobility of a positive ion. The calculations were in agreement with experimental results.

16.5. *Jaffe's Theory.*   In 1931 Jaffe [15, 16] presented a general theory of ionization in liquids which was based on Langevin's earlier work [70]. Jaffe's theory rests on the assumption that the ions produced during the irradiation of a liquid with α-particles or protons are not uniformly distributed in space. The ions are grouped along the track of the ionizing particles inside columns. The track of the particle is the axis of symmetry of the column whose diameter is initially very small but which increases in the course of diffusion. This interpretation of the phenomenon of ionization in liquids is confirmed by photographs of the tracks of ionizing particles in Wilson's cloud chambers, in photographic emulsions and in bubble or diffusion chambers.

The effect of the recombination of ions is, of course, of considerably greater importance when the distribution of ions is non-uniform than when it is uniform (cf. §13.1). Langevin and other scientists observed preferential recombination in gases under high pressure (10 or more atm.). Preferential recombination is only of essential importance in very short periods of time (a small fraction of a second) immediately after ionization. The effect of diffusion later separates the electrons from their parent ions. This phenomenon is important for all types of radiation (α, β and γ-radiation).

The axial symmetry of ion distribution is of decisive importance in columnar recombination (Jaffe). The basic principles of this theory are briefly described below from the point of view of the application of the theory to the conduction mechanism in dielectric liquids.

Immediately after the process of ionization (for instance, as a result of an α-particle) the ions are arranged along a column (fig. 16.8) and have a lateral density n_0:

$$n_0 = \frac{N_0}{\pi b^2} \exp\left(-\frac{r^2}{b^2}\right),   (16.34)$$

where N_0 denotes the linear density (i.e. the number of ions per cm of the column length), $r^2 = x^2 + y^2$ is the radius of the column and b is related to the mean distance $\bar{r}_0$ of an ion from the centre of the column in accordance with the formula:

$$\bar{r}_0 = b\left(\frac{\pi}{4}\right)^{1/2}.   (16.35)$$

It is convenient to assume that the electric field operates perpendicularly to the axes of the columns. The following equation describes the changes in time of the density n of ions of both signs (caused by the processes of recombination and diffusion and by the removal of the ions by the electric field):

$$\frac{dn}{dt} = D\left(\frac{d^2 n}{dx^2} + \frac{d^2 n}{dy^2}\right) + uE\frac{dn}{dx} - \alpha n_+ n_-, \qquad (16.36)$$

where α, D and u denote the coefficients of the recombination, diffusion and mobility of ions, respectively. It was necessary to assume that $\mathrm{div}\, E = 0$.

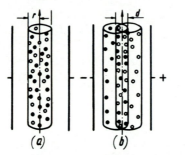

Fig. 16.8. Ion distribution along the path of the ionizing particle: (a) without electric field, (b) with electric field. The full circles denote positive ions, open circles negative ions. The ions were separated a distance $d = 2uEt$ in the cylindrical field. Ref.: Adamczewski [100].

Certain mathematical difficulties were involved in a solution of eqn. (16.36). For this reason Jaffe found an approximate solution by first considering only the phenomenon of diffusion and secondly, only the phenomenon of recombination. The equation found for ion density was as follows:

$$n_\pm = \frac{N}{\pi(4Dt + b^2)}\exp\left[-\frac{(x \pm uEt)^2 + y^2}{4Dt + b^2}\right]. \qquad (16.37)$$

This equation represents two ion columns of opposite signs (fig. 16.8) drifting apart with a relative velocity of $2uE$. N denotes a momentary linear density changing with times as a function of the field:

$$\frac{N_0}{N_t} = 1 + \frac{\alpha N_0}{2\pi}\int_0^t\left\{\exp\left[-\frac{2u^2 E^2 t^2/(4Dt + b^2)}{4Dt + b^2}\right]\right\} dt. \qquad (16.38)$$

An approximate solution of this problem is obtained when the above value is transferred to formula (16.37) and when the value n is transferred from formula (16.37) to (16.36). This solution is characterized by the fact that with an increasing t the value of N_t does not tend to 0 but to a certain boundary value N_∞. The latter is a function of the electric field strength E and represents the number of ions removed from the column by the field.

The solution of this problem necessitates the use of derivatives of cylindrical functions which are difficult to calculate. It is therefore useful to apply certain combinations of theoretical formulae and experimental data on the basis of which it is possible to define the value of N_∞. In particular it is possible to specify the ratio of the number of ions which are removed by the field from the column and which obey the law of normal recombination. For this purpose the following relationship is used:

$$\frac{N_\infty}{N_0} = \frac{1}{1 + (\alpha N_0/8\pi D) f(x)}, \qquad (16.39)$$

where

$$f(x) = x^{-1/2} \int_0^\infty e^{-s} \left[s\left(1 + \frac{s}{x}\right) \right]^{-1/2} ds, \qquad (16.40)$$

$$s = \frac{2u^2 E^2 (y - b^2)^2}{16 D^2 y}; \quad y = 4Dt + b^2; \quad x = \frac{b^2 u^2 E^2 \sin^2 \varphi}{2D^2}. \qquad (16.41)$$

For high values of x (i.e. for the high field strength E) the integral in eqn. (16.40) tends rapidly to 1, $f(x)$ tends to 0 and $N_\infty = N_0$; when $E = 0 \ N = 0$.

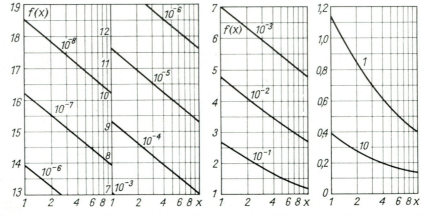

Fig. 16.9. Curves of $f(x)$ against x. Ref.: Zanstra [82]; see also Adamczewski [68].

Since the ratio N_∞/N_0 determines the ratio of the measured charge to the saturation charge, eqn. (16.39) can be replaced by

$$\frac{1}{i} = \frac{1}{i_s} + \frac{\alpha N_0}{8\pi D} \frac{f(x)}{i_s}. \qquad (16.42)$$

The above formula shows a distinct dependence of the current intensity on the electric field strength in view of the fact that the function $f(x)$ contains E and for high values of E $f(x) = A/E$ where the constant $A = (\pi/2)^{1/2} D/ub$.

On the basis of the tables and diagrams calculated and plotted by Zanstra [82] (cf. also Adamczewski [43]) it is possible to calculate the values of the function $f(x)$ for small values of E (several thousand volts per cm). The relationship $1/i = \Phi[f(x)]$ is represented by a straight line in the diagram (fig. 16.9). It is possible to evaluate the specific ionization N_0 and the values of the saturation current i_s from the slope of this straight line.

Jaffe's theory was accepted for a long time with very few reservations. It was not only used to explain the processes of ionization caused both in gases under high pressure and in dielectric liquids by heavy ionizing particles (such as α-particles or protons), but it was also used to interpret the effects of ionization induced in these media by β, x and γ-radiation. In the case of ionization by heavy ionizing particles the existence of ion columns along the tracks of the particles is quite definite but it is doubtful if such columns exist when ionization is induced by x or γ-rays.†

In 1952 Kramers [184] modified Jaffe's theory. Lea's assumptions were applied (cf. § 16.6) by Kramers, who placed a different value on the rôles of preferential recombination and ion diffusion in the first moment after ionization. However, Kramers's theoretical results in high fields were similar to Jaffe's.

In 1953 Richardson [83] published the results of his experiments on ionization in dielectric liquids irradiated by collimated beams of α-particles. The beams were placed at various angles to the direction of the electric field in a capacitor. It was expected that if the external electric field was perpendicular to the axis of an ion column, the number of ions freed by the field from the influence of preferential recombination would be much greater than if the field had been operating parallel to the column. Because the experiment failed to reveal such a dependence on the direction of the field, Richardson assumed that the main rôle was played by δ electrons whose energy was about 1000 ev. Richardson also calculated that if the column had a diameter of 0·05 μ and if the ionization potential was assumed to be about 30 ev then the electric field inside this column should be as high as 10^7 v/cm. This means that the effects produced by external fields of the order of 10^5 v/cm would be so insignificant in the measured current (due to the extraction of ions from the column) that it would be impossible to detect them during experiments.

16.6. *Lea's Theory.* In 1934 Lea [80] put forward a theory to explain the lack of saturation in ionized gases and in gases under high pressure. Lea maintained that while a substance was being ionized by x or γ-rays, the secondary electrons δ (with an energy of 100 to 200 ev) played a decisive rôle. These electrons emerge along the tracks of the photo- and Compton electrons at very small intervals and in turn produce small groups of ions (two to three ion pairs). It is mainly preferential recombination which occurs in these groups. Initially the ion pairs separate as a result of diffusion to form columns and after a long period a uniform distribution of ions takes place.

Lea's theoretical considerations differed from Jaffe's in that the cylindrical symmetry of the columns was replaced by the spherical symmetry of ions. The following formula is obtained for the ion density n at a distance r from the centre of a group of ions:

$$n = \frac{\nu_0}{\pi^{3/2} b^3} \exp(-r^2/b^2), \qquad (16.43)$$

where ν_0 denotes the number of ions in the group.

† See also Januszajtis [336], Terlecki [330], Chybicki [337].

In place of eqn. (16.37) the following is obtained from (16.36):

$$n_{\pm} = \frac{\nu}{\pi^{3/2}\,4(Dt+b^2)^{3/2}} \exp\left[-\frac{(x\pm uEt)^2 + y^2 + z^2}{4Dt+b^2}\right], \qquad (16.44)$$

where

$$\nu = \frac{\nu_0}{1 + (\alpha\nu_0/8\pi^{3/2}\,bD)\,P(x)} \qquad (16.45)$$

denotes the number of all the ions pairs which remain in the group after the time τ. While the function $f(x)$ increases to $2\log 1\cdot5/\sqrt{x}$, in the case of small values $P(x)$ tends to the constant value $k = 1\cdot8$–$2\cdot0$ which is determined by

$$k = \frac{(4D\tau+b^2)^{1/2}}{b}. \qquad (16.46)$$

This differs from Jaffe's relationships in so far as the constant linear density N_0 is replaced by $(\nu/\nu_0)\,N_0$. In order to test the validity of his theory Lea conducted a series of experiments in gases under various pressures (nitrogen and hydrogen). Samples of radium $(B+C)$ and radium E (with radon) were used as sources of ionization. When a lead filter 3 cm thick was used the mean energy of γ-radiation was 1·5 Mev and that of the Compton electrons was 0·75 Mev. When γ-rays were filtered through a 2 mm layer of iron their energy was 0·25 Mev and that of the Compton electrons was an average of 0·06 Mev. Lea concluded from his measurements of the degree of saturation as a function of voltage and gas pressure that approximately 70% of the ions remained in groups and were produced by secondary δ electrons. This conclusion was supported by the slow decrease in the number of ions for 'weak' radiation for which the value of N_0 should have been higher. With increasing pressure, group recombination grows rapidly. However, it has only a slight dependence on the field $(E \sim 500 \text{ v cm}^{-1})$ which, in the case of columnar ionization, should play a greater rôle.

In 1939 Adamczewski [43] attempted to test Lea's theory in liquids irradiated by x-rays of various energies (26·8–40 kev). Hexane was ionized by means of short ionization pulses at constant intervals of 0·04 sec and about 0·001 sec, while the electric field $(E$ up to 3200 v cm$^{-1})$ was being applied the whole time. Adamczewski obtained a number of experimental relationships for $Q = f(E)$ (some of which are given in fig. 16.10). It was possible to calculate Q_0 from the relationship $1/Q = f[f(x)]$ (fig. 16.11) and N_0 could be estimated on the basis of the following formula:

$$\frac{1}{Q} = \frac{1}{Q_0} + \frac{1}{Q_0}\frac{\alpha\nu_0}{8\pi^{3/2}\,bD}P(x) + \frac{1}{Q_0}\frac{\alpha N_0}{8\pi D}f(x). \qquad (16.47)$$

The coefficients of the functions $P(x)$ and $f(x)$ were determined from plots of the experimental results. It was possible to investigate whether the energy of radiation changed these coefficients and in what way (figs. 16.12, 16.13 and 16.14).

Adamczewski's research tended to suggest that the linear densities of the ions N_0 varied with the mean quantum energy of radiation (from $9 \cdot 2 \times 10^4$ for $26 \cdot 8$ kev to $3 \cdot 32 \times 10^4$ for 40 kev). This would indicate that the columnar theory (fig. 16.14) was correct. The value of N_0 was not influenced by a change in the intensity of the x-rays at a constant voltage.

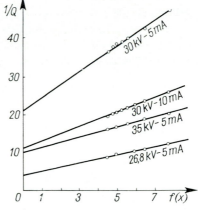

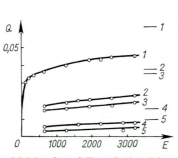

Fig. 16.10. $Q = f(E)$ relationship for various x-ray beam strengths at $t \leqslant 0 \cdot 001$ sec. The figures on the right-hand side show the theoretically predicted saturation values, Q is in e.s.u. cm^{-3}. Ref.: Adamczewski [68].

Fig. 16.11. Curves of $1/Q$ against $f_1[f(x)]$ with $t_2 = 0 \cdot 06$ sec for various x-rays beams. Ref.: Adamczewski [68].

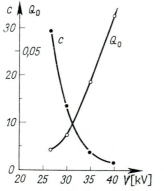

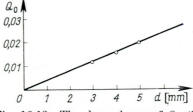

Fig. 16.12. C and Q_0 plotted as functions of U. Ref.: Adamczewski [68].

Fig. 16.13. The dependence of Q_0 (in e.s.u. cm^{-3}) on electrode separation d at 35 kv, 5 mA. Ref.: Adamczewski [68].

In fields of 1000–3000 v cm^{-1}, Adamczewski estimated the percentage of the collected ion charge to be between 50 and 60. A further increase of saturation following an increase in field strength was very slow. Adamczewski estimated the full saturation of the current (i.e. a 100% collection of the ion charge produced during ionization) to be about 10^8 v cm^{-1}.

Figure 16.15 shows both theoretical diagrams in accordance with formula (16.48) and also Adamczewski's experimental points for two different pulses of ionization:

$$Q = Q_\infty \left[1 - \frac{\alpha Q_\infty}{2e(u_+ + u_-)E} \right]. \tag{16.48}$$

Adamczewski's experiments still retain their importance in view of the present state of knowledge on this subject. His theoretical interpretation, however, is problematic and requires further analysis on the basis of the data now available. The value suggested by Adamczewski in 1939 for the 'saturated field' is still accepted in more recent works (Richardson [83]).

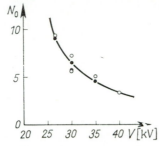

Fig. 16.14. N_0 as a function of U for t_1 (open circles) and t_2 (full circles). Ref.: Adamczewski [68].

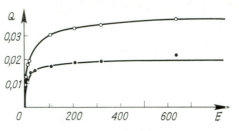

Fig. 16.15. Theoretical curves of $Q = f(E)$ from eqn. (16.48) and experimental results [43]. Open circles: 30 kv, 5 mA at $t_1 = 0.04$ sec; full circles: 40 kv, 5 mA at $t_2 = 0.001$ sec; Q is in e.s.u. cm^{-3}. Ref.: Adamczewski [68].

16.7. *Plumley's Theory*.† On the basis of Onsager's theory [64], Plumley [21] suggested an explanation for the increase in the intensity of the current in high fields. In Plumley's opinion even those molecules of the liquid whose dielectric constant was very low (2–4) are likely to dissociate in higher electric fields. It is possible to conclude, therefore, that neutral molecules generally form pairs of positive and negative ions which may dissociate into two ions in an electric field. Thus, for heptane:

$$2\,C_7H_{16} \rightleftarrows C_7H_{15}^- + C_7H_{17}^+.$$

If U_0 denotes the normal value of the ionization energy, this energy would be reduced by

$$\Delta U_x = -\frac{q^2}{\varepsilon r_0} - qEr_0 \tag{16.49}$$

† Plumley, Frenkel, Małecki and other scientists studied the problem of systematic increases of the electric current in liquids influenced by an electric field (and even under natural conditions, i.e. without ionization). Plumley's theory discussed a new mechanism for the production of ions in high electric fields. The basic principles of this theory will be briefly discussed in view of the fact that this mechanism can also occur in any form in ionized liquids.

in the presence of the electric field E (r_0 denotes the separation of the ions at the maximum of the potential barrier and equals $(q/\varepsilon E)^{1/2}$, q denotes the ion charge and ε—the dielectric constant of the liquid). Hence it can be calculated that

$$\Delta U_x = -2q^{3/2}\left(\frac{E}{\varepsilon}\right)^{1/2}. \tag{16.50}$$

The ionization energy of a molecule is $U_x = U_0 + \Delta U$; therefore,

$$U_x = U_0 - 2q^{3/2}\left(\frac{E}{\varepsilon}\right)^{1/2}. \tag{16.51}$$

Consequently the dissociation constant $K(E)$ determined from the equation $dn/dt = nK(E)$ (n denotes the number of non-dissociated molecules per cm^3) is:

$$K(E) = C\exp\left(-\frac{U_x}{kT}\right) = C\exp\left[-\frac{U_0 + 2q^{3/2}(E/\varepsilon)^{1/2}}{kT}\right]$$

$$= K(0)\exp\frac{2q^{3/2}(E/\varepsilon)^{1/2}}{kT}. \tag{16.52}$$

Also taken into consideration was the following known relationship taken from the theory of statistical mechanics:

$$\frac{d\ln K(E)}{dt} = \frac{U_x}{kT^2}. \tag{16.53}$$

The measured intensity of the current in the region of saturation is:

$$i = nqK(E)\,dS, \tag{16.54}$$

that is, when the value of $K(E)$ is taken from eqn. (16.52) the final formula is:

$$i = C\exp\frac{2q}{kT}\left(\frac{qE}{300\varepsilon}\right)^{1/2}, \tag{16.55}$$

where C is a constant. The relationship

$$\ln i = AE^{1/2} \tag{16.56}$$

is an important conclusion of this theory and can be tested by experiments. The graph of $\ln i = f(E^{1/2})$ is thus linear.

The general relationship (16.56) is confirmed by experiments and particularly those conducted in high electric fields. Plumley [21] (fig. 16.16), Reiss [22] and Pao [3] compared this relationship with their results. It reflects the qualitative character of the dependence of the current on both the temperature and the dielectric constant (cf. fig. 18.3). The quantitative results, however, were not in accordance with experimental results. The validity of this theoretical interpretation is open to question, especially in the case of low fields and it is generally assumed that the consistency between the final general formula of the theory and the experimental results may be purely accidental. The theory of conduction itself may indicate a similar dependence of the current on both the temperature and the field while the mechanism of conduction may be quite different. Certain dependences taken from the theory of electronic semiconductors are similar and in this case the relationship $\log i = f(E)$ is very often linear.

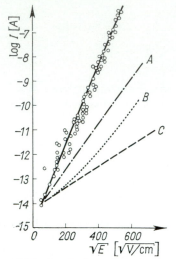

Fig. 16.16. $\log I$ plotted against E for heptane with $d = 0 \cdot 0055$ cm. A, Plumley's theory; B, Onsager's theory; C, thermionic emission theory. Ref.: Plumley [21].

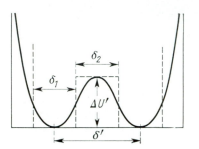

Fig. 16.17. Potential distribution in the vicinity of a molecule (see text). Ref.: Frenkel [101].

16.8. *Frenkel's Theory.* Frenkel [101] based his theory of electrical conduction in dielectric liquids on his own general theory of liquids. According to this theory the molecules of a liquid are bound together by the forces of cohesion but owing to thermal motion they widely oscillate around their position of equilibrium. There is a distinct probability that in the course of oscillation a molecule might acquire such a high kinetic energy that it would be able to separate itself from its neighbouring molecule and travel a distance equivalent to its dimensions and then stop in a new position of equilibrium. The number of travelling molecules is determined by the difference U_0 between the potential energy U_2 of a molecule in an excited state and the potential energy U_1 of a molecule in a normal state ($U_0 = U_2 - U_1$). The number of molecules also depends to a great extent on the kinetic energy of a molecule in its random thermal motion, i.e. it depends on the temperature T of the liquid. Figure 16.17 shows the distribution of the potential in the vicinity of a single molecule or ion in the liquid. The distribution is surrounded by the wall of the potential. $\Delta U'$ denotes the height and δ_2 the width of the wall. The width of the well is denoted by δ_1 and the distance between the two neighbouring positions by δ'.

When an ion is influenced by an external force, the energy of the potential is reduced as it moves towards the direction of the action of the force. The distribution of the potential is shown in fig. 16.18 by the dotted line which contrasts with the normal state represented by a continuous line. The frequency of the molecule's free vibration when in a

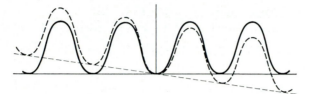

Fig. 16.18. Potential distribution in the vicinity of an ion without a field (continuous line) and in a field (broken line). Ref.: Frenkel [101].

static position is denoted by ν (frequency is the number of possible jumps over the wall of the potential U_0). The number n of molecules which surmount the wall in unit time and in a given direction is:

$$n = \frac{n_0}{6} \nu \exp\left(-\frac{U_0}{kT}\right). \tag{16.57}$$

Number 6 is in the denominator because one-third of the molecules moves in one of three perpendicular directions and a half of this third turns in one direction. In the absence of external forces the activated molecules of the liquid move at random in various directions and surmount the potential wall arriving at new positions of equilibrium. This description of thermal motion in liquids explains various phenomena including the effect of temperature on the viscosity of the liquid.

If a non-polar liquid contains a number of ions, their motion in the absence of an electric field differs only slightly from the thermal motion of molecules. After the application of an electric field, however, the ions acquire additional velocity towards the field and thus gain extra kinetic energy $\Delta U = qE\delta/2$ (q denotes the ion charge, E—the electric field strength and δ—the length of the free path of an ion (fig. 16.18)). ΔU is thus the work of the electric field along half of the length of the free path. The additional number of ions crossing over the potential wall as a result of the action of an electric field is:

$$\Delta n = \frac{n_0}{6} \nu \left[\exp\left(-\frac{U_0-\Delta U}{kT}\right) - \exp\left(-\frac{U_0+\Delta U}{kT}\right)\right]. \tag{16.58}$$

The following expression is obtained for lower fields $\Delta U \ll kT$, when the functions $\exp(-\Delta U/kT)$ and $\exp(\Delta U/kT)$ are expanded:

$$\Delta n = \frac{n_0}{3} \nu \exp\left(-\frac{U_0}{kT}\right)\frac{\Delta U}{kT} = \frac{n_0 qE\delta\nu}{6kT} \exp\left(-\frac{U_0}{kT}\right). \tag{16.59}$$

The number of additional jumps made by one ion per second towards the field is determined by the expression:

$$\frac{\Delta n}{n_0} = \left(\frac{q\delta\nu}{6kT}\right) E \exp\left(-\frac{U_0}{kT}\right).$$

The velocity of the transfer of ions towards the field is:

$$v_E = \frac{\Delta n}{n_0}\delta = \frac{q\delta^2\nu}{6kT} E \exp\left(-\frac{U_0}{kT}\right) = \frac{\Delta U\delta\nu}{3kT} \exp\left(-\frac{U_0}{kT}\right) \tag{16.60}$$

and the mobility of ions in the liquid is:

$$u = \frac{v_E}{E} = \frac{q\delta^2\nu}{6kT} \exp\left(-\frac{U_0}{kT}\right) = \frac{\Delta U\delta\nu}{3kTE} \exp\left(-\frac{U_0}{kT}\right). \tag{16.61}$$

It is clear that the mobility of ions in a liquid depends on the length of the 'free' path of an ion, the temperature of the liquid and also on the work necessary to separate an ion from its neighbouring molecules. In weak fields ($\Delta U \ll kT$) this mobility does not depend on the field strength. The intensity of the ion current in the region below saturation point is formulated as follows:

$$i = n_0 q v_E = n_0 q u E,$$

i.e.

$$i = \frac{n_0 q^2 \delta^2 \nu}{6kT} \exp\left(-\frac{U_0}{kT}\right) E, \tag{16.62}$$

and the electrical conductivity of the liquid is:

$$\sigma = \frac{i}{E} = n_0 q u = \frac{n_0 q^2 \delta^2 \nu}{6kT} \exp\left(-\frac{U_0}{kT}\right). \tag{16.63}$$

In this equation n_0 denotes the number of ions per cm³.

According to Frenkel the mechanism of electrical conduction in dielectric liquids is as follows: an ion produced by an external agent attaches itself to a molecule thus forming an entity. As a result of thermal motion, however, the ion is able to separate itself from the molecule. In this process the ion loses energy when overcoming the coupling forces, i.e. when generating activation energy. After separating from the molecule the ion travels along a certain 'free' path, whose length is comparable to the dimensions of the molecule, and attaches itself to another molecule. The mean length of the free path δ depends both on the structure of the molecule and on the moving ion.

The forces coupling ions and molecules together in the gas are very small and therefore the mobilities of the gaseous ions largely depend on the time of their free transition and on the ratio of the ion charge to the ion mass. The ions in a liquid may be surrounded by neutral molecules (this is characteristic for polar liquids). For this reason the ions in a liquid have similar values of mobility (10^{-4} cm² sec⁻¹ v⁻¹) whereas the mobilities of gaseous ions may be of different orders of magnitude.

It should be stressed that the velocity v_E of ions moving towards the electric field (cf. (16.60)) is not synonymous with the general velocity c of the movement of ions. The latter is determined by the number and length of the jumps made by one ion per second in any direction:

$$c = \frac{n\delta}{n_0} = v_E \frac{n}{\Delta n} = v_E \frac{3kT}{\Delta U} = \frac{qE}{6kT}\delta. \qquad (16.64)$$

For $\Delta U \ll kT$, c is much higher than v_E. It follows from formula (16.63) that the conductivity of ions in a liquid should increase rapidly with an increase in temperature:

$$\sigma = A \exp\left(-\frac{B}{T}\right), \qquad (16.65)$$

where A equals $n_0 q^2 \delta^2 \nu/6kT$ and $B = U_0/k$. In comparison with the term $\exp(-U_0/kT)$ the value A changes very slowly with increasing temperature. To a first approximation, therefore, it can be assumed that the relationship $\ln \sigma = f(B/T)$ should be linear and this is generally consistent with experimental data. In the case of high fields Frenkel failed to take into account the dependence of the ionization current on the strength of the electric field, i.e. when $\Delta U \sim kT$ or $\Delta U > kT$ and when the simplifications used in formulae (16.58) and (16.59) were no longer valid. When formula (16.58) is fully incorporated into eqn. (16.63) there should be a further increase of the current in proportion to the increasing strength of the electric field.

There are certain concepts which try to explain the mechanism of electrical conduction in insulating liquids by ascribing the increase in current to non-linear changes of the potential existing between the electrodes of a capacitor. Silver [245] published an explanation along these lines and based his calculations on Thomson's theory. It can, however, be argued, on the basis of experimental data, that the distribution of the potential is linear in pure dielectric liquids.

16.9. *Adamczewski's Theoretical Conception.* Adamczewski [270] has recently found a general formula for electrical conduction in ionized dielectric liquids. This formula reflects with considerable precision the dependence of measured ionization currents on the following parameters: the strength of the electric field, the distance between the electrodes, temperature, the dose rate and the density of the liquid both in low fields (above saturation point) and very high fields. Adamczewski's qualitative description of the process is consistent with the results of experiments conducted by other scientists. In the case of quantitative analysis certain minor corrections are necessary for the constant values included in the formula.

The formula is based the on generally accepted assumption that the mechanism of charge transfer in dielectric liquids has the following characteristics:

(1) The probability of an ion being displaced in a liquid free from an electric field is in proportion to $\exp(-W/kT)$ (W denotes the activation

19

energy necessary to surmount the potential barrier of the forces binding the charge carrier to its environment and T is the temperature of the liquid).

(2) The influence of the electric field results in the reduction of the activation energy in this process by γdE (i.e. $W_i = A - \gamma dE$).

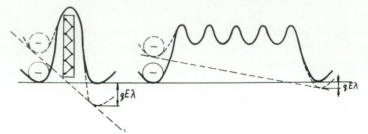

Fig. 16.19. Diagram of the influence of an electric field on the variation of ionic activation energy during motion in a liquid. Ref.: Adamczewski [270].

This is significant because the electric field reduces the activation energy by $eE \Delta x$ (Δx represents a section of the electron's free path). It is possible to state that

$$W = A - eE \Delta x. \tag{16.66}$$

A simplified diagram of this process is shown in fig. 16.19. The length of the molecule through which the charge carrier passes is marked in the diagram for the energy of activation. After many experiments, Adamczewski concluded that a certain linear dependence between the energy of activation and the length of the molecular chain exists in paraffin hydrocarbons.

This subject is closely related to the principles of the theory of natural conduction in both gases and liquids and a detailed analysis made it possible to formulate the following general relationship:

$$I = \frac{a n_0 q d}{kT} \exp \left\{ \left[-\frac{(A - \gamma dE)}{kT} \right] - \exp\left(-CE \right) \right\}, \tag{16.67}$$

where $\gamma = 2 \cdot 5 \times 10^{-6} e$, $C = (0 \cdot 693 u / \alpha n_0 d) \approx 10^{-2}$ cm/v, n_0 denotes the volume density of the free electrons and d—the distance between the electrodes. The last component in the curly brackets, $\exp(-CE)$, is important only for low values of E (from 0 to the saturation region) and it corresponds to the volume recombination.

Equation (16.67) can be written in another form, namely,

$$I = \frac{a n q d}{kT} \exp \left[-\frac{(A - \gamma dE)}{kT} \right], \tag{16.67a}$$

where

$$n = \frac{n_0}{1 + \alpha n_0 (d / uE)} \tag{16.67b}$$

and then the phenomenon of volume recombination is expressed by eqn. (16.67b).

The relationship γdE is obtained by integrating the work of the electron $eE\,dx$ along the line of the electric field which is situated between the plates of the liquid-filled capacitor. The number of traps encountered by the electron on its way from one electrode to another is included in the expression γd. In accordance with Adamczewski's formula published in [1] and in [36] of Part IV.

The results of the application of Adamczewski's general formula under different experimental conditions are given below. Relevant experimental data supplied by other scientists are also taken into account.

At a constant temperature T the general relationship is:

$$I = I_0 \exp(\gamma' dE), \qquad (16.68)$$

where $I_0 = Cned$ and $\gamma' = \gamma/kT$. For small values of γ', when $\gamma' dE < 1$:

$$I = I_0(1 + \gamma' dE). \qquad (16.69)$$

Adamczewski suggested this formula in 1934 for several hundred current–voltage characteristics with different values of n_0, d and E.

This formula was given in its most general form by Jaffe in his earlier works as:

$$I = I_0 + cE. \qquad (16.70)$$

Although formula (16.70) is confirmed by many contemporary scientists, opinions differ as to whether dE contains d and whether $\gamma' = \text{const.}/d$.

Figure 16.20 illustrates various characteristics (taken from Adamczewski's previous publications) and also formulae corresponding to these characteristics.

In high fields from 20 to 30 kv/cm there is an exponential growth of $\exp(\gamma' dE)$ and high voltage characteristics of the type $I = I_0 \exp(\gamma' dE)$ are obtained. These characteristics have been explained in terms of collision ionization and avalanche processes which occur in gases (cf. Nikuradse [2] and House [11] in Part IV). A number of characteristics for high fields supplied by Nikuradse, House, Green and others are given in fig. 18.2 and in figs. 18.11 and 18.12 together with Adamczewski's numerical formulae which correspond to the general expression (16.67).

In fields of up to 20 kv/cm the current increases following an increase in temperature $(T \neq \text{const.})$. The system of curves obtained under such conditions is consistent with the experimental data found by Pao, Jachym, Coelho and others. The application of this formula at very low temperatures is of special interest. Gaeta's data [131] for liquid helium ionized by rays from a ^{147}Pm source ($E_m = 0\cdot229$ мev) were used to determine the following relationship for a system of straight lines at temperatures of $0\cdot83°$к and $1\cdot925°$к:

$$I = \frac{2\cdot1 \times 10^{-17}}{kT} \exp\left[-\frac{(7\cdot2 \times 10^{-5} - 9\cdot5 \times 10^{-7} E)}{kT}\right]. \qquad (16.71)$$

This formula refers to the curves with a minus sign (i.e. the β-electrons are returned to the electrode which emitted them). However, when the direction of the voltage changed, i.e. when the β-electrons have passed through the whole volume of the chamber, the current $n_\beta e$ should be added to I (n_β denotes the number of β-electrons emitted per second in the liquid).

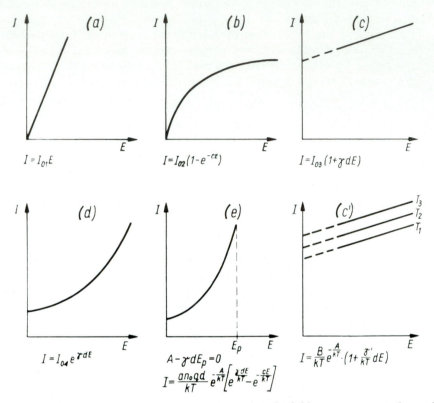

Fig. 16.20. Plots of eqn. (16.17) for various electric fields (at constant electrode separations). Ref.: Adamczewski [270].

A comparison of all the experimental parameters is impossible because they are not all precisely defined. It seems certain, however, that none of the theories of electrical conduction in ionized liquids is so consistent with experimental results as that of Adamczewski. The latter's theory was the most consistent with regard to a wide range of such parameters as the dose rate, the strength of the electric field, the distance between the electrodes, temperature, the molecular structure of the liquid, etc. The value of γ in Adamczewski's formula is proportional to the mean free path of the electron in the liquid:

$$\gamma = \frac{M}{d(n-1)\,\rho N}, \qquad (16.72)$$

where ρ denotes the density of the liquid, M—the molecular mass and N—Avogadro's number. As early as 1934 Adamczewski discovered that when $5 \leqslant n \leqslant 9$, γ decreases with growing density in saturated hydro-carbons. Jachym [239] confirmed this discovery for a wide range of data (see fig. 17.6). Attention must be drawn to the important fact that the general formula (16.67) indicates the criterion of electrical breakdown, i.e. the lowest threshold value for the breakdown field strength (E_{bd}) at which this phenomenon can occur.

It is evident from the above discussion that none of the theories quoted in this chapter explain satisfactorily all the phenomena connected with the flow of an electric current in ionized dielectric liquids. These theories do, however, contribute several interesting suggestions which can be utilized to help solve this problem. The validity of these theories can only be decided by further detailed investigations on both natural and ionization-induced conduction in dielectric liquids at various temperatures. These investigations should concentrate especially on the current–voltage characteristics in both low and high electric fields and on the parameters characterizing the charge carriers (ions or electrons). In short, attention must be paid to the coefficients of mobility, recombination and diffusion.

16.10. *Summary of the Formulae.* A summary of the formulae referring to the dependence of the current on the strength of the electric field in the so-called saturation region at constant temperature is given below:

For gases under normal pressure: $i = i_s = \text{const.}$

An approximate relationship both for gases under high pressure and for dielectric liquids: $i_s = i_0 + cE$

In accordance with Jaffe's theory: $\dfrac{1}{i} = \dfrac{1}{i_s}\left(1 + \dfrac{A}{E}\right)$

In accordance with Plumley's theory: $\ln i = A_1 + B_1 E^{1/2}$

In accordance with the theory of electronic semiconductors: $\ln i = A_2 + B_2 E$

The dependence of the current on temperature in accordance with the theories of Plumley and Frenkel and also the theory of electronic semiconductors: $\ln i = A_3 - \dfrac{B_3}{T}$

Formulae (11.6) and (17.4) (Adamczewski [1] and [45]): $i = i_0(1 + \gamma dE)$

and the general formula (16.67) Adamczewski [270]: $i = \dfrac{anqd}{kT}\exp\left[-\dfrac{(A - \gamma dE)}{kT}\right]$

$$(16.73)$$

A, A_1, A_2, A_3, B_1, B_2 and B_3 denote constants in the above equations. Adamczewski has also suggested another way to explain the mechanism of the increase of the ionization current in the saturation region. The increase was attributed to the influence of molecules excited to a state of great energy either in the initial ionization process or in a secondary process as a result of preferential recombination. Molecules in such a state are prone to dissociation into ions and even to ionization caused by collisions with electrons of low energy or, possibly, by the action of the electric field itself. Adamczewski's calculations of effective cross sections in such processes resulted in values of 10^{-16} cm² which were generally consistent with results for similar processes in gases. It should be emphasized, however, that this explanation requires both detailed examination and theoretical formulation on the basis of more recent experimental data. In any discussion of electrical conduction in dielectric liquids at very high electric fields it is necessary to take into account (*a*) the theory of the emission of electrons from the cathode, (*b*) the theory of the reduced work function in the process of electron emission from a metal into a liquid, (*c*) the avalanche multiplication of the number of free electrons or ions in a liquid as a result of collision ionization and (*d*) the distribution of spatial charge in a liquid.

These problems will be discussed at length in Part IV, where conduction at very high fields and electric breakdown will be analysed.

16.11. *The Possibilities of Applying the Theory of Semiconductors to Explain the Mechanism of Conduction in Dielectric Liquids.* In the last few years a number of published works have indicated the possibility of explaining electrical conduction in solid dielectrics in terms of the band theory of electronic semiconductors and dielectrics. It is well known that the theory of the mechanism of electronic conduction in metals, semiconductors and insulators is based on the model of energy bands.† These bands are produced by the superimposition on one another of single energy levels of valence electrons in atoms when the electrons are concentrated into a crystal lattice. According to Pauli's rule, bands formed by the superimposition of N levels can contain a maximum of $2N$ electrons. When the highest energy level is fully occupied by electrons (*A*) and when the nearest, empty and next highest energy band (*C*) is at a distance of ΔE, a substance can, depending on the value of ΔE, be an insulator ($\Delta E \gg kT$; fig. 16.21(*a*)), electronic semiconductor ($\Delta E \sim kT$; fig. 16.21(*b*)) or a metal (fig. 16.21(*c*)). According to this model the conductivity band is only partially occupied in those substances which are good conductors of electricity (e.g. metals) (fig. 16.21(*c*)). For this reason an electric field can easily increase the energy of free electrons in such substances.

It is understood that an increase of conductivity in dielectrics caused by irradiation with ionizing rays is a result of the electrons, situated in the fully occupied band, being supplied with an energy higher than ΔE. Thermal motions can be the source of such energy in electronic semiconductors when ΔE is about 0·03 ev. Diagrams of the bands (cf. fig. 16.21)

† Electronic semi-conductors are treated in detail by Mott and Gurney [85], Joffe [139], Shockley [87] and Kittel [86].

are valid for pure and homogeneous substances. Impurities or even defects and irregularities in the crystal lattice can produce additional local levels in a good insulator. These levels considerably facilitate the energetic transition of electrons from the occupied zone A (fig. 16.22) to

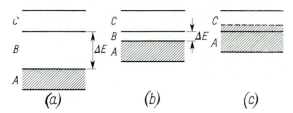

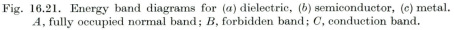

Fig. 16.21. Energy band diagrams for (a) dielectric, (b) semiconductor, (c) metal. A, fully occupied normal band; B, forbidden band; C, conduction band.

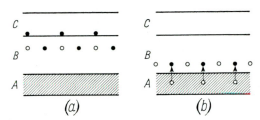

Fig. 16.22. Energy band diagrams for (a) semiconductor with donor additives, (b) semiconductor with acceptor additives. A, fully occupied normal band; B, forbidden band; C, conduction band.

the conductivity zone C. Two different processes should be distinguished: (a) when the atoms of an impurity easily lose their electrons (fig. 16.22(a)), additional local bands concentrate closer to the conductivity band and form a donor-type semiconductor n, and (b) when the atoms of an alien substance capture the electrons, additional local levels (fig. 16.22 (b)) are formed closer to the valance band A facilitating the capture of electrons from this band. These semiconductors are known as acceptor-type semiconductors p. Electronic conduction plays the major part in donor semiconductors. In acceptor semiconductors, however, the main rôle is taken by so-called 'hole conduction', i.e. the movement of the empty spaces (left by the electrons) in an opposite direction to the movement of the electrons.

Theories explaining the most important properties of electronic semiconductors are based on these band models. The properties include the value of conductivity $\sim 10^{-5}$–10^{-12} $\Omega^{-1}\,\mathrm{cm}^{-1}$, the considerable influence of impurities on conductivity, the exponential dependence of conductivity on temperature, the mechanism of photoconduction (i.e. the increase in conductivity as a result of irradiation with ultra-violet, x or γ-rays).

Some scientists have recently considered the analogies between phenomena in the mechanism of electrical conduction in electronic semiconductors and ionized dielectrics in order to analyse the possibilities of applying the band theory to explain phenomena occurring in dielectrics and the mechanism of electrical breakdown in particular. Although these scientists concentrated their attention on such solid dielectrics as polyethylene, polytetrafluoroethylene, polystyrene, mica, amber, etc., the similarities between certain phenomena in both solid and liquid dielectrics suggest the possibility of useful comparisons.

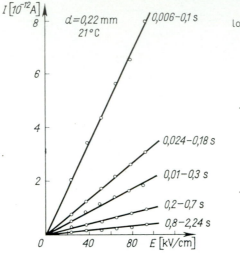

Fig. 16.23. Dependence of initial currents I in solid paraffin on the electric field strength E for different charge collecting times t. Ref.: Ścisłowski [138].

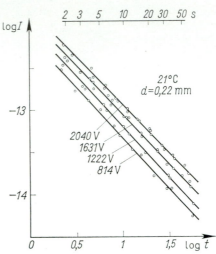

Fig. 16.24. Relationship $I = At^{-n}$ plotted on a log–log scale. The time t is in the range 2–60 sec. Ref.: Ścisłowski [138].

The following works should be mentioned: Fowler's [104] investigations on x-ray-induced conduction in insulating materials, several less important experiments made by Fowler and Farmer [129], theoretical works by Fröhlich [105], Simpson [89] and others. From 1934–39 Ścisłowski [138] conducted the first systematic investigations of electrical conduction in solid paraffin ionized by x and γ-rays at various temperatures. He used very thin layers of paraffin which were completely degassed in their hot state and solidified in a vacuum ionization chamber. He found that the electric current observed in paraffin was the sum of both the conduction and absorption currents. The conduction current rapidly decreased when irradiation ceased and the absorption current I satisfied the law $I = At^{-n}$, where t denotes the time and the coefficient A is in proportion to the applied voltage. Figures 16.23 and 16.24 show a plot of $I = f(E)$ for the initial current and a plot for the relationship $I = f(t)$ is given in a logarithmic plot for various voltages. It is clear from figs. 16.23 and 16.24 that the initial currents in solid paraffin are fully in accordance with Ohm's law while the absorption currents obey the law $I = At^{-n}$ (where

n is almost 1). When the temperature is increased from 17°c to 46°c there is an increase in both currents; this increase, however, was different in each case. The extensive experimental data collected by Ścisłowski is still valuable for new theories of electrical conduction in both solid and liquid dielectrics

In 1966 Jachym [281] investigated the electrical conductivity in the two phases: liquid and solid in the same substance (in octadecane n-$C_{18}H_{38}$) ionized by x-rays. He has found that in the moment of freezing the electrical conductivity rapidly jumped down about fifty times. The ionization current in the liquid varied according to the law $I = I_0 \exp(-W_i/kT)$, the activation energy W_i decreasing with increasing electric field from the value of 0·2 ev (for 0·5 kv/cm) to about 0·15 ev (for 6 kv/cm). Therefore one can conclude that both natural and induced conductivity of dielectric liquids follow the relationship given by the author (16.67).

Fowler suggested in [104] and [129] that conduction in solid dielectrics is of two kinds: static conduction similar to that in semiconducting crystals and induced conduction caused by an ionizing agent. Static conduction, like semiconductors, is characterized by the following dependence on temperature:

$$\sigma = \sigma_0 \exp\left(-\frac{W}{kT}\right), \tag{16.74}$$

where W denotes the distance between the occupied band and the conduction band. The static current is then determined by the equation:

$$I = \sigma E = qunSE, \tag{16.75}$$

where u denotes the mobility of the electron (of the order of 10^{-3} $cm^2\,v^{-1}$ sec^{-1}), n—the number of electrons in cm^3 and E—the electric field strength. The other symbols have already been explained.

Ionization conductivity is determined by (a) the number n_0 of electrons transferred to the conduction band, (b) the number of positive electron 'traps' and (c) the recombination coefficient α which is defined by the equation:

$$n_0 = \alpha nm, \tag{16.76}$$

where $\alpha = Qv$ (v denoting the mean velocity of the electron and Q—the effective cross section for the electron's collision with the trap which produces recombination); m denotes the number of traps per cm^3 and n—the number of electrons per cm^3 at a given moment. The mechanism of conduction is also determined by the structure of a given substance, i.e. whether it is completely or partially crystalline or amorphous.

Although solid hydrocarbons have no defined crystalline structure they often exhibit a molecular regularity (from 60 to 80%) which is dependent on temperature. In Fowler's opinion the energy levels in such substances should be arranged as shown in fig. 16.25. The system of levels for single atomic groups is shown in (a) and a diagram for a substance whose structure is distorted is given in (b). The diagram shows the significance of electron traps in a single group and also indicates the way in which the conduction levels of one atomic group form traps for another. Fowler also suggested that the electron traps produced by ionization can be distributed in the gap between the bands either in a uniform (fig. 16.26 (*a*)) or exponential way (fig. 16.26 (*b*)). The kind of recombination of electrons with traps and the duration of ionization in the substance are determined by the type of distribution within the gap.

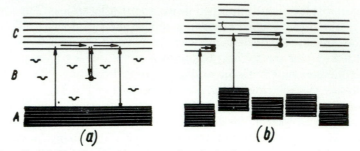

Fig. 16.25. Level distribution in energy bands (*a*) for one group of atoms, (*b*) for a substance with a disarranged structure. Ref.: Fowler [104].

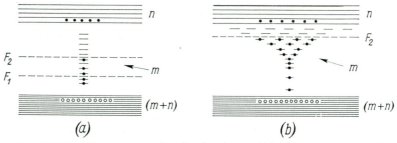

Fig. 16.26. Electron trap distribution in a solid dielectric (see text).
Ref.: Fowler [104].

These theoretical suppositions can be illustrated by the results of two series of measurements taken for polyethylene at temperatures of 80°c and 20°c (cf. Fowler [104, 129]). Figure 16.27 shows plots of the relationship $\log \Delta i = f(t)$, where $\Delta i = i - i_D$, i denotes the ionization current, i_D— the 'dark' current and of the relationship $1/\Delta i = f(t)$. The substance is irradiated by an x-ray beam for 10 min. and the irradiation is switched off at $t = 0$ after which the current decreases with time. It is evident that this decrease is different for various substances. Fowler attributed this difference to various distributions of the electron traps at two different temperatures. When the distribution of the traps is uniform ($M_E = \text{const.}$) the density of the electrons is reduced according to the law:

$$\frac{1}{n} = \frac{1}{n_0} + \frac{\alpha}{kT}(E - E_0)t, \qquad (16.77)$$

where E_0 denotes the level of the lowest trap.

When it is an exponential distribution $M_E = A \exp(E_0/kT)$, then

$$\frac{1}{n} = \frac{1}{n_0} + \frac{\alpha T_1}{T} t. \qquad (16.78)$$

Figure 16.28 shows plots of the dependence of electrical conductivity σ_x on the reciprocal of T in a semilogarithmic plot. E_0 and E_x denote the

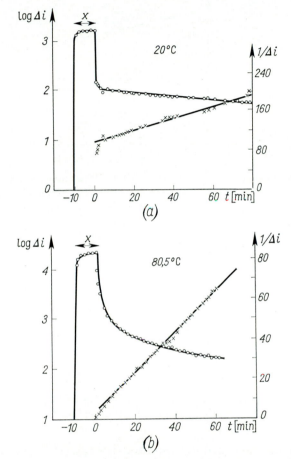

Fig. 16.27. Dependence of ionization current Δi on time t in polyethylene; $\log \Delta i = \log (i - i_2) = f(t)$; i, ionization current; i_2, dark current (unit of current is 2×10^{-14} A), time -10–0 is the irradiation time, time $t \geqslant 0$ is the recombination time: (a) at 20°c, (b) at 80·5°c. Ref.: Fowler [104].

activation energies of these processes. The plot of the relationship $n_0/n = f(t)$ is given in fig. 16.29. Both formulae resemble the relationship for the recombination of ions in liquids (cf. (13.2)).

The experiments described above can be put into practice in many different ways, e.g. the construction of new detectors for research in atomic radiation. Certain substances irradiated with a dose of up to 10^6 R/min still retain their increased conductivity for a period of 20 to

30 hours after irradiation has ceased. Much work has been done on the effects of recombination in electronic semiconductors because of their significance in the process of conduction in semiconductors. The latter are widely applied in technology, especially for crystal diodes and transistors (see bibliography on this subject in Hannay [142], Chapter 11).

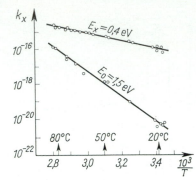

Fig. 16.28.   Temperature dependence of conductivity σ_x in a solid dielectric. E_0 and E_x are activation energies. Ref.: Fowler [104].

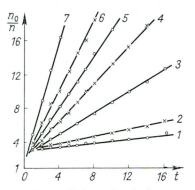

Fig. 16.29.   Plots of n_0/n against T in polyethylene for a number of different radiation doses. Ref.: Fowler [104].

This problem will not be fully discussed because it is only of subsidiary importance for conduction in dielectric liquids. The so-called collision recombination to which Sosnowski [143] drew attention in 1955 is given in outline below.

An energy of about 1 ev is released in the general process of recombination. This energy can be divided into three separate processes: (1) emission of a photon (this process prevails in the case of phosphorus), (2) vibrations of the crystal lattice (related to the emission of a large number of phonons whose maximum energy is $k\theta$, where θ denotes Debye's temperature and $k\theta$ is of the order of 0·01 ev) and (3) the transfer of energy to another electron in the so-called collision recombination process. The energy of excitation is transmitted to another electron in the form of

kinetic energy. This is, therefore, the reverse process of collision ionization in the course of which an electron gives its kinetic energy to another
electron bound to an atom or a molecule and increases the excitation of
the molecule to a higher level of energy. In radiative and phonon recombination the probability of a given process occurring is proportional to the
number of electrons (or traps). In collision recombination this probability
is proportional to the square of this number. Sosnowski was able to show
that the dominant rôle of collision recombination can be proved by
experiment.

Much of this discussion is valid for the problem of conduction in liquids.
The organization of the molecules in liquids, however, is both less perfect
and less permanent and the positive ions ('holes' and electron 'traps')
drift in the electric field.

A great deal of research has recently been conducted on the problem of
radiation-induced conduction in plastics (RIC) using very high doses.
Compton *et al.* [279] irradiated polyethyleneterephthalate with 30 MeV
electron pulses from a linear accelerator and obtained a conductivity of
about $\sim 10^{-8} \, \Omega^{-1} \, cm^{-1}$ at 5×10^9 rad sec^{-1} (Mylar). Two kinds of fast and
delayed carriers were discovered. The number of carriers was investigated
as a function of time, temperature, dose rate, type of electrodes, etc.
Compton *et al.* analysed the mechanism of this phenomenon and concluded that in ionized matter carriers of one kind are immediately trapped,
whereas carriers of the second type remain free until recombination.

In 1965 O'Dwyer [280] discussed in theoretical terms the current–
voltage characteristics in thin dielectric films from the point of view of
quantum mechanics. For various models of the mechanism of this process, dependence of $I = f(U)$ was established.

In addition to modern theories of natural and induced conduction in
dielectrics attention should also be paid to recent research on the distribution of electrons in molecules of various substances, including their
significance and relation to the structures of absorption spectra and to
conduction in semiconductors and dielectrics. It is clear from this research
that apart from a group of electrons which are strongly bound together
with individual atoms (so-called σ electrons), there also exist in chemical
molecules groups of electrons which are less closely bound to atoms and
are, therefore, able to move quite freely within certain atomic groups or
even within a whole molecule (see § 1.5). The latter group of electrons
includes π electrons which exist in the molecules of both non-saturated
organic compounds (with double or coupled bonds) and aromatic compounds, as well as n electrons which exist in atoms other than C and H
(e.g. in O, N, etc., atoms—the so-called heteroatoms). π electrons move
freely within a space limited by the external dimensions of the molecule.

The electrical properties of insulating substances should be related to
both the existence and number of these quasi-free electrons. Research
on this problem will probably contribute a great deal to the understanding
of the mechanism of electrical conduction in some dielectric liquids.

Works published on electric breakdown in dielectric liquids, and on the dependence of breakdown strength value on the chemical structure of the molecules, have greatly contributed to elucidating the problem of the mechanism of slowing down the electrons in a liquid in high electric fields. This research has, therefore, also helped to explain certain problems of conduction in low electric fields. Part IV of this monograph will deal with the problems of electric strength and breakdown in insulators. These are very important for physics and technology and it will also be possible to apply the results of these experiments to further investigations in the field of dielectric liquids in general.

The practical application of liquid-filled ionization chambers to the dosimetry of ionizing radiation will be discussed in the following chapter. Adamczewski and his coworkers, as well as Blanc and his coworkers, have for a long time been concerned with the great importance of this problem for all fields related to dosimetry, including medicine and radiology and radiobiology in particular.† Although much time has been spent on experiments the problem remains unsolved because of great difficulties in the theoretical interpretation of those phenomena which occur when the liquid is ionized. For this reason, the experimental aspects of this problem will be discussed only after its theoretical principles have been given. Some of the basic problems of both dosimetry and liquid dosimetry were discussed in Chapter 8. In view of the significance of this problem, the present state of knowledge in this field and the results of the present author and other scientists will be presented in detail below.

† See also new papers [310–339].

Chapter 17

The Application of Liquid-filled Ionization Chambers to Problems of Dosimetry

17.1. *The Necessity of Introducing Liquid-filled Ionization Chambers.* It is generally recognized that the problem of the dosimetry of ionizing radiation plays an important rôle in the field of contemporary physics, technology, medicine, military science, etc. The mechanism of ionization and electrical conduction in both gases and liquids is one of the most important problems in physics related to dosimetry. This was discussed in Part II. In this chapter attention will be given to the practical application of experimental and theoretical results to the construction of a dosimeter, designed on the basis of an ionization chamber filled with a dielectric liquid. It has already been shown in previous sections that there is a considerable difference between the phenomena of ionization and the conduction of a current in gases and liquids. For this reason it is of paramount importance to use a dielectric liquid and not a gas in the chamber. It would, therefore, be a mistake to apply the readings from a gaseous dosimeter to the radiobiological effects in the tissue. An average energy E_p, necessary to produce one pair of ions in the gas, has been accepted as one of the basic dosimetric values in gaseous dosimetry. This energy is the result of the total of all the energies used for the above-mentioned processes in a volume of gas, divided by the number of ion pairs counted in the same volume. This quotient is 34 ev. (for air).

$$E_p = \frac{\sum_i E_i}{N} = 34 \text{ ev/ion pair.} \qquad (17.1)$$

There is a significant difference between the average value 34 ev/ion pair and that of the ionization energy measured in individual processes in the same gas. Depending on the kind of gas used, the latter is from 12 to 17 ev. The surplus of energy, therefore, is used for other processes which do not produce pairs of ions. The following problems arise when dosimetry is applied to radiobiology, health physics and medicine:

(1) Are those processes which are not reflected in the number of ion pairs of greater biological importance than those of the ionization itself?

(2) Is the energy absorbed in liquid and tissue evenly distributed between all the processes mentioned above? Is the number of ion pairs produced in cm^3/sec a criterion of biological interaction?

(3) Is the dependence of radiation sensitivity on energy, observed in some dosimetric substances, the same for all biologically important processes?

(4) Are the effects of non-uniform ionization, which may be of great significance only in certain small areas of a cell and which have not been observed in gases, of decisive importance in radiobiology?

Apart from these general remarks describing the difference in the course of the phenomena of ionization in gases and liquids, particular attention should be given to the action of a gaseous or liquid ionization chamber working under so-called normal conditions (a standard ionization chamber).

A standard ionization chamber must have the following characteristics (cf. Attix [158]):

(1) Charge particle compensation, i.e. at the boundary of the region under investigation in a given substance, each particle of a defined energy traversing the boundary in one direction corresponds to a particle of the same energy crossing it in an opposite direction.

(2) Integral charge particle equilibrium, i.e. the energy dispersed inside a defined volume of this region by charged particles is such, as to appear that all the charged particles formed in that volume had lost their kinetic energy in the volume. This condition can be expressed in another way, i.e. the integral energy transfer should be equal to the integral absorbed dose. The energy transfer of x or γ-radiation at any point in the irradiated region is the total of kinetic energies transferred to the charged particles plus the energy directly to the matter in the region, the latter being divided by the mass of the volume.

It is difficult to satisfy these two conditions in gas-filled ionization chambers. The chamber must be so constructed and the ionizing beam inside the chamber must be adjusted in such a way that (*a*) the photoelectrons emitted from the window of the vessel cannot reach the active volume of the ionization chamber, (*b*) the electrons emitted from the active volume perpendicularly to the electrodes cannot reach the electrodes before they have reached the end of their path in the matter, (*c*) the ionizing beam should not produce photoelectrons from the metal in the electrodes and (*d*) the electric field in the chamber should be completely uniform.

Many scientists have been concerned with the investigation of phenomena occurring in standard gaseous ionization chambers. In 1960 Attix [158] published the results of his theoretical research and experimental measurements.† These were carried out in a special chamber with sliding walls in order to establish the influence of the range of the photoelectrons produced in gas. Figure 17.1 shows the phenomenon which takes place when a dose, exposed at point P in the active volume V', is determined. It is necessary to determine the total ionization produced by those electrons formed within the volume V' and to eliminate the action of B electrons which are formed outside the volume. In order to make this possible certain experimental conditions are required which would provide for an equilibrium of electrons entering into and departing from a given

† See also [314].

region of the active volume. Such conditions would guarantee both the compensation and integral equilibrium of charge particles.

These phenomena explain changes in the ratio of energy transfer to the absorbed dose. It can be seen from fig. 17.2 that when the absorbing layer

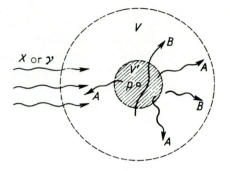

Fig. 17.1. Diagram of the phenomena involved in the definition of an x-ray or γ-ray dose at the point *P*. *A* denotes electrons escaping from the active area *V'*, *B*, electrons from outside of this area (*V*). Ref.: Attix [158].

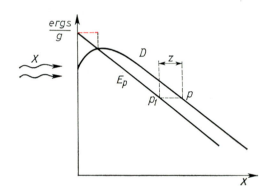

Fig. 17.2. The dependence of the absorbed dose *D* and transfer energy E_p on the thickness of the absorbing layer. Ref.: Attix [158].

has a large thickness x, both the quantities vary in a similar way (the curves are parallel). On the other hand, for smaller thicknesses of x the dose is smaller than the energy transfer because some of the secondary electrons produced in the investigated layer in the course of ionization move forward outside the boundary of the layer. There is an equilibrium for the thickness d of the absorbing layer.

The following relationship exists between the absorbed dose D_a and the energy transfer E_p:†

$$D_a = AE_p(1 + \mu z), \qquad (17.2)$$

† The concept of the energy transfer E_p is equivalent to that of K.E.R.M.A. recently put into practice by I.C.R.U. (cf. footnote, Chapter 8, p. 85).

20

where A is a constant, μ denotes the attenuation coefficient of the incident beam and z—the distance between the points P' and P in fig. 17.2.

Since the range of the photoelectrons depends on the quantum energy of radiation, it is obvious that the dimensions of a standard gaseous ionization chamber must increase with an increase in the energy of ionizing radiation. In this way, for x-rays of 6–25 kev, the collector electrode of such a chamber is 10 mm along the beam and 50 mm perpendicular to it. The plates of the guard electrodes should have the dimensions of 60 mm × 50 mm with an opening cut out for the collector electrode. The input aperture should be placed at a distance of 65 mm from the centre of the collector electrode. When x-rays of a greater penetration are used the dimensions of the chamber should be increaseed accordingly or the pressure inside the chamber should be increased.

In the case of liquid-filled ionization chambers the problem changes because photoelectrons in liquids have a range from a few microns to about 10 mm (cf. fig. 6.1 for higher quantum energies). In view of this fact the dimensions of liquid-filled ionization chambers can be small.

17.2. *Chemical Liquid Dosimeters.*   Investigations on the radiolysis of water and other substances (cf. §§ 9.4 and 9.5) have supplied new material on the influence of ionizing radiation in liquids and the results have been applied to the construction of chemical dosimeters with liquids. The principle of the action of chemical dosimeters consists in making use of changes in the colours of some dyes in solutions, and changes in pH value, electrical conductivity, the concentration of some kinds of ions, absorption of light, etc. These changes are induced in various chemical substances by their absorption of a dose of ionizing radiation.

Chemical dosimeters can be divided into three different groups depending on their range of action. To the first group belong dosimeters for large energetic doses greater than 1000 rads (10^5 erg/g), to the second group—dosimeters sensitive to doses from 46·5 to 560 rads (i.e. from doses able to induce weak symptoms of radiation sickness to lethal doses) while the third group includes chemical dosimeters for neutrons.

The so-called Fricke dosimeter will serve as an example of a dosimeter from the first group. It is the aqueous solution of ferrous sulphate in a solution of sulphuric acid. The most frequent concentrations used are 4×10^{-5} mole $FeSO_4$ and 0·4 mole H_2SO_4 in pure distilled water. Under the influence of irradiation, bivalent ferrous ions Fe^{2+} are oxidated to trivalent ferrous ions Fe^{3+}. The amount of Fe^{3+} effused in an air-saturated solution is in proportion to the dose (up to $4·7 \times 10^4$ rads). The concentration of trivalent iron is determined by means of an ultra-violet spectro-photometric method. The radiation yield G, i.e. the number of molecules or ions produced as a result of the absorption of 100 ev of radiation energy, is 15·5 Fe^{3+} ions/100 ev for the region of 0·6–20 мev quantum energy.

The mechanism of this reaction is as follows: as a result of radiolysis (cf. §9.4) molecules of the OH group and H_2O_2 molecules are produced in the solution and they act on bivalent ferrous ions:

$$\left.\begin{array}{c} Fe^{2+} + OH \rightarrow Fe^{3+} + OH^-, \\ H + O_2 \rightarrow HO_2, \\ H_2O_2 + Fe^{2+} \rightarrow Fe^{3+} + OH + OH^-, \\ H^+ + Fe^{2+} + HO_2 \rightarrow H_2O_2 + Fe^{3+}. \end{array}\right\} \quad (17.3)$$

In this way the number of trivalent ferric ions grows linearly when the radiation dose is increased.

The dependence on temperature of dosimeters of this type is relatively small (from 5°c to 54°c). The magnitude of the dose D in rads is determined by means of the following formula as a function of the density d of an irradiated substance and of the concentration C (mol/m³) of trivalent ferric ions:

$$D = 0.964 \times 10^6 \frac{C}{Gd} \quad \text{(rad)}, \qquad (17.4)$$

where G denotes the radiation yield defined above. Dosimeters of this kind when carefully made from very pure chemical substances can give readings of the dose with a maximum error of a few percent.

Table 17.1. Radiation yields of chemical liquid dosimeters

Type of radiation	Quantum (particle) energy	G (Fe^{3+})	G (Ce^{3+})
γ-radiation (^{60}Co)	1·25 мev	15·5	2·3
X-radiation	(8–18) kev	~ 12·2	—
X-radiation	(50–60) kev	13·9	3·2
	220 kev	15·0	2·7
β-electrons (^{32}P)	1·7 мev	15·5	—
β-electrons (^{3}H)	18 kev	12·0	4·95
Electrons	2 мev	15·5	2·3
Deuterons	15 мev	10·6	2·8
Deuterons	5 мev	8·0	3·2
α-particles (^{210}Po)	5·3 мev	5·3	—
B fragments (n, α) Li	—	—	4·4

The second type of liquid dosimeter is the cerium dosimeter based on the reduction of Ce^{4+} to Ce^{3+} in an aqueous solution. This reaction occurs as a result of the irradiation of an acid aqueous solution of cerous and ceric salts. The radiation yield in this dosimeter is much lower than that in the Fricke dosimeter. For this reason the former dosimeter is only useful for measuring larger doses of the order of 10^9 erg/g, i.e. about 10^7 r for γ-radiation. Table 17.1 gives a list of the radiation yields of both types of dosimeters.

17.3. *Liquid-filled Ionization Chambers as Dosimeters.* The results of experimental and theoretical works on the mechanism of ionization and conduction in dielectric liquids have facilitated the practical application of liquid-filled ionization chambers to the dosimetry of ionizing radiation and other problems can be solved by experiments in radiobiology and radiochemistry.

Adamczewski's works [1, 94, 182] suggest that the dependence of the ionization current on the density of the medium can be expressed in the following form:

$$I = C\dot{D}Sd(\rho + 0.4)(1 + \gamma U). \tag{17.5}$$

This relationship was established by experiments with a very narrow density range of liquids belonging to the saturated hydrocarbon group ($\rho_5 = 0.63 - \rho_9 = 0.75$). Formula (17.5) indicates that the ratio of ionization effects in hexane and in air is only $(0.668 + 0.4)/0.4 = 2.65$ instead of $\rho_h/\rho_p = 0.668/0.001293 = 515$ times as expected from the ratio of densities. This small increase in the measured effect of ionization in a liquid has been corroborated in experiments conducted by Terlecki [33, 201, 310], Gzowski [31] and more recently by Januszajtis [47, 183]. In a standard gaseous ionization chamber they obtained currents only a few times smaller than those in a liquid-filled chamber. In non-standard air-filled chambers where strong photoelectric wall effects were evident, the currents were even larger than in the liquid.

If the energy $W_h = 26$ ev (cf. table 8.1) is necessary to produce one pair of ions in hexane, the efficiency of the collection of ions in hexane in low fields (the so-called efficiency of the chamber) does not exceed 2.5 per cent. In short, the rest of the radiation energy absorbed in the liquid does not contribute to the ionization currents under observation. This may be the result of either initial recombination or other physical and chemical processes which do not affect the number of collected ions. When the total energy absorbed in the liquid is divided by the measured number of ion pairs, an average energy of approximately 1700 ev is obtained. This is the energy required to produce one pair of registered ions in the liquid. This value corresponds to that of 34 ev for air.

On the basis of a more general relationship (Adamczewski [270]):

$$I = \frac{anqd}{kT}\exp\left(-\frac{A - \gamma dE}{kT}\right), \tag{17.6}$$

ıt is possible to calculate the strength of the radiation dose at different temperatures and for different distances between the electrodes in different electric fields.

It is clear that the dosimetry of ionizing radiation in liquids requires further experimental and theoretical research. Investigations are being conducted by some members of our Institute and the results obtained so far promise an early solution of this problem. Several papers concerning this problem were published between 1955 and 1966: Adamczewski [94–96,

160, 182], Januszkiewicz [48], Przybylak [92], Terlecki [201, 267], Januszajtis [47, 160, 183], Schütz [199] and Schmidt [198], and between 1962–66 by the French group Blanc, Mathieu, Vermande and others [159, 166–68, 208–11] and between 1965–66 by the Italian group Ladu, Pellicioni, Rocella, Piccioti and Ramacciotti [300–3].†

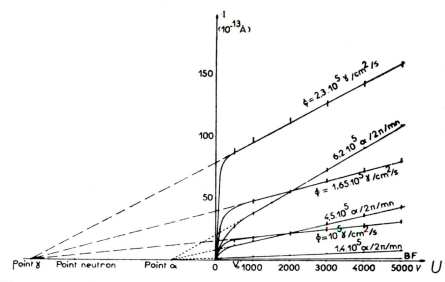

Fig. 17.3. (a) Current–voltage characteristics for various radioactive sources. Ref.: Blanc *et al.* [211].

Some of the Polish publications have already been discussed and the others will be dealt with below. Although the French scientists have partly repeated experiments previously published by Poles, they have also applied them to quantitative experiments with x, γ-rays and neutrons. Figures 17.3 and 17.4 show the current–voltage characteristics (after Blanc *et al.* [211]), for various types of ionizing radiations (γ, α, n) and different values of radiation flux. The authors of these works have determined the voltage values defining the intersection points of the $i = i_0(1 + \gamma U)$ lines with the axis of abscissae . They found that to obtain a current density of 10^{-12} A cm^{-2} in a liquid-filled ionization chamber the following beams of ionizing radiation were necessary:

γ-rays (1·25 мev): $6·4 \times 10^4$ photons cm^{-2} sec^{-1}
neutrons (3 мev): $1·6 \times 10^3 n$ cm^{-2} sec^{-1}
neutrons (14·7 мev): $1·9 \times 10^4 n$ cm^{-2} sec^{-1}

They estimated the lowest detectable beam (N_m) of each radiation to be less than the above values by one order of magnitude. The electrical conductivity was then found to be two to three times greater than the self-conductivity of a liquid which was calculated from current densities of

† In the papers [310–339] there are the results of the investigations in this field from last conferences (up to 1968 year).

5×10^{-14} A cm^{-2}. An accelerator was used as a source of neutrons. When the nuclear reactions were of a D–T type, the neutrons had an energy of 14·7 Mev, but when the reactions were of a D–D type the energy of the neutrons was 3 Mev.

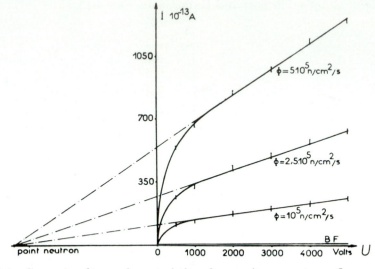

Fig. 17.4. Current–voltage characteristics for various neutron fluxes. Ref.: Blanc *et al.* [211].

In 1962–63 Januszajtis [47, 160] systematically investigated the dependence of the efficiency of a liquid-filled ionization chamber on the energy of ionizing radiation. A specially designed chamber was used for this experiment (cf. § 17.4). On the basis of the ratio of ionization currents in liquids and air, Januszajtis investigated both theoretically and under experimental conditions the ratio of the energies absorbed by these media.

It can be seen from § 7.1 that the attenuation of an x and γ-ray beam, which passes through an absorbent layer of the thickness d and a density ρ, satisfies the formula $I = I_0 \exp(-\mu d)$.

If the absorption coefficient μ/ρ for hexane and air is known, the ratio of the current in the saturation region in hexane to that in air can be formulated as follows:

$$\frac{i_h}{i_p} = \eta \frac{\mu_{ah}\,\rho_p}{\rho_h\,\mu_{ap}} \frac{1 - \exp(-\mu_h x)}{\mu_h x} \frac{W_p}{W_h} \exp(-\mu_h s), \qquad (17.7)$$

where $i_h = i_0 + cE$ denotes the total current measured in hexane at the field strength E, x—the thickness of the absorbent layer in the substance, W_p and W_h denote the average energies required to produce one pair of ions in air ($W_p = 34$ ev) and in hexane ($W_h = 26$ ev). $e^{-\mu s}$ is a corrective for absorption in the layer of the liquid which separates the wall of the chamber from the active volume of the liquid (from where the ions are collected at the collector electrode) and η is the efficiency of the collection of ions in the liquid at a given field strength E.

Since all quantities in eqn. (17.7) can be determined from measurements taken during experiments, with the exception of η and W_h, the ratio η/W_h can be established. The determination of W_h requires either a micro-calorimetric measurement (as in gases) or precise measurements of absorption when the energy of radiation is strictly defined. A more precise evaluation of η necessitates a detailed knowledge of the mechanisms of ionization and conduction in liquids. There is no existing theory to fulfil this need.

On the basis of existing data it is only possible to discuss certain variations of the ratio η/W_h. If 26 ev is assumed to be the value of W_h for hexane vapours, a value of approximately 2·5% (for i_0) is obtained from formula (17.6) for the efficiency of the collection of ions in hexane, i.e. for the efficiency of the liquid-filled ionization chamber. This is in accordance with experimental results. The value of W_h for liquids, however, needs to be more exactly defined by experiments in which the dependence of W_h on the energy must be taken into consideration. Swan [257] and Ullmaier [207] conducted the first experiments of this kind. They obtained 26 ev as the average value for ionization in liquid argon, i.e. the same value as for gaseous argon.

Other investigations relevant to the dosimetry of ionizing radiation in liquids also included research on the dependence of ionization on the density and on the effective atomic number Z_{ef} of the liquid. This number determines the amount of absorbed energy and thus also determines the values of the measured ionization currents. Research has been done in this field by Adamczewski [1], Kao *et al.* [288], Blanc and his colleagues and Jachym [4]. Jachym investigated both decane and carbon tetrachloride and their solutions in a wide range of x and γ-quantum energy (for ^{170}Tm, ^{113}Sn, ^{137}Cs and ^{60}Co). It was established that the ionization current in CCl_4 and its solutions was larger than in decane. Figure 17.5 shows that the largest increase in the ionization current (thirty times) can be seen in CCl_4 when the quantum energy is about 30 kev (in the region of photo-effect).

For isotopes below the quantum energy of 0·4 мev, the ionization current in CCl_4 is approximately twice as large as the ionization current in decane (the region of the Compton effect). Calculations based on an expression similar to formula (17.7), in which the amount of absorbed energy was taken into account, indicated that there was an agreement between the numerical values of the ratio of measured and calculated ionization currents in CCl_4 and decane.

Owing to the very heavy dependence of the ionization current on the effective atomic number of the medium (Z_{ef}), it is possible not only to increase the absorption of the dosimeters of liquid-filled ionization chambers by adding very good insulators to liquid hydrocarbons (CCl_4 or liquid silicones) but also to make solutions of these liquids which have a variable Z_{ef} and corresponding coefficients of absorption. Figure 17.6 shows an increase in the ionization current following an increase (from 5·41 to 16·55) in Z_{ef} for various mixtures as a function of the field strength E for a radiation energy of 28 kev. The logarithmic plot in fig. 17.6 of the dependence of i_0 on Z_{ef} shows that this dependence can be expressed as a simple power dependence $i_0 = CZ_{ef}^n$ where $n = 3 \cdot 15$ and C is constant.

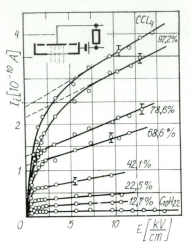

Fig. 17.5. $i = f(E)$ curves for various liquid mixtures, their effective electron number Z_{ef} varying in the range 5·41 to 16·55 (for CCl_4). Ref.: Jachym [239].

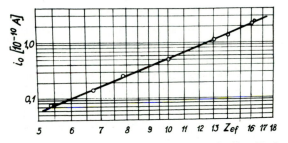

Fig. 17.6. $i_0 = f(Z_{ef})$ relationship plotted logarithmically. Ref.: Jachym [239].

Plots of the dependence of Z_{ef} and the coefficients of absorption on a concentration of CCl_4 in decane (cf. Jachym [239]) facilitate the selection of a liquid with Z_{ef} and μ equivalent to water, tissue or other substances under investigation.

It is well known in dosimetry that the dependence of ionization effects on the quantum energy of electromagnetic radiation is one of the basic problems in measuring the dose of ionizing radiation absorbed by a given substance. This dependence is particularly pronounced in the region of 30 to 250 kev (fig. 17.7), both for gaseous ionization chambers and for films. This kind of dependence can also be observed in other types of dosimeters and presents a great obstacle when measurements are taken because it varies with the kind of dosimetric substance used and therefore requires continuous testing and correcting (fig. 17.8). For this reason it is important for theoretical research to determine the ratio of the absorbed dose (measured in rads) to the exposure (measured in röntgens) as a function of the quantum energy of radiation for certain media and in particular for air and various tissues.

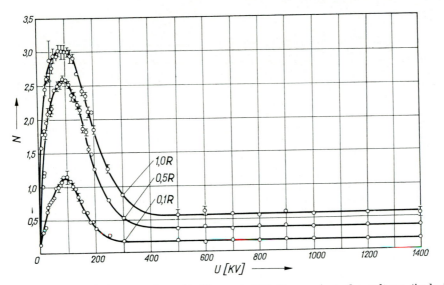

Fig. 17.7. Plot of the capacity of a 'Dupont 552' film against the voltage (in kv) applied to an x-ray tube. Radiation with open window; *r* denotes dose in röntgens. Ref.: Day [103].

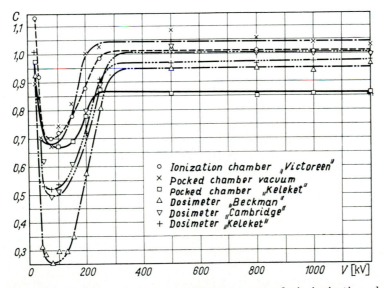

Fig. 17.8. Plot of the correction for different types of air ionization chambers against the radiation energy. *U* is the voltage applied to the x-ray tube. Ref.: Day [103].

Januszajtis [47, 183] conducted investigations in hexane with a view to applying liquid-filled ionization chambers to problems of dosimetry. Radioactive isotopes with a wide range of strictly defined quantum energies were used as a source of radiation. A list of standard radiation sources is given in table 17.2.

Table 17.2. Standard sources of γ-radiation for calibration†

Isotope	$E\gamma$ (Mev)	τ	Isotope	$E\gamma$ (Mev)	τ
^{241}Am	0·05958	458 years	^{54}Mn	0·835	313·5 days
^{57}Co	0·12198	268 days	^{65}Zn	1·114‡	244·4 days
^{141}Ce	0·1455	32·51 days	^{46}Sc	1·118 ⎫	
^{114}In	0·1903	50·00 days		0·892 ⎭	83·9 days
^{203}Hg	0·2791	46·9 days	^{22}Na	1·2736‡	2·58 years
^{51}Cr	0·321	27·8 days	^{60}Co	1·3325 ⎫	
^{198}Au	0·41176	2·7 days		1·1727 ⎭	5·27 years
^{7}Be	0·478	53·4 days	^{88}Y	1·841 ⎫	
^{137}Cs	0·6616	28 years		0·900 ⎭	105 days
^{131}J	0·7239 ⎫	8·066 days	^{124}Sb	2·088 ⎫	
	0·638 ⎪			1·692 ⎭	60·6 days
	0·3645 ⎬		RdTh(ThC″)	2·61425	
	0·2843 ⎪		^{24}Na	2·7535 ⎫	1·91 years
	0·08016 ⎭			1·3679 ⎭	14·87 hours
			Pu—α ⎫		24 400 years
			Ra—α ⎬ ^{9}Be (α,	4·432	1 622 years
			Po—α ⎭ n, γ) ^{12}C		138·4 days

† After Marion, 1960, *Nucleonics*, p. 184.
‡ Plus anihil. 0·51094 Mev.

In the Januszajtis ionization chamber the distance between the electrodes was 2·5 mm (const.). ^{60}Co 3·1 mC was used as a source of ionization and the stream of neutrons from a Po–Be source was 1×10^6 sec^{-1}. The distance between the source of radiation and the ionization chamber was changed from between 7·5 cm to 27·5 cm.

Figure 17.9 shows a graph of the dependence of the ratio of the absorbed dose to the exposure on the energy in hexane for various experimental parameters. It is interesting to note that this curve is very similar to the graph of an analogous relationship in fat tissue (cf. Jaeger [93]).

It is impossible to establish theoretical relationships with any degree of precision in view of an experimental error of almost 10%. Research on this subject is still being carried out. Bearli and Sullivan [30] conducted similar experiments with ionization chambers filled with gas under a pressure of 6 atm. They obtained a series of straight lines for the $I = f(E)$ characteristics up to 30 kv cm^{-1} for α, β and γ-radiation and for neutrons. The equation for the characteristics took the form of $I = kV^n$.

For a given radiation the straight lines intersected the axis E at the same point irrespective of the strength of the dose:

	E_0 (kv cm^{-1})	γ' (v^{-1}) $\times 10^4$
α:	4	2·5
β:	17·1	0·58
γ:	18·0	0·55
n:	13·5	0·74

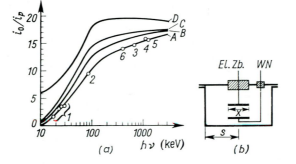

Fig. 17.9. (a) Plot of the ratio i_0/i_p against x and γ-ray energies; i_0 and i_p are ionization currents in the liquid, and air respectively. (b) Quantities x and s appear in eqn. (17.7); *El.Zb.* is the collecting electrode; *WN*, the high-voltage electrode. A, $x = 2$ cm, $s = 3\cdot5$ cm; $B - x = 2$ cm, $s = 2$ cm; $C - x = 2$ cm, $s = 1$ cm; D, $x = 0$, $s = 0$. Ref.: Januszajtis [47].

17.4. *Different Kinds of Dosimetric Ionization Chambers filled with Liquids*. A number of diagrams of liquid-filled ionization chambers used by various scientists have been given in the previous sections of Part III. This section will deal with some special ionization chambers adapted to dosimetric investigations. In recent years these investigations have been extended on a large scale in view of their application to the physics of ionizing radiation and, in particular, to nuclear physics.

Figure 17.10 shows a large ionization chamber whose volume is 1 litre and between 1936 and 1939 Adamczewski [17, 18] used this kind of chamber to investigate the effects of ionization in hexane as a result of cosmic rays (cf. § 11.2). The other types of chambers used by Adamczewski to investigate the dosimetry of x-radiation are shown in figs. 17.11 and 17.12. A cross section of the chamber used by Adamczewski and Przybylak [92] is given in fig. 17.13. The chamber is a parallel plate capacitor whose collector electrode is divided into four separate equal sections which are placed on separate amber insulators. With this kind of chamber it is possible to investigate the effects of absorption in various layers of the liquid along the incident beam. It is also possible to determine the range of photoelectrons and the requirements of the so-called standard ionization chamber.

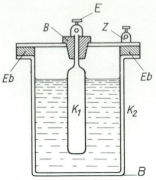

Fig. 17.10. Cross section of a 1 litre liquid ionization chamber. K_1, collecting electrode; K_2, outer voltage electrode (vessel filled with liquid); B, amber; E_b, ebonite; Z, to earth. Ref.: Białobrzeski and Adamczewski [17].

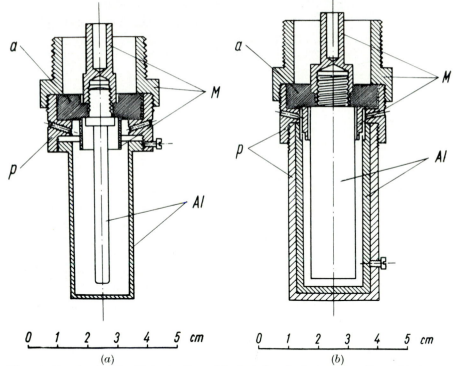

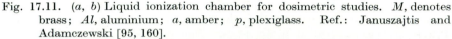

Fig. 17.11. (a, b) Liquid ionization chamber for dosimetric studies. M, denotes brass; Al, aluminium; a, amber; p, plexiglass. Ref.: Januszajtis and Adamczewski [95, 160].

Figure 17.14 gives a cross section of the dosimetric ionization chamber used by Blanc *et al.* [159] who used it in studies concerned with x and γ-rays and neutrons.

Figure 17.15 shows a diagram of a special dosimetric chamber filled with air and liquid. The central part of the chamber is filled with hexane and the external part with air. The electrode which usually contains the liquid is fitted on amber and is connected to an electrometer. The voltages V_1 and V_2 are applied to the central and outer electrodes respectively.

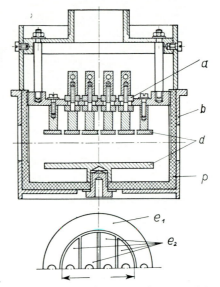

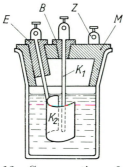

Fig. 17.12. Cross section of a small 50 cm³ ionization chamber. K_1, collecting electrode; K_2, high-voltage electrode; B, amber; E, ebonite; M, brass; Z, to earth. The vessel is a glass beaker with a ground-glass neck. Ref.: Adamczewski [18].

Fig. 17.13. Diagram of a multi-electrode ionization chamber. Ref.: Przybylak [92].

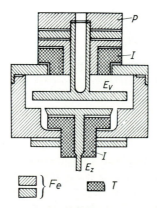

Fig. 17.14. Diagram of a liquid ionization chamber for dosimetric studies. E denotes collecting electrode; E_v, high-voltage electrode; I, insulator; P, cover. Ref.: Blanc *et al.* [159].

Figure 17.16 gives the plots of the chamber's efficiency. Curves are shown for air, liquid and for air and liquid together. As can be seen from the graph, in the case of liquid and air together the dependence of the efficiency of the chamber on the quantum energy of the incident radiation can be reduced to a minimum.

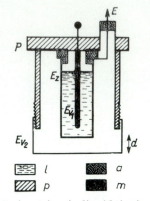

Fig. 17.15. Diagram of a dosimetric air–liquid ionization chamber. E_z, denotes collecting electrode; E_{V_1}, electrode at a potential of V_1; E_{V_2}, electrode at a potential of V_2; E, to electrometer; P, cover; l, liquid; p, perspex; a, amber; m, metal. Ref.: Januszajtis and Adamczewski [160].

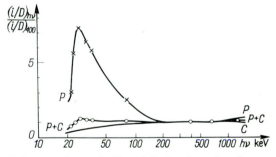

Fig. 17.16. Efficiency of an air–liquid ionization chamber as a function of the radiation energy. P, for air only; C, for liquid only; $P + C$, for air and liquid together. Ref.: Januszajtis [47].

Figure 17.17 shows the special ionization chamber used by Blanc *et al.* [211] for the simultaneous investigation of both natural and ionization currents in the same liquid-filled chamber. The much larger upper electrode was partly covered with α-particles (^{210}Po) in such a way that when the electrode was turned, a stream of α-particles could be directed either inside or outside the chamber.

Figure 17.18 shows a diagram of another chamber recently used for dosimetric research by a group of French scientists under the direction of Blanc. The hexane-filled chamber had three concentric cylinders made of stainless steel and an effective volume of 36 cm³. It was constructed in such a way as to ensure the equilibrium of secondary radiation up to 2 MeV for photons and up to 20 MeV for fast neutrons.

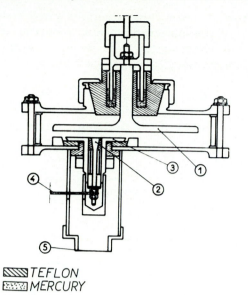

Fig. 17.17. The section through an ionization chamber used for current and pulse measurements: 1, plane electrodes; 2, electrodes for pulse measurements; 3, electrodes for current measurements; 4, connection to an electrometer; 5, closure and connection to the substance. Ref.: Blanc *et al.* [211].

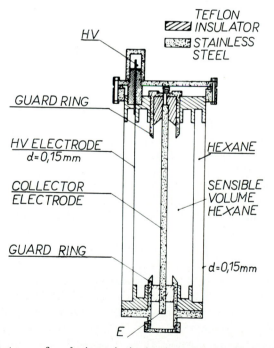

Fig. 17.18. Prototype of a dosimetric ionization chamber for γ-radiation. Ref.: Blanc *et al.* [208].

By means of this chamber these scientists proved the existence of a
linear dependence of the ionization current I on the mean electric field
strength E. In this chamber the field strength is not uniform and is
determined from the formula:

$$E_{\max} = \frac{1}{b-a} \int_a^b \frac{V}{r \log (b/a)} dr, \qquad (17.8)$$

where a and b are the radii of the electrodes and V denotes the applied
voltage. Within the limits of 5 to 25 kv cm^{-1} the linear functions $I = f(E)$

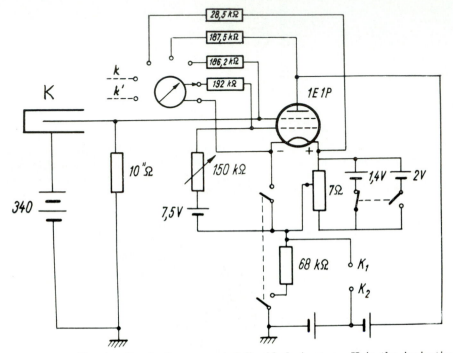

Fig. 17.19. Circuit of a battery-operated liquid dosimeter. K is the ionization
chamber. Ref. Januszkiewicz [48].

intersect the abscissae in the region of 17–18 kv cm^{-1} for different types of
radiation: x-rays (250 kv), γ-rays (^{137}Cs and ^{60}Co) and for radiation from
a betatron (22·5 мev). For fast neutrons the intersection occurs at about
12 kv cm^{-1}.

A circuit diagram of a complete liquid dosimeter with a chamber, valve
electrometer and battery supply is shown in fig. 17.19. Dosimeters of this
type have so far been mainly used to measure doses of x and γ-radiation,
but attempts have been made to detect the effects of cosmic rays.

A diagram of a logarithmic dosimeter constructed by Januszajtis and
Kozłowski [284] is given in fig. 17.20. The collector electrode of the
liquid-filled chamber is connected to the grid of a miniature pentode
placed inside the chamber. This design is quite simple to make and it has

a wide range of applicability within the limits of the strength of the dose from 2 mR/h to 200 mR/h. Compared with a similar apparatus filled with gas under pressure, the logarithmic dosimeter shows a higher sensitivity while the dimensions remain the same. The necessity of applying a relatively high voltage to the chamber is, however, a disadvantage. The main object of this research is to apply these chambers to the dosimetry

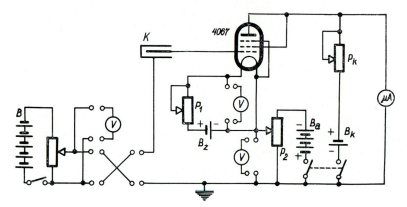

Fig. 17.20. Diagram of a dosimeter with a logarithmic electrometer.
Ref.: Januszajtis and Kozłowski [284].

of electromagnetic radiation in the routine work of radiology, in x-radiation of the energy of the order of 40 kev to 220 kev and in γ-radiation with an energy of several MeV. In addition, these chambers can be used with accelerators for x-radiation up to an energy of 25 MeV (from betatrons), for neutron radiation up to an energy of 14 MeV and for dosimetric investigations of α and β-particles and neutrons.

17.5. *Other Possibilities of Applying Liquid-filled Ionization Chambers.* It was explained in § 11.2 that when an ionization chamber was shielded with thick layers of lead, highly purified dielectric liquids were strongly influenced by cosmic rays. Using this technique, Białobrzeski and Adamczewski [17] used liquid-filled chambers for their investigation of so-called ionization bursts (*Hoffmannstösse*) induced by cosmic radiation (cf. § 11.2).

Liquid-filled chambers can also be used for research on fast neutrons. In such a case dosimetry is based on the indirect action of neutrons. It is a well-known fact that neutrons passing through a substance cannot cause ionization because they do not carry an electric charge. In most cases the neutrons are detected by investigating the atomic nuclei which have been emitted by the neutrons from the absorbent substance. These nuclei have a high velocity and are thus capable of producing ionization. The theory of elastic collisions shows that a nucleus recoiled by a neutron can generally

21

take from the energy of a neutron an energy equal to:

$$\bar{E} = E\,\frac{2M}{(M+1)^2},\tag{17.9}$$

where M denotes the mass of the recoiled nucleus (in units of the neutron mass) and E—the energy of the neutron.

The maximum energy of such a particle can be:

$$E_{max} = E\,\frac{4M}{(M+1)^2}.\tag{17.10}$$

If hydrogen (for which it should be assumed that $M = 1$) absorbs neutrons, then the values of the energies of the recoiled protons will be as follows:

$$E = \tfrac{1}{2}E \quad \text{and} \quad E_{max} = E.\tag{17.11}$$

Their mean energy is therefore half of the energy of the neutron and their maximum energy may be equal to the total energy of the neutron. If $I(v)\,dv$ is the intensity of the stream of neutrons per cm²/sec in the region of the velocity of the neutrons v and $v+dv$, the ionization current in the chamber, whose effective volume U is filled with a substance whose density is ρ and which has a molecular mass A, is:

$$i = \frac{NUQ}{A}\int_{v_{min}}^{v_{max}} I(v)\,\sigma(r)\,dv,\tag{17.12}$$

where N denotes Avogadro's number and $\sigma(r)$—the active cross section of a collision between a neutron and a proton. For neutrons with an energy of 0·4–4 MeV this cross section oscillates between 0·5 and 1·5 barns (1 barn = 10^{-24} cm²). The velocity at which a proton has a sufficiently high energy to cause a detectable ionization pulse is denoted by the lower boundary of an integral and the maximum velocity at which the ionization pulse of a proton is still detectable is denoted by the upper boundary of the integral. It should be noted that protons whose energy is higher than an energy corresponding to v_{max} can pass through the chamber without losing the amount of energy necessary to produce a detectable ionization pulse. The ionization effects induced by the atomic nuclei (which are heavier than protons) present in the dielectric liquid itself are neglected in formula (17.12), as are the effects of ionization induced by those atomic nuclei emitted from the walls and electrodes of the chamber.

The influence of fast neutrons on the electrical conduction of dielectric liquids was established in the course of experiments by Schütze [199], Gazda (cf. Adamczewski [96]), Blanc *et al.* [159] (cf. §17.3), Schmidt [198] and other [310, 312, 318–320, 330, 339].

In 1966 Jahns and Jacobi [212] discussed the mechanism of ionization and electrical conduction in liquid hexane from the point of view of Onsager's theory of the recombination of ions and Jaffe's theory of columnar ionization. They found the radiation yield G for hexane to be equal to 0·13 for ions which had avoided initial recombination. This value corresponds to an average probability of avoiding recombination of 3% and to the mean initial distance 80 nm between the two ions forming a pair.

On the basis of their calculations Jahns and Jacobi conducted a number of experiments on the possibility of applying hexane ionization chambers

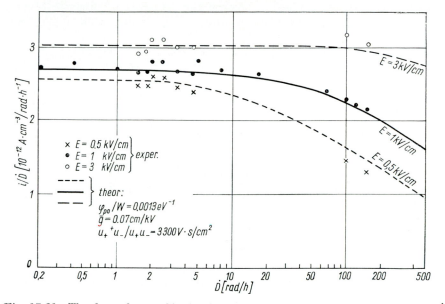

Fig. 17.21. The dependence of ionization current (i) in hexane on the dose rate ($\dot{D}$) per rad/hour (at 20°c). Comparison of experimental and theoretical data. Ref.: Jahns and Jacobi [212].

to dosimetry. They found the following expression for the dependence of the ionization current i on the strength of the dose $\dot{D}$:

$$\frac{i}{i_e} \approx \frac{2 \cdot 5 \times 10^{-12} d\dot{D}}{\sigma_e} \left(\frac{1}{E} + 7 \times 10^{-5} \right), \qquad (17.13)$$

where E is measured in v cm^{-1}, $\dot{D}$—in rads/h, σ_e (natural conductivity)—in Ω^{-1} cm^{-1} and d—in cm. The plots of this dependence for the fields $0 \cdot 5$ kv cm$^{-1} < E < 3$ kv cm^{-1}, given in fig. 17.21 indicate that the dependence $i/\dot{D} = f(\dot{D})$ is constant for doses up to about 100 rads/h when $E = 3$ kv/cm. The region over which $i/\dot{D}$ is constant decreases with decreasing strength of the electric field E.

Figure 17.22 shows plots of the dependence of the ionization current on the strength of the dose for a stream of neutrons in hexane and in tissue.

Particular attention should be given to the possibility of applying liquid-filled ionization chambers to the investigation of heavy particles (protons, deuterons and α-particles) accelerated to high energies of the order of 10^8–10^{10} ev by modern accelerators. Research on these fast particles artificially produced in large numbers has recently been carried out in big laboratories and is of great importance in the physics of the atomic nucleus. Since particles of such a high energy have a range of 100 km in air, investigations using a detector with a stopping power a hundred times greater than air under normal pressure should significantly increase the accuracy of the sensitivity of measuring techniques.

In 1955 Adamczewski [96] pointed out the possibility of applying ionization chambers to investigations of the processes of photodisintegration in the nuclei of carbon atoms induced by γ-radiation of 17·6 MeV

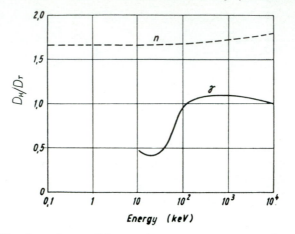

Fig. 17.22. The dependence of dose ratio on quantum energy for a neutron flux in hexane and in tissue. Ref.: Jahns and Jacobi [212].

energy from p-Li reaction. Such investigations conducted by other scientists have so far revealed that this kind of radiation results in the disintegration of the carbon nuclei into three particles, whose total energy is about 10·2 MeV. Such experiments were mainly carried out by means of the photographic emulsion technique. Despite the advantages of this technique, disturbances are caused by the photodisintegration of other nuclei (mainly ^{16}O) and another disadvantage is the difficulty in precisely determining the total energy of the α-particles ejected in this photo-reaction with carbon. The application of a liquid ionization chamber filled with a saturated hydrocarbon (e.g. hexane) in these investigations should separate the effect of the photodisintegration of carbon nuclei in the dielectric (which contains only carbon and hydrogen atoms) from other processes of this type.

Adamczewski calculated that when the energy of γ-rays (produced in the reaction of protons, whose energy is about 441 keV and intensity about 40 μA, with lithium) is 17·6 MeV and when the average intensity of the beam is about 2×10^6 quanta/sec, two to three disintegrations of carbon nuclei should occur per second in cm^3 of hexane inside a solid angle unit of the beam. Each such process should be accompanied by an ionization pulse corresponding to the formation of about $0·3 \times 10^6$ ion pairs. The measuring apparatus should register this phenomenon without any difficulty.

It should be stated in conclusion that despite great technological difficulties in obtaining highly purified dielectric liquids and maintaining them in this state for the long time required for measurements, the ionization chamber filled with a dielectric liquid can play an important rôle in very difficult and delicate problems of measurement in contemporary physics and technology. Although such a chamber cannot provide full saturation of the ionization current it makes it possible to determine precisely the value of the ionization current at a given strength of the electric field and at a given dose rate. It is also possible to evaluate the number of heavy ions produced in a liquid in a state of equilibrium in the region of the apparent saturation of the current. Therefore, its response satisfies the conditions which exist in ionized liquids (or tissues) after the effects of preferential recombination cease to be of importance. For the purposes of determining doses in a given region, a liquid-filled ionization chamber is as accurate as dosimeters of other types.

References to Part III

[1] ADAMCZEWSKI, I., 1934, *Acta phys. pol.*, **3**, 235; 1935, *Ibid.*, **4**, 427.

[2] PRICE, W. J., 1958, *Detection of Nuclear Radiation* (New York: McGraw-Hill).

[3] PAO, C. S., 1943, *Phys. Rev.*, **64**, 60.

[4] JACHYM, B., 1966, *Acta phys. pol.*, **29**, 21.

[5] ROSSI, B., and STAUB, H., 1949, *Ionization Chambers and Counters* (New York: McGraw-Hill).

[6] CAMPBELL, J., and O'CONNOR, D., 1956, *Basis of Measuring Methods using Radioactive Isotopes* (Warsaw: Polish Scientific Publishers).

[7] AGLINTSEV, K. K., 1965, *Applied Dosimetry* (London: Iliffe) (translated from Russian).

[8] BIERLEYEV, T. I., 1953, *Czech. J. Phys.*, **10**, 59.

[9] BONCH-BRUYEVICH, A. M., 1957, *Application of Electronic Valves in Experimental Physics* (Warsaw: Polish Scientific Publishers) (in Russian).

[10] SCHINTLMEISTER, J., 1942, *Elektronenröhre als physikalische Messgerät* (Vienna: Springer).

[11] DU BRIDGE, L., and BROWN, S., 1934, *Rev. scient. instrum.*, **4**, 532.

[12] GZOWSKI, O., and TERLECKI, J., 1959, *Acta Phys. pol.*, **18**, 191.

[13] THOMAS, D. G. A., and FINCH, H. W., 1959, *Electronic Eng.*, **22**, 395.

[14] JAFFE, G., 1910, *Ann. Phys.*, **32**, 148.

[15] JAFFE, G., 1913, *Ann. Phys.*, **42**, 303.

[16] JAFFE, G., 1929, *Ann. Phys.*, **1**, 977.

[17] BIAŁOBRZESKI, C., and ADAMCZEWSKI, I., 1935, *Bull. Acad. pol. Sci.*, **A120**.

[18] ADAMCZEWSKI, I., 1936, *Nature, Lond.*, **137**, 994.

[19] ROGOZIŃSKI, A., 1941, *Phys. Rev.*, **60**, 148.

[20] CURIE, P., 1903, *Comp. Rend.*, **134**, 420.

[21] PLUMLEY, H. J., 1941, *Phys. Rev.*, **59**, 200.

[22] REISS, K. H., 1936, *Ann. Phys.*, **20**, 325.

[23] NIKURADSE, A., 1934, *Das flüssige Dielektrikum* (Berlin: Springer).

[24] ADAMCZEWSKI, I., and BIAŁOBRZESKI, C., 1935, *Nature, Lond.*, **136**, 109.

[25] HOFFMANN, G., 1922, *Physik. Z.*, **13**, 480; 1922, *Ibid.*, **13**, 1029; 1927, *Ibid.*, **28**, 729.

[26] COMPTON, A. H., WOLLAN, E. O., and BENNETT, R. D., 1934, *Rev. Sci. Inst.*, **5**, 415.

[27] TELEGDI, V. L., and ZÜNTI, W., 1950, *Helv. phys. Acta.*, **23**, 745.

[28] TELEGDI, V. L., and EDER, M., 1952, *Helv. phys. Acta.*, **25**, 55.

[29] STEINKE, E. G., 1933, *Z. Phys.*, **85**, 210.

[30] BEARLI, J., and SULLIVAN, A. M., 1963, *Report CERN*, **63**, 1.

[31] GZOWSKI, O., 1961, Doctorate Thesis, Technical University of Gdańsk, Poland.

[32] GZOWSKI, O., 1962, 1962, *Nature, Lond.*, **194**, 173.

[33] TERLECKI, J., 1961, Doctorate Thesis, Technical University of Gdańsk, Poland.

[34] TERLECKI, J., 1962, *Nature, Lond.*, **194**, 172.

[35] BLANC, D., MATHIEU, J., and BOYER, J., 1961, *Nuovo Cim.*, **9**, 932.

[36] GIBAUD, R., 1958, *J. Phys.*, **19**, 175.

[37] IVANOV, W. I., 1960, *Izwiestija WUZ*, **1**, 115; 1960, *Ibid.*, **2**, 134; 1959, *Atomn. Energ.*, **7**, 73.

[38] STACEY, F. D., 1959, *Aust. J. Phys.*, **12**, 105; *Ibid.*, **11**, 158.

[39] WILLIAMS, R. L., and STACEY, F. D., 1957, *Can. J. Phys.*, **35**, 928.

[40] ADAMCZEWSKI, I., 1936, Doctorate Thesis, University of Warsaw, Poland.

[41] ADAMCZEWSKI, I., 1937, *Ann. Phys.*, **8**, 309.

[42] ADAMCZEWSKI, I., 1937, *Acta phys. pol.*, **6**, 432.

[43] ADAMCZEWSKI, I., 1939, *Acta phys. pol.*, **8**, 31.

[44] ADAMCZEWSKI, I., 1959, *Phys. Verh.*, **4–6**, 34.

[45] ADAMCZEWSKI, I., 1961, *Atompraxis*, **9**, 327.

[46] JACHYM, B., 1963, *Acta phys. pol.*, **24**, 243.

[47] JANUSZAJTIS, A., 1963, *Acta phys. pol.*, **24**, 809.

[48] JANUSZKIEWICZ, E., 1962, Diploma Thesis, Technical University of Gdańsk, Poland.

[49] MORANT, M. J., 1961, *Nature, Lond.*, **187**, 48.

[50] TERLECKI, J., and GZOWSKI, O., 1962, *Acta phys. austr.*, **15**, 337.

[51] SWAN, D. W., 1960, *Proc. phys. Soc.*, **76**, 36.

[52] LE PAGE, R. W., and DU BRIDGE, L. A., 1940, *Phys. Rev.*, **58**, 61.

[53] BIJL, H., 1913, *Verh. dt. phys. Ges.*, **15**, 210.

[54] MEYER, L., and REIF, E., 1958, *Phys. Rev.*, **110**, 279.

[55] TYNDALL, A., and POWELL, C., 1930, *Proc. R. Soc.*, **A129**, 162.

[56] SATO, T., NAGAO, S., and TORIYAMA, Y., 1956, *Br. J. appl. Phys.*, **7**, 297.

[57] LeBLANC, O. H., 1959, *J. chem. Phys.*, **30**, 1443.

[58] WIZEVICH, J., and FRÖHLICH, H., 1934, *Ind. Engng Chem.*, **26**, 269.

[59] DAVIDSON, N., and LARSH, A. E., 1950, *Phys. Rev.*, **77**, 706.

[60] VON HIPPEL, J., 1937, *J. appl. Phys.*, **8**, 815.

[61] DEWHURST, H. A., 1957, *J. phys. Chem.*, **61**, 466; *Ibid.*, **62**, 15; 1956, *Ibid.*, **24**, 1254.

[62] ŚWIĘTOSŁAWSKA-ŚCISŁOWSKA, J., and ADAMCZEWSKI, I., 1937, *Acta phys. pol.*, **6**, 425.

[63] ADAMCZEWSKI, I., 1947, *Bull. Acad. pol. Sci.*, **A145**, 145.

[64] ONSAGER, L., 1934, *J. chem. Phys.*, **2**, 509.

[65] SUTHERLAND, W., 1909, *Phil. Mag.*, **18**, 341.

[66] LLEWELLYN-JONES, F., 1957, *Ionization and Breakdown in Gases* (London: Methuen).

[67] BROWN, S. C., 1959, *Basic Data of Plasma Physics* (New York: John Wiley).

[68] ADAMCZEWSKI, I., 1939, *Acta phys. pol.*, **8**, 45.

[69] GAZDA, E., 1964, *Acta phys. pol.*, **25**, 17.

[70] LANGEVIN, P., 1903, *Ann. Phys.*, **28**, 243.

[71] FISHMAN, E., 1955, *Twin. Journ.*, **59**, 469.

[72] UMIŃSKI, T., 1964, Doctorate Thesis, Technical University of Gdańsk, Poland.

[73] DOUGLASS, D. C., and McCALL, D. W., 1958, *J. chem. Phys.*, **62**, 1102.

[74] HAHN, K. L., 1950, *Phys. Rev.*, **80**, 580.

[75] NICHOLAS, J. F., 1959, *J. chem. Phys.*, **31**, 922.

[76] MACHE, H., 1932, *Phys. Z.*, **33**, 43; 1934, *Ibid.*, **35**, 296.

[77] HORNBECK, J. A., and WANNIER, G. H., 1951, *Phys. Rev.*, **82**, 458.

[78] FROST, L. S., 1959, *Phys. Rev.*, **105**, 354.

[79] CROWE, R. W., 1956, *J. appl. Phys.*, **27**, 156.

[80] LEA, D. E., 1934, *Proc. Camb. Phil. Soc.*, **30**, 80.

[81] EYRING, H., 1936, *J. chem. Phys.*, **4**, 283.

[82] ZANSTRA, H., 1935, *Physica*, **2**, 817.

[83] RICHARDSON, E. W. T., 1953, *Proc. phys. Soc.*, **A66**, 403.

[84] SCHUMANN, W. O., 1932, *Z. Phys.*, **76**, 707.

[85] MOTT, N. F., and GURNEY, R. W., 1957, *Electronic Processes in Ionic Crystals* (London: Oxford University Press).

[86] KITTEL, C., 1960, *Introduction to Solid State Physics Theory* (New York: John Wiley).

[87] SHOCKLEY, W., 1950, *Electrons and Holes in Semiconductors* (New York: Van Nostrand).

[88] BROWN, S. C., and ROSE, D. J., 1952, *J. appl. Phys.*, **23**, 711, 719, 1028.

[89] SIMPSON, J. H., 1950, *Proc. phys. Soc.*, **A63**, 86.

[90] DORNTE, R. W., 1940, *Ind. Engng Chem.*, **32**, 1529.

[91] SMOLUCHOWSKI, R., 1959, *Radiation Research*, **1**, 26.

[92] PRZYBYLAK, I., 1962, Diploma Thesis, Technical University of Gdańsk, Poland.

[93] JAEGER, T., 1965, *Principles of Radiation Shielding Protection Engineering* (New York: McGraw-Hill) (translated from German).

[94] ADAMCZEWSKI, I., 1961, *Selected Topics in Radiation Dosimetry*, I.A.E.A., 191.

[95] ADAMCZEWSKI, I., 1961, *A Report from Moscow Conference*: "Dosimetry and Application of Radioisotopes".

[96] ADAMCZEWSKI, I., 1955, *Zeszyty Nauk. P. G. Chem.*, **1**, 25.

[97] UMIŃSKI, T., and DERA, J., 1963, Conference on Electronic Processes in Dielectric Liquids, Durham.

[98] VON ENGEL, A., 1956, *Handbuch Phys.*, **11**, 504.

[99] HASSE, H. R., 1926, *Phil. Mag.*, **1**, 39.

[100] ADAMCZEWSKI, I., 1958, *Postępy Fiz.*, **9**, 49, 261.

[101] FRENKEL, J., 1947, *Kinetic Theory of Liquids* (Oxford: Clarendon Press).

[102] BATES, D. R., and KINGSTON, A. E., 1961, *Nature, Lond.*, **189**, 652.

[103] DAY, F. H., 1951, *X-ray Calibration of Radiation Survey Meters, Pocket Chambers and Dosimeters*, Circ. Nat. Bur. Stand., No. 507.

[104] FOWLER, J. F., 1956, *Proc. R. Soc.*, **236**, 464.

[105] FRÖHLICH, H., 1947, *Proc. R. Soc.*, **A188**, 521; 1958, *Theory of Dielectrics* (Oxford: Clarendon Press).

[106] LANDOLD, H., and BÖRNSTEIN, R., 1956, *Physikalisch-chemische Tabellen* (Berlin: Springer).

[107] MEYER, L., 1961, *Proc. 7th Int. Conf.* (University of Toronto Press).

[108] CARERI, G., 1961, *Physica*, **26**, 89.

[109] CARERI, G., FASOLI, U., and GAETA, F. G., 1960, *Nuovo Cim.*, **10**, 15, 774.

[110] CARERI, G., and GAETA, F. G., 1961, *Nuovo Cim.*, **10**, 20, 152.

[111] CARERI, G., SCARAMUZZI, F., and THOMPSON, J. O., 1959, *Nuovo Cim.*, **10**, 13, 186.

[112] CHONG, P., and INUISHI, Y., 1960, *Techn. Reports Osaka Univ.*, **10**, 545.

[113] CHONG, P., KAWARABAYASHI, T., and INUISHI, Y., 1960, *Techn. Reports Osaka Univ.*, **10**, 25.

[114] CHONG, P., YMANAKA, C., and SUITA, T., 1959, *Techn. Reports Osaka Univ.*, **9**, 17.

[115] ATKINS, K. R., 1959, *Phys. Rev.*, **116**, 1339.

[116] ABE, R., and AIZU, K., 1961, *Phys. Rev.*, **123**, 10.

[117] MEYER, L., DAVIS, H. T., RICE, S. A., and DONNELLY, R. J., 1962, *Phys. Rev.*, **126**, 1927.

[118] DAVIS, H. T., RICE, S. A., and MEYER, L., 1962, *J. chem. Phys.*, **39**, 947.
[119] FORSTER, E. O., 1962, *J. chem. Phys.*, **37**, 1021.
[120] RICE, S. A., and ALLNATT, A. R., 1961, *J. chem. Phys.*, **34**, 2144.
[121] KAHAN, E., 1964, Doctorate Thesis, University of Durham.
[122] NOWAK, W., 1963, Doctorate Thesis, Technical University of Szczecin, Poland.
[123] ZINOWIEWA, K. N., 1958, *Zh. eksp. teor. Fiz.*, **34**, 609.
[124] CARR, H. Y., and PURCELL, E. M., 1954, *Phys. Rev.*, **94**, 630.
[125] GAZDA, E., 1963, *Acta phys. pol.*, **24**, 209.
[126] GAZDA, E., 1963, *Nature, Lond.*, **200**, 767.
[127] NERNST, W., 1904, *Ann. Phys. Lpz.* (*Boltzmann Festschrift*), p. 504.
[128] ADAMCZEWSKI, I., 1963, Conference on Electronic Processes in Dielectric Liquids, Durham.
[129] FOWLER, J. F., and FARMER, F. T., 1955, *Nature, Lond.*, **175**, 516, 590, 648.
[130] CHYBICKI, M., and GZOWSKI, O., 1966, *Z. Phys. Chem.*, **233**, 117.
[131] GAETA, F. G., 1962, *Nuovo Cim.*, **26**, 1173.
[132] BOAG, J. W., 1950, *Br. J. Radiol.*, **23**, 601; 1952, *Ibid.*, **25**, 649.
[133] BOAG, J. W., and WILSON, T., 1952, *Br. J. appl. Phys.*, **3**, 222.
[134] LOEB, L. B., 1956, *Handbuch Phys.*, **11**, 471.
[135] LEOB, L. B., 1955, *Basic Processes of Gaseous Electronics* (Berkeley: University of California Press).
[136] MOULINIER, P., 1941, *Comp. Rend.*, **213**, 802.
[137] NICHOLAS, J. F., 1959, *J. chem. Phys.*, **31**, 922.
[138] SCISŁOWSKI, W., 1935, *Acta phys. pol.*, **4**, 123; 1937, *Ibid.*, **6**, 403; 1939, *Ibid.*, **7**, 127; 1939, *Ibid.*, **7**, 214.
[139] JOFFE, A. F., 1960, *Physics of Semiconductors* (New York: Academic Press).
[140] ROSIŃSKI, K., 1954, *Postępy Fiz.*, **5**, 305; 1955, *Ibid.*, **6**, 66.
[141] PIGOŃ, K., 1956, *Wiadomości Chemiczne*, **9**, 454.
[142] HANNAY, N. B., 1959, *Semiconductors* (New York: Reinhold Publishing Corporation).
[143] SOSNOWSKI, L., 1956, *Postępy Fiz.*, **7**, 107.
[144] BASSET, J. (cited by 'Lambda' in *Hydrodynamics*, 1945).
[145] ROSSI, C., and BIANCHI, E., 1961, *Nature, Lond.*, **189**, 822.
[146] SWAN, D. W., 1961, *Nature, Lond.*, **190**, 904.
[147] FORSTER, E. O., 1963, Conference on Electronic Processes in Dielectric Liquids, Durham.
[148] FORSTER, E. O., 1964, *J. chem. Phys.*, **40**, 86.
[149] FORSTER, E. O., 1964, *J. chem. Phys.*, **40**, 91.
[150] RICE, S. A., and CHOI, L., 1962, *Phys. Rev.*, **8**, 410.
[151] DAVIS, H. T., STUART, S. A., and MEYER, L., 1962, *J. chem. Phys.*, **37**, 1521.
[152] CREMER, E., 1955, *Advances in Catalysis*, **7**, 75.
[153] RUETSCHI, P., 1951, *Z. Phys. Chem.*, **14**, 277.
[154] MORANT, M. J., 1963, Conference on Electronic Processes in Dielectric Liquids, Durham.
[155] SLETTEN, A. M., and LEWIS, T. J., 1963, *Br. J. appl. Phys.*, **14**, 883.
[156] HEYLEN, A. E. D., and LEWIS, T. J., 1962, *Proc. Phys. Soc.*, **80**, 422.
[157] CHONG, P., SUGIMOTO, T., and INUISHI, Y., 1960, *J. phys. Soc. Japan*, **15**, 1137.
[158] ATTIX, F. H., 1961, *NRL Rept.* 5646. U.S. Naval Res. Lab., Washington, D.C.
[159] BLANC, D., MATHIEU, J., and VERMANDE, P., 1963, *Nuclr Instrum. Meth.*, **18**, 777.
[160] JANUSZAJTIS, A., and ADAMCZEWSKI, I., 1963, Conference on Electronic Processes in Dielectric Liquids, Durham.
[161] KORPALSKA, I., and ADAMCZEWSKI, I., 1963, Conference on Electronic Processes in Dielectric Liquids, Durham.
[162] ADAMCZEWSKI, I., 1965, *Br. J. appl. Phys.*, **16**, 759.

[163] ADAMCZEWSKI, I., and JACHYM, B., 1966, *Acta phys. pol.*, **30**, 767.

[164] ROESCH, W. C., and DONALDSON, E. E., 1956, *Proc. of the International Conference on the Peaceful Uses of Atomic Energy*, **14**, 172.

[165] NAGHIZADEH, J., and RICE, S. A., 1962, *J. chem. Phys.*, **36**, 2710.

[166] VERMANDE, P., 1963, Doctorate Thesis, University of Toulouse, France.

[167] BLANC, D., MATHIEU, J., and VERMANDE, P., 1962, *Nucl. Electronics*, **285**, 1.

[168] BLANC, D., MATHIEU, J., and BOYER, J., 1961, *Nuovo Cim.*, **19**, 929.

[169] COELHO, R., 1962, *Bull. Soc. Fr. Electr.*, **25**, 13.

[170] BOYER, J., 1963, Doctorate Thesis, University of Toulose, France.

[171] MATHIEU, J., 1963, Doctorate Thesis, University of Toulouse, France.

[172] SUGITA, K., 1960, *Br. J. appl. Phys.*, **2**, 536.

[173] GUIZONNIER, R., 1954, *Rev. gén. Élect.*, **8**, 489; 1958, *Ibid.*, **67**, 637.

[174] GUIZONNIER, R., 1956, *J. Phys. Rad.*, **17**, 121A.

[175] GUIZONNIER, R., 1961, *J. electrochem. Soc.*, **6**, 108.

[176] GUIZONNIER, R., 1961, *Comp. Rend.*, **253**, 807.

[177] GUIZONNIER, R., 1961, *J. electrochem. Soc.*, **108**, 519.

[178] GUIZONNIER, R., and THOMAS, C., 1959, *J. Phys. Rad.*, **20**, 153A.

[179] GIBAUD, R., 1962, Doctorate Thesis, University of Bordeaux, France.

[180] MORANT, M. J., 1963, *Br. J. appl. Phys.*, **14**, 469.

[181] WATSON, P. K., 1963, *Nature, Lond.*, **199**, 646; 1965 (October) Conference on Electrical Insulation, Pennsylvania.

[182] ADAMCZEWSKI, I., 1964, *Sur le Mécanisme de l'Ionisation et de la Conductibilité électrique dans les Liquides diélectriques* (Paris: Acad. Pol. Sci.).

[183] JANUSZAJTIS, A., 1964, Doctorate Thesis, Technical University of Gdańsk, Poland.

[184] KRAMERS, H. C., 1952, *Physica*, **18**, 10.

[185] GUIZONNIER, R., and AGUIRRE, P., 1962, *Comp. Rend.*, **255**, 294.

[186] GUIZONNIER, R., and MOUCLIER, M., 1962, *J. Phys. Rad.*, **23**, 155A.

[187] GUIZONNIER, R., and BESNARD, M., 1959, *J. Phys. Rad.*, **249**, 684.

[188] VERMEIL, C., MATHESON, M., LEACH, S., and MULLER, F., 1964, *J. Chim. phys.*, **61**, 596.

[189] STAUDHAMMER, P., and SEYER, W. F., 1957, *J. appl. Phys.*, **28**, 405.

[190] JACHYM, B., 1964, Doctorate Thesis, Technical University of Gdańsk, Poland.

[191] JACHYM, B., 1963, *Acta phys. pol.*, **24**, 785.

[192] JACHYM, B., 1964, *Acta phys. pol.*, **25**, 385.

[193] JACHYM, B., 1964, *Acta phys. pol.*, **26**, 1061.

[194] GAZDA, E., 1963, Doctorate Thesis, University of Toruń, Poland.

[195] CRAGGS, J. D., and TOZER, B. A., 1958, *Proc. R. Soc.*, **247**, 337.

[196] FREEMAN, G. R., 1963, *J. chem. Phys.*, **39**, 988; 1964, *Ibid.*, **40**, 907.

[197] ZAKY, A. A., and HOUSE, H., 1963, *J. appl. Phys.*, **34**, 3194.

[198] SCHMIDT, W. F., 1964, *Atompraxis*, **3**, 157.

[199] SCHÜTZE, W., 1955, Patentschrift 924226, Klasse 219, Gruppe 1801, Bund. Rep. Deutch.

[200] GZOWSKI, O., 1962, *Z. phys. Chem.*, **221**, 288.

[201] TERLECKI, J., and KASZTELAN, S., 1965, *Zesz. Nauk. W. S. P. Mat. Fiz.*, **5**, 17.

[202] MAŁECKI, J., 1964, *Worns of Kom. Mat. Przyr. Poz. Tow. Przyj. Nauk (Dielectric Physics)*, **11**, 113, 125 .

[203] GRAY, E., and LEWIS, T. J., 1965, *Br. J. appl. Phys.*, **16**, 1049.

[204] HUMMEL, A., ALLEN, A. O., and WATSON, F. H., 1966, *J. chem. Phys.*, **44**, 3431.

[205] HUMMEL, A., and ALLEN, A. O., 1966, *J. chem. Phys.*, **44**, 3426.

[206] WATSON, P. K., and CLANCY, T. M., 1965, *Rev. Scient. Instrum.*, **36**, 217.

[207] ULLMAIER, H. A., 1966, *Phys. Med. Biol.*, **11**, 95.

[208] BLANC, D., MATHIEU, J., and PATAU, J. P., 1966, *Health Physics*, **12**, 1589.

[209] MATHIEU, J., PATAU, J. P., and BLANC, D., 1966, *J. Physique*, **27**, 495.

330   *III. Electrical Conduction in Dielectric Liquids at Low Fields*

[210] MATHIEU, J., BLANC, D., CARTAILLAC, D., and TORRES, L., 1966, *J. Phys.*, **27**, 246.
[211] BLANC, D., MATHIEU, J., TORRES, L., and VERMANDE, P., 1965, *L'Onde electrique*, **457**, 1; 1965, *Health Physics*, **11**, 63.
[212] JAHNS, A., and JACOBI, W., 1966, *Z. Naturf.*, **21**, 1400.
[213] ZAJĄC, H., 1967, Doctorate Thesis, Technical University of Gdańsk, Poland.
[214] BÄSSLER, H., 1964, *Phys. kondens. Matérie.*, **2**, 187.
[215] BÄSSLER, H., and RIEHL, N., 1965, *Z. Naturf.*, **20a**, 227, 587.
[216] VOGEL, H., and BÄSSLER, H., 1964, *Z. Naturf.*, **19a**, 1070.
[217] BÄSSLER, H., MAYER, P., and RIEHL, N., 1965, *Z. Naturf.*, **20a**, 394.
[218] BÄSSLER, H., 1966, *Z. Naturf.*, **21a**, 447, 454.
[219] BÄSSLER, H., and DRIES, B., 1966, *Z. Naturf.*, **21a**, 441.
[220] BÄSSLER, H., 1964, *Z. Naturf.*, **19a**, 1389.
[221] HEYLEN, A. E. D., 1958, *J. chem. Phys.*, **29**, 813.
[222] HEYLEN, A. E. D., 1957, *Nature, Lond.*, **180**, 703.
[223] HEYLEN, A. E. D., and LEWIS, T. J., 1956, *Br. J. appl. Phys.*, **7**, 411.
[224] VON ENGEL, A., 1955, *Ionized Gases* (Oxford: Clarendon Press).
[225] SECKER, P. E., 1965, *Br. J. appl. Phys.*, **16**, 1527.
[226] COE, G., HUGHES, J. F., and SECKER, P. E., 1966, *Br. J. appl. Phys.*, **17**, 885.
[227] SECKER, P. E., JESSUP, M., and HUGHES, J. F., 1961, *J. scient. Instrum.*, **43**, 515.
[228] SECKER, P. E., and LEWIS, T. J., 1965, *Br. J. appl. Phys.*, **16**, 1649.
[229] KUFFEL, E., and RADWAN, R. O., 1956, *Proc. I.E.E.*, **113**, 1863.
[230] ADAMCZEWSKI, I., and KOZŁOWSKI, K., to be published.
[231] MOLNAR, J. P., 1951, *Phys. Rev.*, **83**, 933.
[232] LAUER, E. J., 1952, *J. appl. Phys.*, **23**, 300.
[233] ALLIS, W. P., 1956, *Handbuch Physik.*, **21**, 383.
[234] BHALLA, M. S., and CRAGGS, J. D., 1960, *Proc. Phys. Soc.*, **76**, 369.
[235] KAVADA, A., and JARNAGIN, R. C., 1966, *J. chem. Phys.*, **44**, 1919.
[236] KALLMANN, H. P., KRASNANSKY, Y. J., and ORENSTEIN, A., 1964, *J. chem. Phys.*, **40**, 3740; 1965, *Ibid.*, **43**, 1931.
[237] PITTS, E., TERRY, G. C., and WILLETS, F. W., 1966, *Trans. Faraday Soc.*, **62**, 2858.
[238] KALINOWSKI, J., 1967, *Acta phys. pol.*, **31**, 3.
[239] JACHYM, A., 1967, *Zesz. Nauk. P. G. Fiz.*, **1**, 51.
[240] FREEMAN, G. R., 1964, *J. chem. Phys.*, **40**, 907.
[241] FREEMAN, G. R., 1963, *J. chem. Phys.*, **38**, 1022.
[242] FREEMAN, G. R., 1963, *J. chem. Phys.*, **39**, 3537.
[243] RICE, S. A., and JORTNER, J., 1965, The Theory of Ionic and Electronic Mobility in Liquids, *Progress in Dielectrics*, Vol. 6, 183.
[244] EDELES, H., and CRAGGS, J. D., 1963, *The Coefficients of Thermal and Electrical Conductivity in High Temperature Gases*, Vol. 5, 187.
[245] SILVER, M., 1965, *J. chem. Phys.*, **42**, 1011.
[246] PAPOULAR, R., 1965, *Electrical Phenomena in Gases* (London: Iliffe).
[247] CARERI, G., DUPRE, F., and MAZZOLDI, P., 1966, *Quantum Fluids*, p. 305.
[248] REIF, F., and MEYER, L., 1960, *Phys. Rev.*, **19**, 1164.
[249] MAZZOLDI, P., 1966, *Supplemento al Nuovo Cimento*, **4**, 1.
[250] CUNSOLO, S., 1961, *Nuovo Cim.*, **21**, 76.
[251] CARERI, G., CUNSOLO, S., and MAZZOLDI, P., 1964, *Phys. Rev.*, **136**, 303.
[252] DI CASTRO, C., 1966, *Nuovo Cim.*, **42**, 251.
[253] CARERI, G., CUNSOLO, S., and MAZZOLDI, P., 1964, *Phys. Rev.*, **136**, A303.
[254] CARERI, G., CUNSOLO, S., MAZZOLDI, P., and SANTINI, M., 1965, *Phys. Rev. Lett.*, **15**, P564, 1–5.
[255] SCHNYDERS, H., RICE, S. A., and MEYER, L., 1965, *Phys. Rev.*, **15**, N512, 1–4.

[256] SCHNYDERS, H., and RICE, S. A., 1965, *Phys. Rev. Lett.*, **15**, 187.

[257] SWAN, D. W., 1965, *Proc. Phys. Soc.*, **85**, 1297.

[258] BRUSCHI, L., MAZZOLDI, P., and SANTINI, M., 1966, *Phys. Rev. Lett.*, **17**, M542, 1–3.

[259] BRUSCHI, L., MARAVIGLIA, B., and MAZZOLDI, P., 1966, *Phys. Rev.*, **143**, 84.

[260] MEYER, L., and REIF, F., 1960, *Phys. Rev. Lett.*, **5**, 1.

[261] DAVIES, D. B., and MATHESON, A. J., 1966, *J. Chem. Phys.*, **45**, 1000.

[262] SWAN, D. W., 1963, *Proc. phys. Soc.*, **82**, 74.

[263] SWAN, D. W., 1964, *Proc. phys. Soc.*, **83**, 659.

[264] ZESSOULES, N., BRINKERHOFF, J., and THOMAS, A., 1963, *J. appl. Phys.*, **34**, 2010.

[265] PICKARD, W. F., *Nature, Lond.*, **201**, 283.

[266] ADAMCZEWSKI, I., and JACHYM, B., 1968, *Acta phys. pol.*, **34**. (in print)

[267] TERLECKI, J., 1966, *Acta phys. pol.*, **29**, 743.

[268] CHYBICKI, M., 1967, Doctorate Thesis, University of Toruń, Poland.

[269] JACHYM, A., and JACHYM, B., 1967, *Acta phys. pol.*, **31**, 733.

[270] ADAMCZEWSKI, I., 1968, to be published.

[271] JACHYM, A., and JACHYM, B., 1968, *Acta phys. pol.*, **31** (in print).

[272] DRUYVESTEYN, M. J., 1936, *Physica*, **3**, 65.

[273] STARODUBCEW, C. B., and TURIEKII, M., 1964, *Mon. A. N. Uz. S.S.R. Fiz. Mat.*, **6**, 83.

[274] STARODUBCEW, C. B., and GENERAŁOWA, W. W., 1963, *Uz. S.S.R. Panlar. A. N. Fiz. Mat.*, **1**, 46.

[275] BEVERLY, W. B., and SPIEGEL, V., JR., 1964, *J. Res. natn. Bur. Stand.*, **A68**, 1.

[276] MICKIEWICZ, M. K., and PROTOPOPOW, O. O., 1965, *Ukr. Fiz. Ż.*, **10**, 342.

[277] PROTOPOPOW, O. O., 1963, *Ukr. Fiz. Ż.*, **8**, 1039.

[278] POPOW, W., 1962, *Dokl. Bułg. Ak. Nauk.*, **15**, 821.

[279] COMPTON, D. M. J., CHENEY, G. T., and POLL, R. A., 1965, *J. appl. Phys.*, **36**, 2434.

[280] O'DWYER, J. J., 1966, *J. appl. Phys.*, **37**, 599.

[281] JACHYM, B., 1966, *Acta phys. pol.*, **30**, 1049.

[282] BLANC, D., MATHIEU, J., PATAU, J. P., FRANÇOIS, H., and SAUDAIN, G., 1966, *Health Physics*, **12**, 1589.

[283] SCHMIDT, W. F., 1964, *Atompraxis*, **3**, 157.

[284] JANUSZAJTIS, A., and KOZŁOWSKI, K., 1967, *Zesz. Nauk. P. G.*, **105**, 131.

[285] AUKERMANN, L. W., 1954, *Phys. Rev.*, **93**, 929.

[286] KALINOWSKI, J., 1967, *Zesz. Nauk. P. G. Fiz.*, **105**, 69.

[287] YAHAGI, K., KAO, K. C., and CALDERWOOD, J. H., 1966, *J. appl. Phys.*, **37**, 4289.

[288] KAO, K. C., PHAHLE, A. M., and CALDERWOOD, J. H., 1965, *Proc. Sixth Elect. Insul. Conf., New York*, p. 242.

[289] THEOBALD, J. K., 1953, *J. appl. Phys.*, **24**, 123.

[290] FELICI, N., 1959, *C. R. Acad. Sci., Paris*, **249**, 1196.

[291] BRIERE, G., *J. Phys. Radium*, **2**, 500.

[292] BRIERE, G., and FELICI, N. J., 1964, *C. R. Acad. Sci., Paris*, **259**, 3237.

[293] FELICI, N. J., 1964, *Br. J. appl. Phys.*, **15**, 801.

[294] FELICI, N., 1960, *C. R. Acad. Sci., Paris*, **250**, 3960.

[295] BARRET, S., BRIERE, G., FELICI, N., and PIERRE, G., 1965, *C. R. Acad. Sci., Paris*, **261**, 4400.

[296] LOWER, S. K., and EL-SAYED, K. M. M. A., 1966, *Chem. Rev.*, **66**, 199.

[297] RÜPPEL, H., and WITT, H. T., 1958, *Z. phys. Chem.*, **15**, 321.

[298] CROWE, R. W., 1966, *J. appl. Phys.*, **37**, 1515.

[299] PRASSAD, A. N., and CRAGGS, J. D., 1960, *Proc. phys. Soc.*, **76**, 223; 1961, *Ibid.*, **77**, 385.

[300] LADU, M., PELICCIONI, M., and ROCCELLA, M., 1965, *Nucl. Instrum. Meth.*, **34**, 178.

[301] LADU, M., PELICCIONI, M., PICIOTTI, E., and RAMACCIOTTI, D., 1965, *Rev. Scient. Instrum.*, **36**, 1241.

[302] LADU, M., and PELLICCIONI, M., 1966, *Nucl. Instrum. Meth.*, **39**, 339.

[303] LADU, M., PELLICCIONI, M., and ROCCELLA, M., 1965, *Nucl. Instrum. Meth.*, **37**, 318.

[304] LADU, M., and PELLICCIONI, M., 1967, *Nucl. Instrum. Meth.*, to be published,

[305] LADU, M., PELLICCIONI, M., and ROCELLA, M., 1967, *Nucl. Instrum. Meth.*, **3040**, 1.

[306] LADU, M., PELLICCIONI, M., and ROCCELLA, M., 1967, *Neutron Monitoring*, p. 145 (Vienna: I.A.E.A.).

[307] TERLECKI, J., 1968, *Z. phys. Chem.*, **237**, 314.

[308] MAGEE, J. L., FUNABASHI, H., and MOZUMBER, A., 1964, A.E.C. Document No. C00-38-378.

[309] GAZDA, E., 1965, *Acta phys. pol.*, **27**, 881.

[310] ZIELCZYŃSKI, M., 1966, Conference in Vienna, I.A.E.A.

[311] BIRKS, J., 1963, *J. Phys. Chem.*, **67**, 2199.

[312] PSZONA, S., ZIELCZYŃSKI, M., and ŻARNOWIECKI, K., 1966, *I.R.P.A.*, *Congress Rome*, No. 110 (in press).

[313] CRAGGS, J. D., and TOZER, B. A., 1960, *Proc. R. Soc.*, **254**, 229.

[314] ATTIX, F. H., and ROESCH, W. C., 1966, *Radiation Dosimetry* (New York, London: Academic Press).

[315] ULLMAIER, H., 1964, *Z. für. Phys.*, **178**, 44.

[316] EBERT, H. G., and KOEPP, R., 1964, *Z. phys. Chem.*, **43**, 5.

[317] LENKEIT, S., and EBERT, H. G., 1964, *Naturwissenschaften*, **10**, 5237.

[318] MATHIEU, J., BLANC, D., CAMENADE, P., and PATAU, J. P., *J. Chim. Phys.* (in print).

[319] BLANC, D., MATHIEU, J., PATAU, J. P., FRANÇOIS, H., and SOUDAIN, G., 1966, Conference in Vienna, I.A.E.A.

[320] BLANC, D., MATHIEU, J., DELARD, R., FRANÇOIS, H., and SOUDAIN, G., 1967, Health Physics Society Annual Meeting, Washington, D.C.

[321] CHARALAMBUS, S., 1967, *Nucl. Instrum. Meth.*, **48**, 181.

[322] BLANC, D., 1966, *Les Radioéléments* (Paris: Masson).

[323] BHALLA, M. S., and CRAGGS, J. D., 1961, *Proc. Phys. Soc.*, **78**, 438.

[324] SCHMIDT, C. W., KAO, K. C., CALDERWOOD, J. H., and McGEE, J. D., 1966, *Nature, Lond.*, **210**, 192.

[325] LLEWELLYN JONES, F., 1957, *Ionization and Breakdown in Gases* (London: Methuen).

[326] KUFFEL, E., and RADWAN, R. O., *Proc. I.E.E.*, 1966, **113**, 1863.

[327] HEYLEN, A. E. D., and LEWIS, T. J., 1963, *Proc. R. Soc.*, **271**, 531.

[328] LEWIS, T. J., 1965, *Br. J. appl. Phys.*, **16**, 1049.

[329] BLANC, D., MATHIEU, J., PATAU, J. P., FRANÇOIS, H., and SOUDAIN, G., 1966, Conference in Regensburg.

[330] TERLECKI, J., 1968, *Problems of Application of Liquid-filled Ionization Chambers to Dosimetry of Photon and Neutron Radiation* (Gdańsk: G.T.N.).

[331] ADAMCZEWSKI, I., 1968, *Les Phénomènes d'Ionisation et de Conduction dans les Diélectriques Liquides* (Paris: Masson et Cie).

[332] ADAMCZEWSKI, I., 1968 (March), Conference in Toulouse: *Journées d'Électronique*.

[333] BLANC, D., MATHIEU, J., PATAU, J. P., FRANÇOIS, H., and SOUDAIN, G., 1968, Conference in Toulouse: *Journées d'Électronique*.

[334] IVANOV, W., 1968, Conference in Toulouse: *Journées d' Électronique*.

[335] ADAMCZEWSKI, I., 1968 (September), Conference in Grenoble: *Phénomènes de Conduction dans les Liquides isolants* (in print).

[336] JANUSZAJTIS, A., 1968, Conference in Grenoble (in print).

[337] CHYBICKI, M., 1968, Conference in Grenoble (in print).

[338] MATHIEU, J., BLANC, D., PATAU, J. P., 1968, Conference in Grenoble (in print).

[339] TERLECKI, J., 1968, Conference in Grenoble (in print).

[340] BRIERE G.. CAUQUIS, G., and ROSE, B., 1968, Conference in Grenoble (in print).

[341] BUTTLE, A., and BRIGNELL, J. E., 1968, Conference in Grenoble (in print).

[342] FORSTER, E. O., and LANGER, A. W., 1968, Conference in Grenoble (in print).

[343] GUIZONNIER, R., 1968, Conference in Grenoble (in print).

[344] HUYSKENS, H., 1968, Conference in Grenoble (in print).

[345] JACHYM, A., and JACHYM, B., 1968, Conference in Grenoble (in print).

[346] ATTEN, P., and GOSSE, J. P., 1968, Conference in Grenoble (in print).

[347] KALINOWSKI, J., 1968, Conference in Grenoble (in print).

[348] KAHAN, E., and MORANT, M. J., 1965, *Br. J. appl. Phys.*, **16**, 943.

[349] COHEN, M. H., and LEKNER, J., 1967, *Phys. Rev.*, **158**, 305.

[350] MILLER, L. S., and SPEAR, W. E., 1967, *Phys. Letters*, **24A**, 47.

[351] LEKNER, J., 1967, *Phys. Rev.*, **158**, 130.

[352] MILLER, L. S., HOWE, S., and SPEAR, W. E., 1968, *Phys. Ber.*, **166**, 871.

[353] GHOSH, P. K., and SPEAR, W. E., 1968, *J. Phys. C. (Proc. Phys. Soc.)*, **1**, 1347.

[354] HUCK, J., 1968, *Colloques Intern.* (Grenoble: C.N.R.S.).

[355] HILL, C. E., and HOUSE, H., 1968, *Colloques Intern.* (Grenoble: C.N.R.S.).

[356] SHARBAUGH, A. H., and BARKER, R. E., 1968, *Colloques Intern.* (Grenoble: C.N.R.S.).

[357] ESSEX, V., and SECKER, P. E., 1968, *Brit. J. Appl. Phys.*, **1**, 63.

[358] BLANC, D., MATHIEU, J., and PATAU, J. P., 1967, *Proceedings of the Conference*: "Microdosimetry" (Ispra).

[359] ADAMCZEWSKI, I., 1968, *Colloques Intern.* (Grenoble: C.N.R.S.), II.

[360] ADAMCZEWSKI, I., and JACHYM, B., 1969, *Acta Phys. Polon.*, **34**, 53.

[361] HALPERN, B., LEKNER, J., 1967, *Phys. Rev.*, **156**, 351.

[362] BLOOR, A. S., and MORANT, M. J., 1968, *Colloques Intern.* (Grenoble: C.N.R.S.).

[363] POLLARD, A. F., and HOUSE, H., 1968, *Colloques Intern.* (Grenoble: C.N.R.S.).

[364] FELICI, N. J., 1968, *Colloques Intern.* (Grenoble: C.N.R.S.).

C.N.R.S.: Centre National de la Recherche Scientifique.

PART IV

The Electrical Conductivity of Dielectric Liquids at High Fields and Electrical Breakdown

Chapter 18

Electrical Conductivity at High Fields

18.1. *Introduction*. In the three preceding parts of this book the results of experimental and theoretical work mainly concerning the electrical conductivity of dielectric liquids at low fields have been presented. Part IV contains the collected results of investigations on liquid dielectrics subjected to electric fields from 100 to 2500 kv/cm and includes electrical breakdown. In the last few years these studies have advanced considerably due to modern measuring methods applied in research with high-purity liquids. Particulary noteworthy are the investigations undertaken at Queen Mary College, London, under the direction of Tropper and Lewis, work carried out by Macfadyen in Birmingham, and research conducted by the American group (Crowe, Sharbaugh, Bragg and others). These investigations have been discussed in papers by Lewis [1, 129], Watson [126], Morant [125] and Sharbaugh and Watson [132], on which Part IV of this book is partly based. A monograph by Nikuradse [2] contains results of work carried out before 1932, and work later than 1932 may be found in the monograph by Skanavi ([3], Chapters IX, X and XI) and those by Kok [4] and Balygin [127]. Results from several hundred original papers published before 1966 are also quoted in this book.

The problem of conductivity in dielectric liquids in high fields is so complicated that many processes not yet fully explained are involved in this range of electric fields. Experimental work is here divided into two groups: the study of applications, and physical investigations of a fundamental character. The first group covers usual low-purity liquids with a natural conduction of the order of 10^{-10} to 10^{-13} Ω^{-1} cm^{-1}, the ones used for industrial purposes. Here the most important factor is the degree of liquid contamination, and the scatter in results is between 70% and 50%. Nevertheless, these liquids are of considerable practical use as they can be used in industry without treatment and for a variety of purposes. Liquids for physical investigations must be of the highest purity (10^{-18} to 10^{-19} Ω^{-1} cm^{-1}), the electrodes must be suitably prepared, and the presence and

effect of dissolved gases taken into consideration. There is a high reproducibility of results and scatter can be as low as 5%. Investigations in
which short impulses, of the order of a microsecond, were applied for very
high fields (a million volts per centimetre), are of particular interest and
they have thrown much light on the mechanism of electrical breakdown
in dielectric liquids. Interesting results have also been obtained in studies
of prebreakdown impulses (micro-breakdown) in liquids.

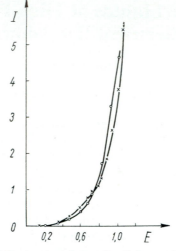

Fig. 18.1. Dependence of the current intensity I (in $A \times 10^{-10}$) on the electric field
strength E (in MV cm^{-1}) in hexane for two different polarities. Ref.: House [11].

18.2. *Current–voltage Characteristics.* Natural conduction in dielectric
liquids in high fields has been investigated by many workers including
Baker and Boltz [5], Nikuradse [2], Le Page and Du Bridge [6], Dornte [7],
Plumley [8], Pao [15], Goodwin and Macfadyen [9], Green [10], and more
recently House [11], Lewis [1, 12–14], Morant [125], Kahan [179], Kao
and Calderwood [153, 160, 199, 200, 205], Watson and Sharbaugh [132,
161] and others. Most works show that above a certain value of the
strength of the field (of the order of 100 to 200 kv/cm), an exponential rise
in the current begins and corresponds to the relationship:

$$I = I_0 \exp(\alpha \delta E), \tag{18.1}$$

where I_0 can be a function of temperature and field strength, the coefficient
is a function of field strength, temperature and cathode work function
and δ is the electrode separation. A detailed theoretical discussion of this
relationship can be found in Chapter 20. The graphs in figs. 18.1, 18.2 and
18.3 are typical examples of an experimental dependence of this kind.
From these curves a sharp increase in current, especially above 500–
600 kv cm^{-1} can be observed. Some authors have found a marked
dependence of current on the electrode material (copper, gold, stainless

steel) and the degree of electrode annealing (see fig. 18.4). Detailed investigations have been carried out in order to explain the three main processes responsible for the conduction of current in this range viz.:

(1) Electron emission from metals under the influence of an electric field. This is a surface phenomenon and depends on electrode material

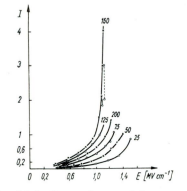

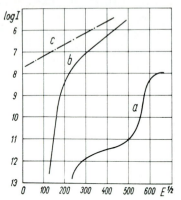

Fig. 18.2. Dependence of the current intensity I (in A × 10^{-9}) on the electric field strength E (in MV cm^{-1}) in hexane at different electrode separations δ (in μm). Ref.: House [11].

Fig. 18.3. The relationship between log I and $E^{1/2}$ in toluene. Platinum and iridium electrodes are coated with (*a*) an oxygen and (*b*) a hydrogen layer; C is the theoretical curve of the Schottky effect. The distance between the electrodes is of the order of 0·07 cm. Ref.: Baker and Boltz [5].

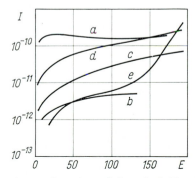

Fig. 18.4. Current I (in A) as a function of electric field strength E (in kv cm^{-1}) in hexane for a number of cathode materials with an electrode separation of 420 μ. *a*, Copper; *b*, gold; *c*, steel; *d*, annealed steel. Ref.: Green [10].

and the method of preparation. The main rôle is played here by the reduction in the magnitude of the work function required for an electron to leave the metal due to the action of an electric field.

(2) The phenomena occurring in the liquid itself which depend on the volume of liquid and hence the electrode separation δ. They include such effects as dependence of conductivity on viscosity, density, temperature

22

and molecular structure, secondary ionization processes in the liquid and the influence of previous electrical breakdown.

(3) Secondary phenomena such as the influence of gases (mainly air and oxygen) adsorbed on the cathode surface and dissolved in the liquid, the effect of inaccurate polishing of electrodes (mainly the cathode), the influence of dust particles, impurities, field distribution and positive ion layer, etc.

(4) The effect of ionizing agents.

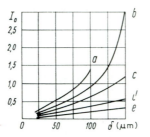

Fig. 18.5. Current I_0 (in A $\times 10^{-9}$) as a function of electrode separation (in μm) (inhexane) for a number of electric field strengths (in MV cm^{-1}): a, 1·2; b, 1·1; c, 1·0; d, 0·8; e, 0·6. Ref.: House [11].

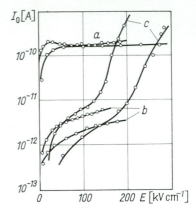

Fig. 18.6. Current I_0 (in A) as a function of the electric field strength E (in kv cm^{-1}) for a number of electrode materials. Each pair of curves is for the same material at two different electrode separations (100 and 420 μ): a, brass cathode; b, gold; c, stainless steel. Ref.: Green [10].

18.3. *Results of Experimental Work.* Results selected from experimental material gathered by various workers in the last ten years illustrating the processes discussed above will now be discussed.

The influence of electron emission from the cathode on current intensity was investigated by evaluating I_0 from eqn. (18.1) from a plot of the relationship $I = f(E, \delta)$. Typical graphs of the function $I_0 = f(\delta)$ can be found in fig. 18.5. These curves illustrate the change in current intensity I_0 with electrode separation and electric field strength.

Green [10] found that the current I_0 is strongly dependent on the cathode surface material; he examined this phenomenon under various experimental conditions and some of his results are presented in fig. 18.6. The relationship between current and electrode separation $I_0 = f(\delta)$ is not a result which is typical of other investigations. Plumley [8], for example, did not find it at all although he used the same liquid as Green (hexane), while Goodwin and Macfadyen [9] and House [11] confirmed it to a greater or lesser extent (see, for example, fig. 18.7).

The relationship between current intensity and temperature has been examined by Le Page and Du Bridge [6]. The results obtained by them are shown in fig. 18.8.

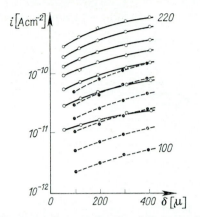

Fig. 18.7. Current density i (in A cm^{-2}) as a function of electrode separation (in μm) for different cathode materials and electric field strengths (from 100–220 kv cm^{-1}): $\bigcirc$, phospher-bronze; $\bullet$, stainless steel. Ref.: Goodwin and Macfadyen [9].

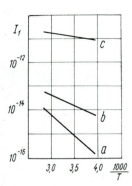

Fig. 18.8. Temperature dependence of current density i (in A cm^{-2}) in toluene from Richardson's equation. The work function is calculated from the gradients of the straight lines: a, 0·33 ev for $E = 10^4$ v cm^{-1}; b, 0·17 ev for $E = 6·25 \times 10^4$ v cm^{-1}; c, 0·049 ev for $E = 2·5 \times 10^5$ v cm^{-1}. Ref.: Le Page and Du Bridge [6].

The influence of collision ionization on current increase in high fields has been the subject of thorough investigations by Nikuradse [2] and Schumann [16] in 1932–34, and more recently by House [11], Goodwin and Macfadyen [9], Lewis [1] and others. The general view is that this phenomenon is similar to that occurring in gases (Von Engel [17]) and that processes such as avalanche ionization and attachment and detachment of electrons to atoms and neutral particles occur.

A number of papers have been published recently (1963–66) dealing with studies of conductivity in liquids subjected to very high fields. A few of them will be mentioned here.

Kao and Calderwood [205] studied the conduction current in hexane, carbon, tetrachloride and transformer oil. They investigated the effect of temperature, hydrostatic pressure, additives, vapour and time. They found two components of conduction current in high fields, one caused by the presence of ions which changes linearly with field, and the other caused by electrode emission from the cathode (the Schottky effect) which has an exponential character. The decrease of current with time was explained by the authors as being due to a decrease in the number of ions and a reduction of their mobility. Figure 18.9 shows a diagram of their liquid chamber adapted for investigations under high pressure and at different temperatures and fig. 18.10 shows their results for the dependence of viscosity and current on temperature for hexane and carbon tetrachloride.

Guizonnier and co-workers [230] studied the conductivity of liquids at voltages ranging from 20 to 100 kv and large electrode separations from

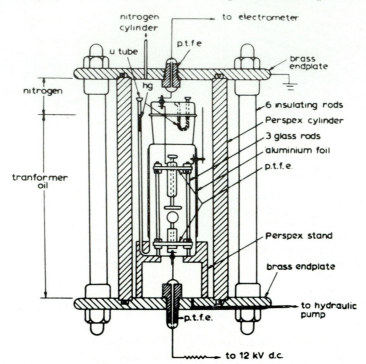

Fig. 18.9.   Pressure chamber, test cell and electrode assembly.   Ref.: Kao and Calderwood [205].

1 to 4 cm. They investigated the effect of impurities, water and temperature and concluded that the relationship between current and time could be expressed in the form $I = I_0 . f(t)$, where $I_0 = A \exp(-W/kT)$ and W has values from 0·27 to 0·41 ev. Their studies showed that the initial current in the liquid can be expressed by

$$I_0 = aV^2 \quad \text{or} \quad I_0 = aV^2 + bV. \tag{18.2}$$

They also suggested that his relationship was due to the water present in the liquid. Their investigation showed that the field distribution between the electrodes is linear for voltages of 20–30 kv and an electrode spacing of about 1·5 cm, while for higher voltages it becomes non-linear when for 100 kv the maximum drop occurs at the electrodes (greater at the anode), within 10% of the space between the electrodes.

As can be seen from Chapter 16.8, part of this relationship can be obtained from the general expression derived by the present author [222] which is based on the transport mechanism of negative carriers (free or solvated electrons) in the liquid. Excitation energy in the transport process is reduced by the field ($W = A - eE\delta$) so that the probability of charge carrier movement increases in proportion to $\exp(-W/kT)$ and for constant temperature is proportional to $\exp(eE\delta)$.

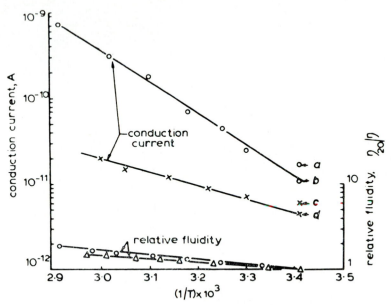

Fig. 18.10. Conduction current and relative fluidity as functions of temperature. ○, Carbon tetrachloride, 100 μm gap, 207 kv/cm; × △, *n*-hexane, 120 μm gap, 240 kv/cm; *a, c*, before raising temperature; *b, d*, after raising temperature. Ref.: Kao and Calderwood [205].

According to this interpretation, the experimental curves obtained by House and Nikuradse can be expressed in the following form:

House:
$$I = 1.04 \times 10^{-10} \exp(1.84 \times 10^{-4} \delta E) \tag{18.3}$$

for
$$25 \leqslant \delta \leqslant 150 \ \mu m, \quad 0.2 \leqslant E \leqslant 1.4 \ \text{mv/cm} \quad \text{(Fig. 18.11)}.$$

Nikuradse:
$$I = 2.43 \times 10^{-19} E^2 \exp(-9.4 \times 10^5/E) \exp(1.94 \times 10^{-4} \delta E) \tag{18.4}$$

(Fig. 18.12).

Both expressions are good general representations of the relationships found experimentally, and also give results in qualitative agreement when the constants are properly evaluated.

Kahan [179] carried out fundamental studies of conduction in hexane at high fields, including breakdown. A chromatographic analysis was made to check the purity of various liquid samples and it was found that the spectroscopic hexane used by many workers consists of up to 20% of hexane isomers and other impurities.

The most interesting results of his work are:

(*a*) The establishment of experimental conditions (parallel electrodes 3 mm in diameter, very pure degassed liquid, and filters) such that even comparatively large currents (10^{-6} A) did not show any fluctuations.

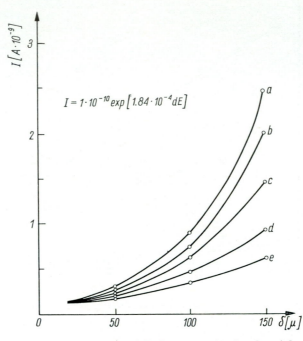

Fig. 18.11. Theoretical curves of $I = f(E)$ from eqn. (18.3) after Adamczewski [222] and experimental results (open circles) after House [11].

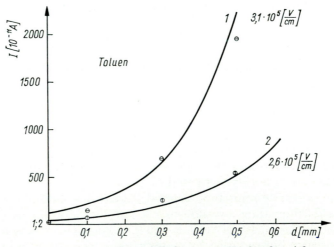

Fig. 18.12. Theoretical curves of $I = f(E)$ from eqn. (18.3) after Adamczewski [222] and experimental results (open circles) after Nikuradse [2].

(*b*) It was found that the relationship $I = f(E)$ can be divided into three main regions: the first from 60 to 150 kv/cm, currents were of the order of 10^{-13} to 10^{-12} A and showed small increase with increasing field and small current fluctuations, the second from 150 kv/cm to 250 kv/cm, currents grew exponentially with field and showed exponential growth of current with more marked fluctuations, and the third from 250 kv/cm to 400 kv/cm, currents reached saturation values of the order of 10^{-7}–10^{-6} A.

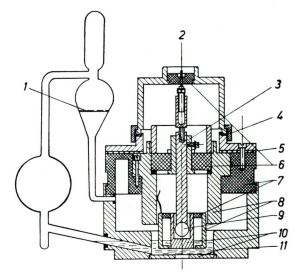

Fig. 18.13. Liquid-filled ionization chamber. 1, Schott filter; 2, to electrometer amplifier; 3, bronze socket; 4, 5, rubber liner; 6, amber insulator; 7, insulators; 8, shielding electrodes; 9, joint, 10, electrodes; 11, investigated liquid. Ref.: Terlecki [244].

Breakdown occurred at about 450 kv/cm, but in liquids containing air the breakdown stress was twice as large. Despite the use of fine filters Kahan observed particles suspended in the liquid, but their effect on conduction current was very small, especially in well-degassed liquids.

In his theoretical considerations Kahan ascribed great effect to the alignment of molecules in high fields caused by the formation of dipoles and their orientation in an electric field. This has a great effect on conduction current and breakdown stress.

In 1965 Terlecki [195] investigated the current and voltage characteristics in a carefully purified and degassed hexane ionized with x-rays in fields up to 150 kv/cm with different filters in the x-ray tube.

The diagram of his chamber is shown in fig. 18.13, and the current and voltage characteristics obtained in fig. 18.14.

The experimental work carried out in this field indicates that there are several phenomena (observed by various investigators) which are common to different liquids and electrode materials and independent of the conditions under which the experiments are carried out. Many phenomena have, however, not been explained satisfactorily, and the results obtained are not always reproducible and may even be contradictory. This disagreement has been ascribed to the influence of gases and water, and to distortion of the electric field.

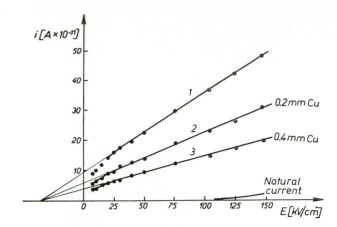

Fig. 18.14. Current–stress characteristics in hexane. Ref.: Terlecki [244].

An interesting addition to these results was contributed by Watson and Sharbaugh in 1960 [18]. They studied electrical conductivity in hexane at high fields using microsecond voltage impulses. The voltage was varied from 1 to 18 kv and the impulse duration time from 1·5 μsec to 10 msec. The electrode separations were from 10 μm to 100 μm. It was found that for fields up to about 1·2 mv cm^{-1}, the increase in number of electrons predicted from earlier theoretical and experimental work did not take place and that for these fields a large rôle is played by the emission of free electrons from the cathode, this emission being largely dependent on the thickness of the oxidized layer on the cathode. The results of these investigations are given in fig. 18.15.

There is some indication of electron multiplication in liquid at fields over 1·3 mv cm^{-1} just prior to breakdown. The results were discussed on the basis of the Fowler–Nordheim and Schottky equations (see § 20.2) and numerical values for cathodic electron emission were evaluated. This work by Watson and Sharbaugh has made a considerable impact on recent theoretical studies, although it has not as yet been confirmed by other researchers.

A very interesting series of investigations was carried out recently by Darveniza and Tropper [19]. They made a detailed investigation of the luminescence effect in dielectric liquids subjected to high electric fields especially just before electrical breakdown (this effect was observed earlier by House [11]). They used chiefly industrial grades of oil which displayed weak luminescence at high fields and pure liquids such as hexane and benzene which displayed no luminscence. However, when anthracene (1 g per litre) or solid scintillator POPOP was added to the pure liquids, marked luminescence was observed.

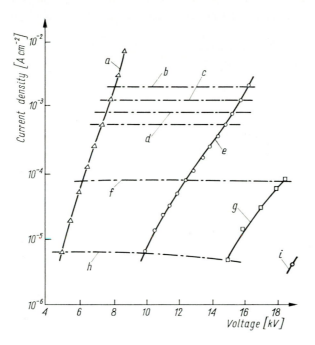

Fig. 18.15. The dependence of conduction current density upon voltage at different electrode separations (in *n*-hexane). Lines of constant field strength are shown as dashed lines. *a*, 2·5 mils; *b*, 3·300 v mil^{-1}; *c*, 3·200 v mil^{-1}; *d*, 3·100 v mil^{-1}; *e*, *d* = 5 mils; *f*, 2·500 v mil^{-1}; *g*, *d* = 7·5 mils; *h*, 2·000 v mil^{-1}; *i*, *d* = 10 mils. Ref.: Sharbaugh and Watson [120].

Figure 18.16 presents the current and light as a function of voltage for oil. As can be seen, light output and current increase with electric field strength in a similar manner. This is even more distinctly shown in fig. 18.17. After several breakdowns, the intensity of the fluorescence in the liquid falls. The luminescence effect begins at field strengths of the order of 500 kv/cm and maximum luminous energy is observed at a wavelength of about 4800 Å. Darveniza and Tropper also investigated the dependence of luminiscence on pressure, the oil distillation temperature and the presence and quantity of additives.

Gosling [116] in his work using purified transformer oil also studied luminescence under various experimental conditions especially with reference to the gas content. Murooka *et al.* [117] carried out a series of investigations in bubble chambers for fields of 1 MV/cm and recorded electrical breakdown with high-speed cameras. They observed that bubbles are formed for the most part near the cathode and that they move towards the anode (see also Krasucki [149]). It was found that a movement of the order of 0·1 mm required about 10 msec.

Some interesting results have been published recently by Jones and Angerer concerning electroluminescence in liquid paraffin and transformer oil [138]. In fig. 18.18 the relationship between electroluminescence

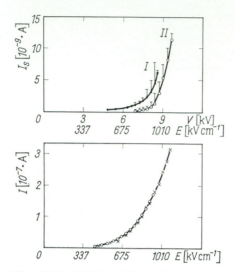

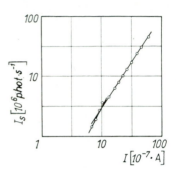

Fig. 18.16. Light–voltage (above) and current–voltage (below) characteristics in oil. Curve I, after one breakdown, curve II, after three breakdowns. Ref.: Darveniza and Tropper [19].

Fig. 18.17. Light intensity I_s as a function of current I (on a log–log plot). Ref.: Darveniza and Tropper [19].

B and conduction current I is shown. The gradient of these curves for transformer oil and liquid paraffin is nearly the same (45°), transformer oil giving more light output for the same conduction current. The presence of POPOP and dissolved nitrogen considerably affected the light emission but a linear relationship of $\log B$ against $\log I$ was observed over several decades. The light output was observed to be lower for ascending values of stress than descending values and when a high-speed oscilloscope was used with a photomultiplier, the light output was found to consist of a series of discrete pulses, even when the liquid was highly stressed. The number of pulses increased with the applied stress.

At Gdańsk Technical University, Kalinowski and Dera [115] carried out a series of measurements on the electroluminescence of various dielectric liquids, both when pure and with additives. The strongest luminescence was obtained in benzene with a small addition of the scintillator POPOP (20 mg/litre) (fig. 18.19 and 18.20). In his later work Kalinowski [193] discussed this phenomenon and presented a hypothesis. He claimed that

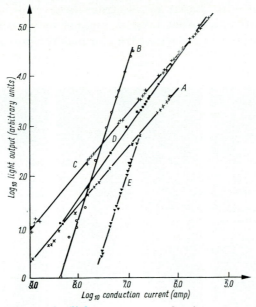

Fig. 18.18. $\log_{10} B / \log_{10} I$ (light output–conduction current) characteristics. *A*, liquid paraffin and transformer oil; *B*, liquid paraffin containing POPOP at a concentration of 2×10^{-5} mole/100 g (28 μm, 51 μm, 81 μm and 125 μm gaps); *C*, transformer oil containing POPOP at a concentration of 2×10^{-5} mole/100 g (125 μm gap); *D*, transformer oil containing POPOP at a concentration of 2×10^{-5} mole/100 g (51 μm gap); *E*, liquid paraffin containing POPOP and dissolved nitrogen at atmospheric pressure (125 μm gap). Ref.: Jones and Angerer [138].

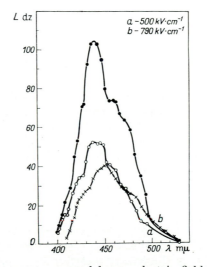

Fig. 18.19. Emission spectrum caused by an electric field in a benzene solution of POPOP and the photoluminescence spectrum (lower curve with crosses) in the same solution caused by a mercury lamp. Ref.: Kalinowski and Dera [115].

electroluminescence is caused by electrons from cold cathode emission which are emitted in a rather random manner and accelerated into the vicinity of the cathode by the strong field formed by a positive space charge.

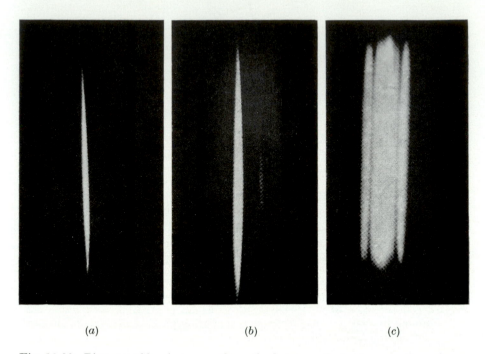

(a)   (b)   (c)

Fig. 18.20. Pictures of luminescence from the between-electrode area of maximum electric field strength $E = 640$ kv cm^{-1} (a) and $E = 900$ kv cm^{-1} (b). Picture of single electric breakdown (c). POPOP solution in benzene (concentration 20 mg/l.). Ref.: Kalinowski and Dera [115].

A number of electrons reach energy levels such that in non-elastic collisions with solvent or additive molecules they cause excitation. Re-emission of energy by the molecules of luminescent material gives the observed luminescence. The intensity of electroluminescence Φ is proportional to the number of re-emission processes in the region near the cathode:

$$\Phi \sim AN \int_{\varphi_0}^{\infty} \gamma(\varphi)\, d\varphi. \tag{18.5}$$

In this expression A is the probability of photon re-emission, N—the density of electrons entering the liquid due to equilibrium cold emission, $\gamma(\varphi)$—function describing the energy distribution of electrons and φ_0—the lowest electron potential for molecular excitation.

Furthermore, assuming that the cold emission could be expressed by the Fowler–Nordheim equations, and that electrons are removed from the region of luminescence chiefly by the electric field, he found an expression for N in terms of the electric field near the cathode E_c. Using the $\gamma(\varphi)$ distribution for gases deduced by Druyvesteyn [194], which takes the effect of an electric field and non-elastic collisions of electrons with molecules into account, he found that the intensity of electroluminescence Φ can be expressed as a function of the electric field near the cathode:

$$\Phi(E_c) = PE_c \exp\left[-\frac{G}{E_c}\right], \tag{18.6}$$

where P and G are constants depending on the electrode material, the condition of its surface, the mobility and free path of electrons in the liquid, the excitation potential of molecules and the probabilities of non-elastic collisions and photon re-emission.

Comparison of eqn. (18.6) with experimental results gave good agreement in the region of applied stresses, under consideration, but there is some deviation for fields near to those for breakdown, and for very low fields.

In connection with studies on dielectric liquids in very high fields, it should be noted here that under these conditions an electrostriction effect was also observed (Schott and Kaghan [164]), i.e. the raising of the liquid level through a distance H in an electric field of strength E according to the equation:

$$H = \left(\frac{E^2}{2\rho g}\right)\left(\frac{\varepsilon}{\varepsilon_0}\right)(\varepsilon - \varepsilon_0), \tag{18.7}$$

where ρ is the liquid density, g—the acceleration due to gravity and ε_0 and ε—the dielectric constants for a vacuum and the liquid, respectively. For example, for polyethylene in a field of 6×10^3 v/cm the experimental value was 1·5 cm. (Theoretical calculations give only 0·043 cm.) The pressure in the liquid increases then by:

$$\Delta P = (\tfrac{1}{2}E^2)\left[\tfrac{1}{2}\varepsilon_0\left(\frac{\varepsilon}{\varepsilon_0}+2\right)\left(\frac{\varepsilon}{\varepsilon_0}-1\right)\right], \tag{18.8}$$

which corresponds to about 0·028 cm hydrostatic pressure.

18.4. *Conclusions from Experimental Work.* On the basis of reproducible experimental results reported by the majority of workers it can be said that:

(1) In low fields up to 100 kv/cm, the volumetric effect dominates in liquid; the current is carried by the ions or electrons of the liquid itself or by impurities present in it. Ions in the liquid may be formed as a result of ionization effects due to natural radiation (cosmic rays, radioactive contamination), artificially produced radiation (x and γ-rays) or as a result of liquid molecule dissociation (spontaneous) or caused by the electric field (see § 16.7).

(2) For fields between 100 and 500 kv/cm the increasing influence of cold electron emission from the cathode surface is observed.

(3) In fields between 500 and 1000 kv/cm the electron emission from the cathode becomes both extremely evident and dominating, and electron multiplication in the liquid develops, which in higher fields leads to electron avalanches and electric breakdown.

Between 300 and 400 kv/cm, sporadic strong ionization current pulses appear but they do not usually lead to breakdown. These pulses, observed in hexane by House [11], were explained by him as being the result of the accidental presence of bubbles of air or oxygen in the liquid or on the electrode surface, or as being due to the presence of particles of impurities that might have accidentally found their way into the liquid. Very often these pulses disappeared completely after a period of time during which the field was acting, which might be explained by spontaneous purification of the liquids. Similar phenomena were observed by Yamanaka and Suita [22] in carbon tetrachloride and mineral oils, in much lower fields, however ($E < 100$ kv/cm), than were used by House.

The various effects of electrode surfaces were explained as being due to different values of the work function for electrons to escape from a metal. Figure 18.4 shows how the shape of the current–voltage characteristic changes for different electrode materials, all other parameters being constant. The work function of an electron for these metals varies according to the series:

$$Fe < Ni < Au < Cu.$$

The effect of ionizing radiation on current intensity has been examined several times. Green [10] found an approximately ten-fold current increase when a radium preparation (9·64 mg Ra in a region between 10^{-12} and 10^{-11} A in fields from 30 to 180 kv/cm) was used. Plumley [8] also investigated the effect of liquid purity, electric field strength and temperature on the conductivity of dielectric liquids. A diagram of the current–voltage characteristics in the form of the relationship $\log I = f/E^{1/2}$ (derived by him for a large number of measurements) is given in fig. 16.16, along with theoretical equations.

An interesting practical application of dielectric liquids in high-voltage engineering was developed by Hughes and Secker [231]. They constructed an electrostatic generator of the Van de Graaff type in which the rubber or silk belt was replaced by a stream of liquid which transfers the electric charge to a collector. Figure 18.21 shows this generator and fig. 18.22 shows curves of the voltage and current generated as a function of the grid-emitter input voltage and conditions under which the liquid is pumped. The maximum voltages obtained using this generator were 70 kv, and the maximum current was of the order of $1·5 \times 10^{-5}$ A. In 1966, Secker *et al.* [240] designed a high-voltage diode with a liquid dielectric (silicones with viscosities of 10·3 and 1·5 cs and transformer oil). This triode is shown in fig. 18.23 and its characteristics given in fig. 18.24.

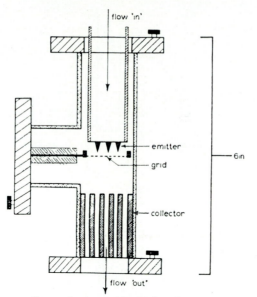

Fig. 18.21. Diagram of an electrostatic high-voltage generator with dielectric instead of the insulating band. Ref.: Hughes and Secker [231].

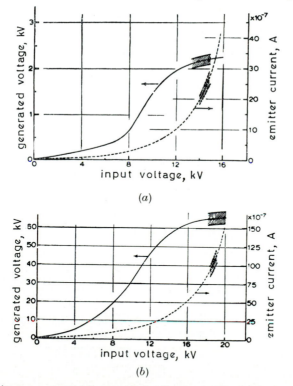

Fig. 18.22. Emitter current and generated voltage as a function of the grid-emitter input voltage. (*a*) No mechanical pumping of the hexane liquid. (*b*) With mechanical pumping giving a flow velocity of approximately 0·5 m/s. Shaded regions indicate the magnitude of experimental scatter. Ref.: Hughes and Secker [231].

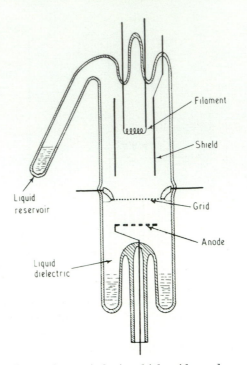

Fig. 18.23. Diagram of two-phase triode, in which grid–anode spacing is 1 cm. The triode is normally operated with the grid earthed, high voltage being applied between the anode and the grid, and the electron-injecting voltage between the filament and the grid. Ref.: Secker *et al.* [110].

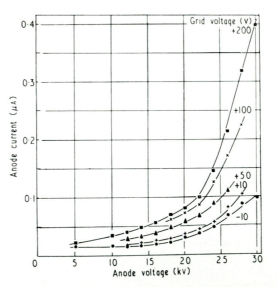

Fig. 18.24. Conduction characteristics of triode filled with 10 centistoke oil for various electron accelerating voltages between filament and grid. Ref.: Secker *et al.* [110].

Chapter 19

Electrical Breakdown

19.1. *Introduction*. Investigations on electrical breakdown in various dielectric liquids and especially in transformer oil have been carried out by many authors since the information obtained from these studies is of great value where commercial application of these liquids is concerned. The influence of electrode materials, impurities and additives in the liquid, gas content, degassing of the liquids and electrodes, the duration of the voltage applied, the rate of increase of the applied voltage and the frequency of alternating voltages, and the effect of temperature and hydrostatic pressure have been carefully studied. The studies were carried out by Inge and Walter [23], Lazarev and Rashektajev [87], Gemant [25], Schwaiger [26], Edler [27], Volkenstein [28], Salvage [29], Peek [51], Skowroński [59], etc. Descriptions of this work can be found in monographs by Skanavi [3], Kok [4], Lewis [1, 129], Sharbaugh and Watson [132], and a short comparison of the most important results is given at the end of this chapter.

19.2. *Breakdown in a Vacuum, in Gases and in Vapours*. Before dealing with the study of breakdown phenomena in liquids, we shall discuss briefly a few fundamental problems related to breakdown in a vacuum, gases and vapours, and in particular pre-breakdown phenomena.

There is a large amount of published work dealing with breakdown in a vacuum and its dependence on cold cathode and thermal emission of electrons from a cathode in low and high fields, the cathode material and electrode curvature, etc.

The problem of electron emission in vacuum has been studied recently (1965, 1966) by Chatterton [224] and Farral and Miller [145, 146], under different experimental conditions. They determined the effects of geometrical factors (distance and curvature of electrodes), electrode material, temperature, and space charge distribution in vacuum on breakdown values and on reproducibility and scatter of the results.

Breakdown in gases and the vapours of various liquids has been studied by many workers who have published interesting experimental results and theoretical calculations. A critical comparison of this work can be found in monographs by:

Loeb [133], Llewellyn Jones [134], Dakin and Berg [135], Meek and Craggs [136] and Papoular [137].

A great amount of work in this field was carried out from 1960 to 1966 by Prasad *et al.* [77–80], Heylen and Lewis [139–43], Crowe [144] and Kuffel [208–10, 237].

Conductivity and breakdown in gases and hydrocarbon vapours have been studied by Heylen and Lewis [139–43]. They determined the electron energy distribution function, transport coefficients and the relationship between breakdown value and molecular structure.

Recently Crowe [144] investigated the effect of time lag on breakdown in electronegative gases particularly in sulphur hexafluoride. This work has been mentioned here as it could find application in the study of dielectric liquids.

Kuffel [209, 210, 237] has investigated the attachment of electrons to oxygen in the air with different humidities and in steam. He has also studied the development of discharges in gases using different experimental conditions and time factors.

A special investigation was carried out on sulphur hexafluoride which is often used in the study of insulating liquids. The presence of this gas results in an increase in breakdown strength. Kuffel and Radwan [208] explained this effect as being due to electron attachment and the formation of heavy ions; they suggested the following model:

(*a*) Resonance capture:

$$SF_6 + e \rightarrow (SF_6^-)^* \rightarrow \frac{SF_6^-}{SF_5^-} + F. \qquad (19.1)$$

(*b*) Excitation and dissociation:

$$SF_6 + e \rightarrow (SF_6)^* + e \rightarrow SF_5 + F^- \rightarrow SF_4 + F + F^-. \qquad (19.2)$$

(*c*) Formation of positive and negative ions:

$$SF_6 + e \rightarrow SF_6^* + e \rightarrow SF_5^+ + F^- + e. \qquad (19.3)$$

(*d*) Electron multiplication:

$$SF_6 + e \rightarrow (SF_6^+)^* + 2e \rightarrow SF_5^+ + F^- + 2e. \qquad (19.4)$$

It should be emphasized that much information about conductivity and breakdown mechanisms has been obtained from studies carried out on solid state. These investigations form a large sphere of scientific research and will not be discussed in this work. Certain results obtained from these studies are of great significance because of similarity of some processes in all three solid, liquid and gaseous phases.

Basic information can be gained from Stratton [188], Mason [189], Dekker [187] and O'Dwyer [192].

Information regarding more recent measuring techniques and results are to be found in references [242–46].

19.3. *Research Methods.* In the following, only recent studies on electrical breakdown in dielectric liquids are discussed. It includes work in which pure liquids were tested under carefully controlled laboratory conditions. The results were reproducible and characteristic of the liquids tested, and not of the impurities they still might contain. Such phenomena as corona, the formation of an electric arc and channel break-down will not be discussed.

This group of studies includes the work carried out in the past ten years by Maksiejewski and Tropper [30], Lewis [1, 12–14, 39], Crowe *et al.* [31–33], Goodwin and Macfadyen [9], Green [10], Edwards [35], Adam-czewski [36] and others. The work carried out using pulse technique

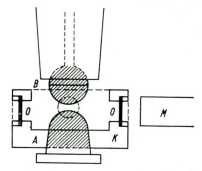

Fig. 19.1. Diagram of a chamber for measuring electric breakdown in liquids. *K*, vessel filled with liquid; *A*, *B*, electrodes; *O*, windows; *M*, microscope. Ref.: Crowe [33].

should be considered carefully, as the results obtained reveal information regarding the value of breakdown strength and the minimum time required for the breakdown to occur. The laboratory conditions in which the tests are carried out greatly affect the accuracy and reproducibility of the results obtained, and therefore a short description of the apparatus used in most of the published work is given below.

In most cases the breakdown tests were carried out in a cell containing two hemispherical electrodes completely immersed in the liquid (fig. 19.1).

The electrodes were made of clean and carefully polished steel, some-times electroplated with phosphor-bronze. The diameter was usually a few centimetres and the distance between electrodes, varying from 0·004 cm to 0·007 cm, was adjusted using a micrometer. Electrodes coated with lacquer were also used and resulted in increased breakdown values. The cell was equipped with two or three windows through which it was possible to illuminate the tested sample of the liquid and observe it through a microscope. Careful examination of each sample was carried out to avoid the presence of small particles or suspended matter in the liquid or on the surface of the electrodes. Impurities so small that they may not be seen even by the naked eye considerably affect the value of breakdown of the liquid and may cause accidental breakdown to occur.

The liquids purchased were of the highest quality and purified in the laboratory, using special, but well-known physicochemical methods. As can be seen from Chapter 4.1, these methods include: the removal of certain compounds (unsaturated hydrocarbons from saturated hydro-carbons), removal of water, filtration using glass-sintered filters (Schott) to remove solid particles, fractional distillation over metallic sodium, degassing of the liquid, and so on.

Morant and Kahan's [250] chromatographic studies indicated that dielectric liquids which are claimed by even the best firms to be spectro-scopically pure may still contain other compounds in proportions such that the actual content of the desired substance is only 80%. A voltage

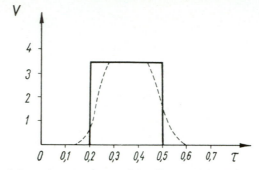

Fig. 19.2. Shape of the voltage pulse used by Crowe [33] (V in kv, τ in μsec). The full lines describe the theoretical shape and the broken line the actual shape.

ranging from a few to several kilovolts was usually obtained from a pulse generator with rectangular pulse shape of variable amplitude and time duration from $0.25\ \mu$sec to $54\ \mu$sec (fig. 19.2). The short impulses eliminated the 'thermal breakdown' which may take place when the electric field is applied for a longer time, which in turn could affect the study of intrinsic breakdown strength of the tested liquid.

The magnitude of breakdown stress was found to depend on the dura-tion of the voltage pulse, at least for the very short pulses (of the order of microseconds), but for longer pulses the increase in pulse duration did not affect the value of breakdown strength significantly (fig. 19.3).

Brignell's experimental technique [171] is worth mentioning here. He developed a method for the automatic measurement of electric strength in liquids by which subjective human errors are eliminated and a large number of statistical experimental results can be obtained in a short period of time (fig. 19.4).

19.4. *Pre-breakdown Phenomena. Current Fluctuation.* In very high fields single large current pulses could be observed before breakdown occurs, as if local breakdown occurred which does not pass the whole way between the electrodes. These are sometimes called 'microbreakdowns'. The frequency of such ionizing pulses and their value depend on field intensity, the type of liquid and its purity, the presence and nature of gas dissolved in the liquid, electrode material, the condition of electrode surfaces and temperature. In some liquids these pulses are sometimes followed by light pulses which can be registered using photomultipliers.

Works in this field have developed considerably recently, mostly because they result in more information about the charge transport mechanism in liquids and about the breakdown mechanism. We shall mention only a few of them here, particularly those which have given new information regarding breakdown mechanism. Huq and Tropper [206] carried out

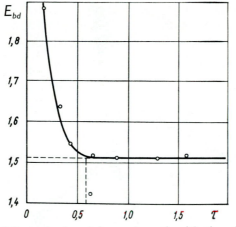

Fig. 19.3. Change of the pulse breakdown strength with time (E_{bd} in MV cm^{-1} and τ in μsec).

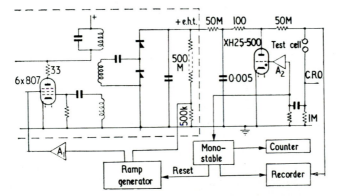

Fig. 19.4. Diagram of an automatic system for making electrical breakdown measurements in liquids. Ref.: Brignell [171].

work in this field using a special apparatus. They investigated the effect of different gases on the frequency and magnitude of current pulses before breakdown occurs, and also the effect of suspended matter, hydrostatic pressure and temperature. Similar work was also carried out by Megahed [176], Nosseir Hawley and Clothier [148] and others.

Garton *et al.* [147, 149–51] showed, using a photographic method, that in viscous liquids electrons and ions may form local gas bubbles in which a small electron avalanche is initiated, resulting in ionizing pulses and discharges. The avalanches develop along arbitrary channels (fig. 19.5) which may reach the cathode and initiate complete breakdown. Krasucki [150] carried out this work in connection with his studies of the electric strength of oil-impregnated paper used for the production of capacitors. He also found that water has a great effect as regards reducing the electric strength of this kind of insulating system.

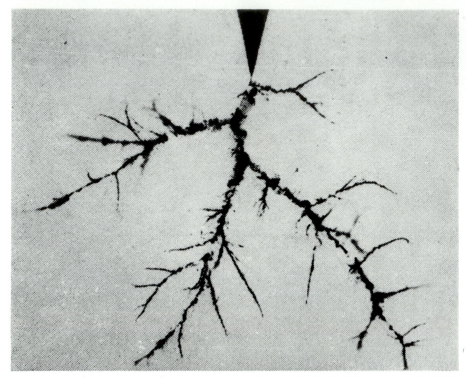

Fig. 19.5. Breakdown channels in hexachlordiphenyl. Ref.: Krasucki *et al.* [150].

Samuel *et al.* [158] also investigated these pulses during the ionization of hexane by γ-rays ^{60}Co (8 rad/h) and explained their occurrence as being due to the existence of free electrons in the liquid, the mobility of which they estimated to be about 10^2 cm^2 v^{-1} sec^{-1}.

Chadband and Wright [152] have studied pre-breakdown phenomena using the Schlieren technique and a high-speed camera (6×10^5 frames per minute). They observed the formation of plasma bubbles on the cathode surface and elongation of these bubbles towards the anode. They claimed that the minimum value of the field in plasma is about five times greater than the mean field in the system, and that the mobility of electrons in this case is of the order of 0.03 cm^2 v^{-1} sec^{-1}.

Calderwood *et al.* [153] observed light pulses in pure hexane at relatively low fields of 24–30 kv/cm near electrodes made from a gramophone needle with a radius of 0.02 mm and a steel sphere 0.6 cm in diameter, with a separation of 1.75 mm. They suggested that electrons emitted from the cathode are accelerated in the plasma region. In the front of the plasma the field is greatly reduced and prevents its propagation.

The statistical nature of breakdown phenomenon, which was pointed out by Lewis, suggests a similar character of various pre-breakdown phenomena. In very high fields, current pulses are observed which are followed by light pulses. These light pulses can easily be detected when anthracene is added to pure hexane. Recently, Gzowski *et al.* [154] found that light pulses are also random if high fields are applied. It is suggested that these light pulses are due to local microdischarges which do not develop into complete breakdowns. A statistical analysis of time delays between the application of a high field, the first and second light pulses, enabled the mean frequency $f(E)$ with which these pulses appear to be determined. It was found that this function depended on the electric field in a similar way as does the function of mean breakdown frequency. Further work by Gzowski [155, 215] indicated the important rôle played by suspended dust and microscopic particles in the gap in the process initiating light pulses. A small particle, when charged, oscillates between the electrodes and could be compared to a large heavy ion which, when approaching an electrode, causes local increase in the field. In this region a microscopic breakdown may appear, registered as light pulse, in very high fields this may develop into a complete breakdown.

19.5. *Experimental Results.* The most significant results obtained in the above studies were:

(1) Determination of the minimum pulse duration τ_0 for which a constant value of the impulse breakdown is obtained.

(2) Investigation of different stages of breakdown phenomenon and in particular:

    (*a*) Field-aided emission of electrons from the cathode.

    (*b*) Mobility of electrons and ions in the liquid and the mechanism by which their motion is hindered.

    (*c*) Formation of electron avalanches and ion avalanches in the liquid leading to breakdown.

(3) Investigation of the effects of liquid properties (density, viscosity, molecular structure, resonance energy) on the value of breakdown strength.

(4) Determination of the effects of electrode material and the electrode surface conditions.

(5) Determination of the effect of temperature.

(6) Determination of the effect of hydrostatic pressure.

Let us consider the most important results obtained in these studies. For very short pulses the breakdown stress depends on the duration of the pulse. A typical relationship is shown in fig. 19.6. $E_{bd} = f(t)$. For a pulse duration smaller than τ_0, the electrical strength depends on pulse duration. Goodwin and Macfadyen [9] have found that for this region the relationship between E_{bd} and pulse duration can be expressed in the form

$E_{bd} = C(\tau - \tau_0)^a$, where C, a and τ are constant for a given liquid. For impulses with time duration longer than τ_0, the breakdown stress does not change with the pulse duration and may be compared with the statistical value of breakdown stress.

There are still some discrepancies in the test results regarding the dependence of τ_0 on the type of the liquid tested, which may alter essentially the existing interpretation of the breakdown mechanism. In the work

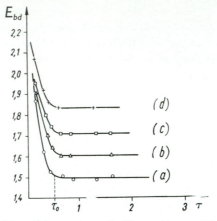

Fig. 19.6. Dependence of breakdown strength E_{bd} on time τ (E_{bd} in MV cm^{-1} and τ in μsec) in saturated hydrocarbons at an electrode separation of $\delta = 6\cdot35 \times 10^{-3}$ cm: a, hexane; b, heptane; c, octane; d, nonane. Ref.: Crowe [33].

of Crowe and co-workers [31–33] and Edwards [35], the value of τ_0 obtained for different liquids is constant (fig. 19.6) and depends only on the stress, but Goodwin and Macfadyen [9] found a systematic change in τ_0 depending on the density or viscosity of the liquids belonging to the same group of saturated hydrocarbons (fig. 19.7). In fig. 19.8 the dependence of E_{bd} on electrode spacing is shown for different liquid hydrocarbons.

It would be of great value to establish which results are acceptable, since the value of τ_0 is of great importance as regards breakdown mechanism. From the value of τ_0 it is possible to estimate the mobility of charge carriers (electrons or ions) causing the breakdown. On this basis, it was estimated that the mobility of these charge carriers is 2×10^{-3}–9×10^{-3} cm^2 v^{-1} sec^{-1} [33], which do not differ markedly from those obtained by the author (Part III [40, 41]), viz. $4\cdot5$–8×10^{-4} cm^2 v^{-1} sec^{-1} for low electric stresses. It is possible that new information from the work of Terlecki [37] and Chong *et al.* [114] will throw some light on this. They have investigated how the mobility of the fast ions changes with change of electric stress in the region 100–300 kv/cm. It was found that in this region of stresses, the mobility of the negative carriers does not change within the limits of experimental accuracy (about 7%).

However, it seems that a coherent interpretation of this phenomenon will be difficult because the transit time for ions is subject to statistical variations (Terlecki [37]). Among the other results emerging from the large source of experimental data which throw light on the mechanism of electrical breakdown, the effect of the liquid itself on the breakdown stress should be mentioned, i.e. the dependence on volume of liquid tested, spacing of electrodes, density and viscosity of the liquid and its molecular structure.

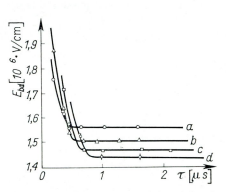

Fig. 19.7. Dependence of breakdown strength E_{bd} on time τ (E_{bd} in MV cm^{-1} and τ in μsec) for hexane with a number of electrode separations δ (in 10^{-3} cm); a, 5·08; b, 6·35; c, 7·62; d, 8·69. Ref.: Goodwin and Macfadyen [9].

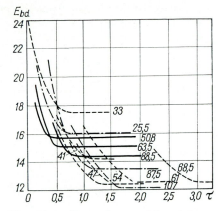

Fig. 19.8. Dependence of breakdown strength E_{bd} on time τ (E_{bd} in MV cm^{-1} and τ in μsec) at a number of electrode separations according to the following workers: ———, Crowe (steel electrodes), 50·8, 63·5, 88·5; – – – –, Goodwin and Macfadyen (phosphor-bronze electrodes), 33, 41, 47, 54, 61, 68; –·–·–·, Edwards (phosphor-bronze electrodes), 25·5, 87·5, 107.

The great amount of experimental work has resulted not only in a large quantity of information, but has also pointed out several characteristic quantitative relationships. The most important are:

(1) Breakdown field stress increases proportionally with increase in the density of the liquid, but the line representing the $E_{bd} = f(\rho)$ relationship does not pass through the origin (fig. 19.9).

(2) The electrical strength of chemical substances with a molecular structure including branched chains (isomers) is lower than the electric strength of similar substances with a straight-chain molecular structure.

(3) Breakdown strength for liquids belonging to the aromatic hydrocarbons is in general greater than the electrical strength of saturated hydrocarbons and also changes linearly with density, but the increase of breakdown stress for these liquids is much greater at increased density than that observed for saturated hydrocarbon liquids.

19.6. *The Effect of Electrode Material and Condition of its Surface on Electrical Breakdown.* Many studies carried out by different workers have shown clearly the effects of electrode material, preparation of the electrode surface and degassing of electrodes on the value of breakdown. The information obtained is by no means consistent and to a great extent depends on the method of electrode preparation, degassing and the type of liquid in which they were immersed. In general, it is claimed that electrical breakdown in liquids depends on two phenomena:

(1) Electron emission from the cathode surface.

(2) The mechanism of the formation of an avalanche of electrons and ions.

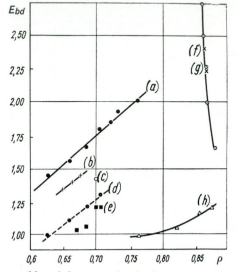

Fig. 19.9. Dependence of breakdown strength E_{bd} (in MV cm^{-1}) on liquid density ρ (in g cm^{-3}) under various experimental conditions: a, normal paraffins, $\tau = 1 \cdot 4\,\mu$sec; b, single branched-chain hydrocarbons, $\tau = 1 \cdot 4\,\mu$sec; c, double branched-chain hydrocarbons, $\tau = 1 \cdot 4\,\mu$sec; d, normal paraffins, direct voltage; e, single branched-chain hydrocarbons, direct voltage; f, g, straight and branched-chain benzene derivatives, $\tau = 1 \cdot 65\,\mu$sec; h, silicons, direct voltage. Ref.: Lewis [1, 39].

Divergence in the experimental results depends mostly on which effect is the predominating one, this depending on the electrode material or purity of the liquid tested. It was possible to show the latter dependence by using a point-plane arrangement of electrodes and changing their polarity. The effect of electrodes is then clearly seen to be large when the point electrode is negative, but when the point electrode is positive the effect of the liquid predominates. In fig. 19.10 the relation between breakdown values and electrode spacing is shown for different electrode materials (Al, Cu, Cr) in hexane. When the point electrode is negative, there is no difference for the three materials, but when the point electrode is positive the breakdown stress increases from aluminium to chromium. Similar results were obtained for transformer oil by Zein El-dine and Tropper [40] using an impulse technique. Studies on the relation between breakdown stress and the conditions of the electrode surface have been carried out by Maksiejewski and Tropper [40], Weber and Endicott [41] and others, but the results differ considerably and point to the statistical nature of the phenomenom.

Recently many new and interesting results have been obtained by Watson and Sharbaugh [18], Swan and Lewis [21] and Ward and Lewis [38] using hexane, and liquid argon, oxygen and nitrogen. In all of these investigations the effect of the cathode surface was emphasized, and Swan and Lewis pointed out the possibility of an anode effect especially

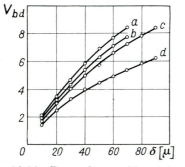

Fig. 19.10. Dependence of breakdown voltage V_{bd} (in kv) on electrode separation δ (in μm) for a number of electrode materials and cathode shapes; *a*, Cr; *b*, Cu; *c*, Al (flat cathodes); *d*, Cr, Cu, Al (point cathodes). Ref.: Lewis [1, 39].

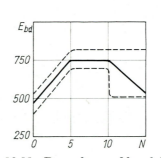

Fig. 19.11. Dependence of breakdown strength E_{bd} (in kv cm^{-1}) on the number of breakdowns N in transformer oil. The broken lines indicate the limits of scatter of experimental results. Ref.: Maksiejewski and Tropper [30].

in the case when negative ions are present in the liquid. The studies on liquid argon are very interesting because in this liquid some authors have detected free electrons. Their mobility is of the order of 2×10^2 cm^2 v^{-1} sec^{-1} (Williams [42], Malkin and Schultz [43]).

19.7. Dependence of Electric Strength on the Number of Breakdowns. Very interesting results have been obtained during studies of the relation between the value of breakdown strength and the number of breakdowns under d.c. conditions. It was found that breakdown values are lower for first breakdowns, increase systematically with number of breakdowns and only after a certain number of breakdowns reach a steady mean value (see fig. 19.11). This phenomenon, known as 'conditioning' has been observed by many authors, and the number of conditioning breakdowns lies in the range from only a few to as many as 20–40. After this conditioning the breakdown stress of the liquid may have increased by 50–100%. This may be explained by assuming that the first breakdowns remove gas bubbles and impurities from the liquid and the surface of the electrodes. The drop in breakdown stress after a large number of breakdowns is ascribed to contamination of the liquid, channels formed in the liquid and damage to electrode surfaces.

19.8. *Effect of Temperature and Hydrostatic Pressure on Electrical Strength.* An increase in the temperature of the liquid usually causes a reduction in the electric strength of the liquid. This was shown by Lewis [1], Salvage [29], Maksiejewski and Tropper [30], Crowe *et al.* [31] and others. A typical relationship between breakdown stress and temperature can be seen in fig. 19.12. It can be seen that outside a certain region close

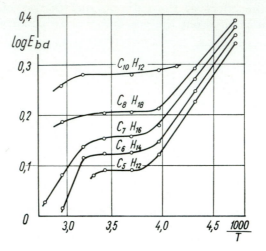

Fig. 19.12. Temperature dependence of the logarithm of breakdown strength in normal saturated hydrocarbons. Ref.: Goodwin and Macfadyen [9].

to room temperature in which the logarithm of breakdown stress does not change with temperature, the stress increases rapidly as the temperature is reduced. Salvage [297] showed that the breakdown stress drops from 900 kv/cm at 5°c to 500 kv/cm at 60°c in hexane under d.c. conditions, and Goodwin and Macfadyen [9] established for several saturated hydrocarbon compounds a general relationship $\log E \sim T^{-1}$.

Decrease of breakdown stress with increase in the temperature of the liquid tested can be explained by the reduction of the density and viscosity of the liquid as the temperature rises, and therefore by an increase in the mobilities of ions and electrons. The relation between breakdown stress and temperature for paraffins and silicon oils is shown in fig. 19.13. In fig. 19.14 the relationship between breakdown stress and density and viscosity is shown. It should be noted that results obtained by different workers here differ, and there are some results published which indicate no relationship between temperature and breakdown stress; a negative coefficient was even found, for example, in the work of Edwards [35] for ethyl alcohol, and in the work of Hoover and Hixson [44] for mineral oils. These discrepancies could perhaps be explained by the relatively little-studied phenomenon of the appearance and disappearance of air bubbles on electrode surfaces.

There is no doubt about the considerable drop in breakdown value close to the boiling point of the liquid (Lewis [39]). Figure 19.15 shows the relationship between the ratio of breakdown stress E_{bd} to the number of molecules per cm³ N and the temperature T for various saturated hydrocarbons according to Edwards [35]. Edwards considered that the three different regions on the curves in figs. 19.14 and 19.15 are due to the change in resistivity of the liquid caused by the vibrations of long chains of hydrocarbon molecules in the pseudo-crystalline liquid structure.

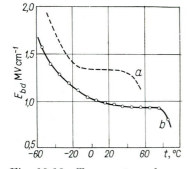

Fig. 19.13. Temperature dependence of breakdown strength for: a, silicons ($\delta = 5 \times 10^{-3}$ cm) and b, paraffins ($\delta = 3 \times 10^{-3}$ cm). Ref.: Lewis [39].

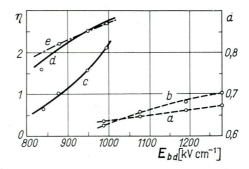

Fig. 19.14. Dependence of breakdown strength E_{bd} on viscosity (in cP) and density d (in g cm⁻³) for silicons (solid lines) and paraffins (broken lines). Ref. Lewis [39].

The studies on changes in breakdown stress in relation to hydrostatic pressure have not given coherent or reproducible results. Maksiejewski and Tropper [30] found a change in E_{bd} from about 200 kv/cm at 155 mm Hg to about 500 kv/cm at 760 mm Hg. A similar dependence was found by Watson and Higham [45] for paraffinic oils which had been carefully filtered and degassed (fig. 19.16). This phenomenon depends on the amount of air dissolved in the liquid and absorbed by the electrodes, and on the duration of applied pulse and on the polarity of the cathode.

From the work of Watson and Sharbaugh [18], it appears that gas bubbles appearing near the cathode or vapour bubbles formed at the cathode due to intensive electron emission just before breakdown have a considerable effect on breakdown stress. They consider that at such a moment a great amount of energy is transferred to the liquid (about 10^4 w/cm³) which may reach as high a value as 10^7 w/cm³ in some regions, leading to evaporation of a small amount of the liquid. The presence of these vapour bubbles results in the dependence of breakdown values on hydrostatic pressure.

The author is of the opinion that in very pure liquids which have been properly degassed, the effect of temperature and hydrostatic pressure could be explained by changes in density of the liquid, and in consequence by a change in the number of C—C bonds which lie in the path of electrons or ions moving in the liquid when subjected to electric field. From eqns. (10.5) and (20.31) in Chapter 20.5, a change in temperature from 0°c to 50°c should result in a reduction of the density of hexane by 6% and the reduction of its electric strength by 7% (see fig. 20.15).

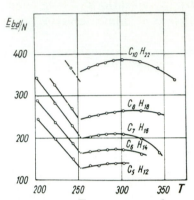

Fig. 19.15. Temperature dependence of breakdown strength E_{bd} (in mv cm^{-1}) for a number of saturated hydrocarbons with an electrode separation of 30 μm. Ref.: Edwards [35], Lewis [39] and Goodwin and Macfadyen [9].

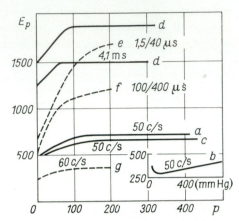

Fig. 19.16. Dependence of breakdown strength E_{bd} (in kv cm^{-1}) on pressure p (in mm Hg) under various experimental conditions: a, b, c, d, for mineral oil with an electrode separation of about 1 mm and alternating voltage; e, f, g, for degassed oil. Curves a, b, c, and d—ref.: Watson and Higham [45]; e, f, g—ref.: Hoover and Hixson [44].

The above considerations may be taken as purely hypothetical since they have been deduced from theoretical speculations for extremely pure liquids and do not take into account the effect of gases dissolved in the liquid and those adsorbed on the surface of the electrodes. They also may refer only to those temperature ranges in which the density changes in a regular manner. It should be noted here, however, that there are many experimental results which support this interpretation of the phenomena described.

19.9. *Dependence of Breakdown Stress on Other Factors.* A few authors have investigated the effect of ionizing radiation on breakdown stress. They have used γ-rays, radium and cobalt-60 with an activity of the order of several millicuries. They did not as a rule detect any significant effect (Lewis [1], Maksiejewski and Tropper [30], Goodwin and Macfadyen [9]) although there were small effects which did not cross the limits of statistical error. Goodwin and Macfadyen found a decrease in τ_0 but it is not known if this effect was not also within the limits of experimental error.

All the above authors suspected that at very high fields the effect of electron emission is of such a magnitude that it overshadows the effect of ionizing radiation. With greater doses of radiation of the order of megarads, which could be obtained using reactors or accelerators, a significant irradiation effect has been observed (Stark and Garton [112]), especially for polyethylene at temperatures above the melting point. Irradiation caused an increase in breakdown stress of about 100–300% with doses of 100–300 megarads (see fig. 9.22). The authors tried to explain this effect by suggesting chemical changes in the compounds, splitting of molecules, and the formation of longer and larger molecules. A similar effect was observed as far as mechanical strength is concerned. No noticeable effect was recorded for ultra-violet radiation on breakdown stress in spite of the fact that this type of radiation was found by Weber and Endicott [47] to be effective in the dissociation of hydrocarbon vapours. It is possible that in liquids the absorption of ultra-violet radiation in outside layers is so great that it does not penetrate the bulk of liquid between the electrodes.

19.10. *The Effect of Additives and Impurities.* The fact that the degree of purity of the liquid tested has a decisive influence on its conductivity and breakdown stress has been accepted by all investigators. Some of them added impurities to very pure liquids in a systematic way in order to determine the quantitative effect of impurities of different kinds on breakdown stress of the liquid. For example, Friese [48], Zein El-dine and Tropper [40] and others added water to very pure oil (up to 200 p.p.m.) and found a considerable decrease in breakdown stress from 700 kv/cm to 100 kv/cm, and also studied the dependence of the breakdown value on the duration of applied d.c. voltage. It was also found that the conduction current increases considerably before breakdown. This effect is well known and explained by the great ability of the water to dissociate and produce a large quantity of ions in the bulk of liquid as well as in the vicinity of the electrodes. A more interesting investigation was carried out by Suzuki and Fuijoka [49] who added different substances which increase the breakdown stress. They found, for example, that the addition of iodine to oil in an amount of 0·01 g per litre increased the value of breakdown stress by 18%, but a greater amount of iodine, of the order of 0·1 g per litre, reduced the breakdown stress by 5%. They found a similar effect for other additives. Ruhle [50] carried out similar work on hydrocarbon liquids but unfortunately the basic liquids tested by him were not sufficiently pure for this type of investigation, and low values of breakdown were obtained (about 350 kv/cm). His results are presented in fig. 19.17. The author suggests that liquids of higher dielectric constant concentrate near to the electrodes, and in effect produce a distortion of the electric field distribution.

The work of Musset *et al.* [53], Booth and Johnson [54] and Angerer [118] should also be mentioned here. Due to potential industrial application, great attention is now being paid to the possibilities of increasing the breakdown stress of liquid insulating materials by adding suitable additives. It was found that the addition of *p*-nitrotoluene to cable oil increases its electric strength by 48%. However, more work will have to to be carried out in conditions which are perfectly controlled and reproducible.

The work of Angerer [118] published in 1963 should be mentioned here. He used different organic and inorganic additives to find their effect on the electric conductivity and breakdown stress of transformer oil and liquid paraffin. The more volatile additives proved to be more effective

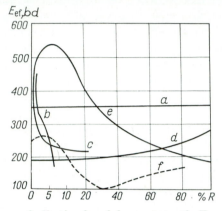

Fig. 19.17. Dependence of effective breakdown strength $E_{\text{ef},bd}$ (in kv cm^{-1}) on the additives concentration (in wt. %) for 50 c/s alternating current: *a*, pure hexane; *b*, hexane with nitrobenzene; *c*, hexane with acetone; *d*, hexane with xylene; *e*, hexane with chlorobenzene; *f*, oil with chlorobenzene. Ref.: Suzuki and Fujioka [49] and Ruhle [50].

than non-volatile ones in liquid paraffin tests. The effect of selenium was very pronounced and was attributed to the possibility of the formation of a protective layer on the electrode surfaces. Two distinct optimum values were found, one for maximum reduction of conduction current and the other for maximum increase in breakdown stress. The two values of additive concentration differ by one order of magnitude, the concentration for minimum conduction current being lower than that for maximum breakdown stress. Angerer proposed a model of the breakdown mechanism in liquids. He suggested that in the first stage of the breakdown mechanism microscopic cavities are formed due to thermal fluctuations enhanced by the presence of the electric field and the mechanical effects of the field, and in the second stage in which electrons moving through the liquid permeated by such microcavities can gain enough energy to bring about dissociation accompanied by gas evolution; here a microbubble can form leading to the third stage in which a gaseous breakdown may take place, when the rate of bubble formation exceeds the rate at which the evolved gas can be absorbed by the liquid. The model gives an explanation for the greater effectiveness of more volatile additives.

19.11. *Effect of Oxygen on Breakdown Stress.* Recently (1957–62) a number of publications have appeared showing the effect of oxygen on the breakdown stress of liquids. This effect is particularly marked in liquid argon. As was mentioned in Chapter 12.6, liquid argon differs from other liquids in that free electrons have been observed with a mobility of the order of 10^2 cm^2 v^{-1}sec^{-1}, which is about 10^6 times greater than the mobility of negative ions in liquid dielectrics. Electron transfer in liquid argon can be described using the classical kinetic theory. Swan [55] calculated that in electric field of intensity of 1 mv/cm the electrons in this liquid can acquire an energy of about 5·5 ev. However, the number of free electrons depends greatly on the presence of oxygen molecules, which exhibit a large electron capture cross section (Biondi [56]). The mean free path of an electron between two collisions with atoms is given by the expression:

$$\lambda = (nNQ \times 10^{-2})^{-1}, \qquad (19.5)$$

where n denotes percentage concentration of oxygen molecules in liquid argon, N—the number of argon atoms per cm^3 ($N = 2 \times 10^{22}$ cm^{-3}) and $Q = 10^{-14}$ cm^2. For $n = 0·001\%$ the mean free path is equal to 5×10^{-4} cm. A different mechanism of electron capture appears at about 7 ev (Hurst and Borther [57]) but has a less significant effect.

Both processes reduce the number of free electrons in the liquid, and therefore stronger fields are required to produce an electron avalanche and electric breakdown. The presence of oxygen in the liquid may also produce a double layer next to the electrodes which reduces the emission from the cathode.

The first studies of this kind on normal liquids were carried out by Sletten [24], Sletten and Lewis [198], Brignell and House [175] and others. Sletten and Lewis showed that in hexane the influence of oxygen causes the electric strength to rise from 0·7 mv/cm to 1·3 mv/cm. Nitrogen, hydrogen and carbon dioxide, on the other hand, give no effect. These results were explained as being due to the electronegativity of oxygen by which it traps electrons and gives rise to the formation of heavy ions with low mobility which hinder the breakdown mechanism. It should be emphasized that they did not claim that the electrodes in any way affected the breakdown value, but rather that the previous action of the electric field had some influence (pre-stressing).

Brignell and House [175] confirmed that oxygen behaves in this manner. For hexane they obtained breakdown values of 1·05 mv/cm in the presence of oxygen, and 0·54 mv/cm for the degassed liquid. (The electrode spacing was 200 μm.) They observed no change in the breakdown value over the temperature range from $-60°$c to $+60°$c. Only at temperatures close to the boiling point ($\sim 69°$c) did the breakdown strength decrease sharply.

19.12. *Other Methods of Studying Liquids Subjected to High Electric Fields.* Among other studies of electrical breakdown in liquids, those which

24

employ photographic methods for the studying of discharge and breakdown should also be mentioned here. Komelkov [58], Skowroński [59],
Liao and Anderson [60] and Sommerman *et al.* [61] have published photographs showing corona discharges, the initiation and development of

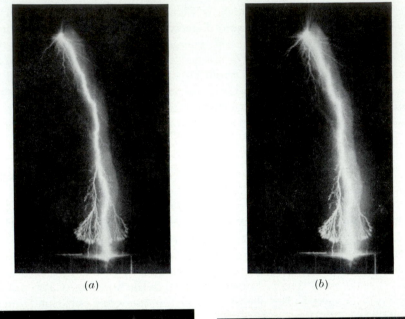

(a)   (b)

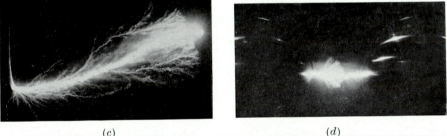

(c)   (d)

Fig. 19.18. Pictures of electric breakdowns in transformer oil: (*a*) and (*b*) rectified
pulsating voltage, plate–point separation 20 mm, preliminary breakdown
and the channels are from anode, main breakdown—from cathode;
(*c*) multiple breakdown; (*d*) breakdown near the point anode. Ref.:
Skowroński [59].

breakdown, and streamer discharges which contain a number of avalanches
taking place in parallel. In fig. 19.18 there are shown several photographs
of a technical grade oil containing small amounts of impurities, traces of
water and gases. As can be seen, the development of a spark is similar to
that which takes place in air. Skowroński considers that electrons
emitted from the cathode play an active part in the development of the
discharge, but that it is initiated in the form of a streamer from the anode
in channels (non-luminous) from which oil has been removed by heavy

positive ions and filled by gases. Discharge takes place when one of the channels reaches the cathode. The so-called bridge discharges are caused by electrons emitted from the cathode or gas bubbles near the anode.

Under the action of the electric field foreign particles are dragged into the region of the spark-gap from the surrounding liquid, concentrate in it and form a chain between the electrodes. For this reason the density of impurities and their nature play an important rôle in the mechanism of

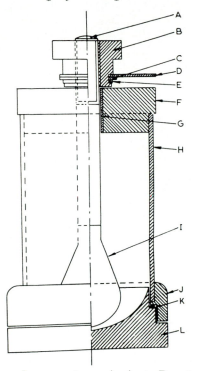

Fig. 19.19. Cup-sphere spark-gap system. *A*, pivot; *B*, nut; *C*, washer; *D*, pointer; *E*, spring; *F*, cover; *G*, cover sleeve; *H*, cylinder; *I*, upper electrode; *J*, ring; *K*, washer; *L*, spherical-cup electrode. Ref.: Skowroński [239].

breakdown. The shape of electrodes and the magnitude of the applied voltage also have an effect on breakdown values. Skowroński emphasizes the effect of the so-called 'electrical wind' which occurs well before the initial discharge (corona) and which may cause an increased breakdown voltage by 'blowing' away the bridge which forms between the electrodes.

Skowroński proposed a different electrode arrangement for testing commercial insulating liquids. A spherical electrode of 20 cm radius inside a hemispherical cup electrode of 40 cm radius should replace the test cell recommended by C.E.J. (two spheres 12·5 cm in diameter at a distance of 2·5 mm) and that by V.D.E. (spherical electrodes with a radius of 25 mm) (see fig. 19.19).

Some authors have studied the effect of gas bubbles and vapour bubbles in the bulk of the liquids and near the electrode surfaces on conduction and breakdown in high fields. It is found that ions produced in these bubbles may lead to the formation of free radicals, which in turn lead to the formation of new gas bubbles of hydrogen or methane.

Basseches and Maclean [63] suggested that unsaturated hydrocarbons are formed in paraffin hydrocarbons by the following reactions:

$$\left. \begin{array}{l} CH_3 + C_nH_{2n+2} \rightarrow CH_4 + C_nH_{2n+1}, \\ 2C_nH_{2n+1} \rightarrow C_nH_{2n+2} + C_nH_{2n}. \end{array} \right\} \qquad (19.6)$$

Buchholz [64] pointed out the chemical changes which lead to increased viscosity and Watson and Higham [65] have investigated the formation of gaps and cavities during breakdown.

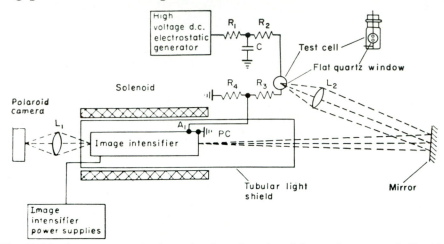

Fig. 19.20. Arrangement for investigation of pre-breakdown phenomena in liquids. Ref.: Smith *et al.* [199].

It seems necessary to carry out further studies in this field using modern methods such as the Schlieren technique, gas bubble chambers (applied mostly in studies of nuclear physics), diffusion chambers and scintillation counters. Some of these studies were commenced some time ago by Nagao [66], Buchholz [64], Bates and Kingston [67], and more recently by Murooka *et al.* [117], Watson and Sharbaugh [122] and Gosling [116]. See also papers presented at the Durham Conference [125].

Between 1963 and 1966 the Schlieren technique was applied in research on electrical breakdown in liquids. Chadband and Wright [152] constructed an apparatus based on this technique, with a high-speed camera operating at 6×10^4–10^5 revolutions per minute and a linear sweep speed of the order of 1 cm/μs. They obtained a number of very interesting photographs of discharge taking place in liquids.

Brown [233], using the Schlieren technique adapted for microscope observations, claimed that there is rotation in the liquid which is dependent on the current intensity (1–500 μA) and the dipole moment. This effect is extremely small in hexane (see figs. 19.20, 19.21, 19.22).

19.13. *Electrical Breakdown in High-frequency Fields.* Studies of the effect of frequency on breakdown strength in dielectric liquids are of great importance in radiotechnology. Unfortunately they have usually

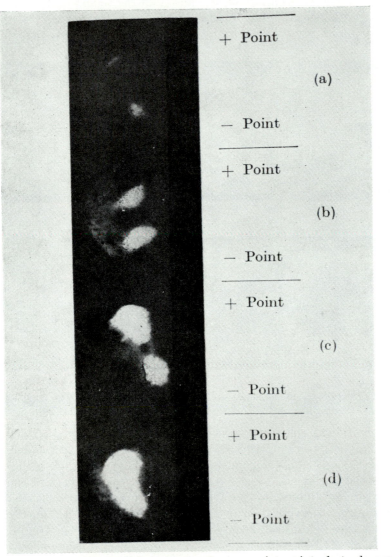

+ Point

(a)

− Point

+ Point

(b)

− Point

+ Point

(c)

− Point

+ Point

(d)

− Point

Fig. 19.21. Pre-breakdown light emission from a point–point electrode system. Ref.: Smith *et al.* [199].

been carried out on liquids of technical grade and in many cases were not reproducible; the results obtained were therefore of no help in reaching useful conclusions. Only the most important work by Koppelmann [68] and Walter and Inge [69] will be mentioned here. For low frequencies

(0–60 c/s) the breakdown strength for hexane and oil increased with frequency by about 60–100%; for high frequencies (0·4–12 Mc/s) the breakdown strength decreased with increased frequency and fell to almost zero. It was supposed that thermal breakdown took place. The result is, however, largely dependent on the purity of the tested liquid.

A very interesting effect was observed in studies on the relationship between breakdown values and the wave form of the applied voltage.

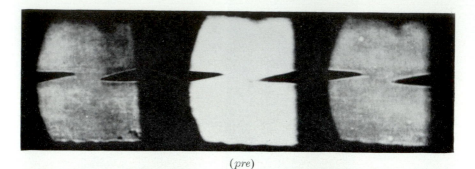

(*pre*)

(*post*)

Fig. 19.22. High-speed Schlieren photographs of pre and post-breakdown phenomena in *n*-hexane. Ref.: Hakim and Higham [238].

In non-purified liquids breakdown strength depended on the r.m.s. value of the applied voltage, while in pure liquids on the peak value of the applied voltage. This dependence may well suggest that in liquids of commercial purity the breakdown mechanism is of a thermal character, and in purified liquids its character depends on the electrical conditions. It was also observed that in the same liquid, initially purified only partially, the breakdown strength increased with the number of breakdowns, initially being dependent on the r.m.s. value of the applied voltage, then on the peak value of the applied voltage. This may well suggest that a purification process occurs during the first series of breakdowns.

The effect of impurities and additives (liquid and gas) is more pronounced at lower frequencies than at high frequencies. For oil, it was found that filtration and removal of water increased the breakdown strength by about three times at a frequency of 50 c/s but by only 1·3 times at frequencies of the order of 10^5 c/s.

The relationship between breakdown strength and temperature shows an optimum in the temperature range between 30–80°C for many liquids, however the maximum value decreases with an increase in frequency (from 0·4 Mc/s to 12 Mc/s). The observed maximum was related to those changes in loss angle (inversely proportional to temperature) for which a minimum occurs at the same temperature. An increase in hydrostatic pressure of from 0–800 mm Hg produces an increase in breakdown strength from 200% to 300%.

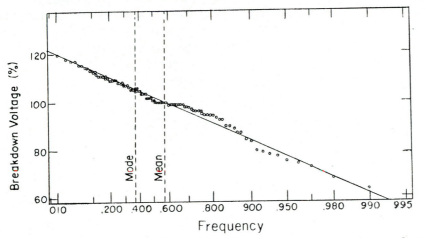

Fig. 19.23. Breakdown voltage distributions for sphere to plane gaps in transformer oil. Ref.: Sletten and Dakin [213].

Ushakow [156] carried out a number of investigations, both experimental and theoretical, on the development of breakdown discharge in polar and non-polar liquids. Out of the first group they used distilled water ($\rho = 3 \times 10^5$ ohm-cm) and technical grade ethyl alcohol and the non-polar liquid was transformer oil of technical grade. Experiments were performed using voltage impulses and point–plane and point–point electrode arrangements. The gap settings were 20, 50 and 165 mm.

Development of the discharge initiated by a leader consisting of a head and a channel was different in polar and non-polar liquids. In polar liquids it developed in steps with light pulses and simultaneous current pulses followed by intervals when no light output was observed, and in non-polar liquids the development of the discharge has a continuous character with light pulses and a continuous weakly luminous channel. They also observed that for a negative-point positive-plate arrangement, the applied voltage at which discharge begins is much greater than for a positive-point negative-plate arrangement. In transformer oil the touching of a plate electrode did not always cause breakdown.

The authors proposed a new approach to the theory of liquid breakdown, and explained the dark intervals and light pulse during the development of a leader as being due to formation and neutralization of the space charge in the front of the streamer head, which has the same polarity as the point electrode. In the case of distilled water they are probably heavy OH$^-$ ions (negative-point electrode), and for oil they are hydrocarbon molecules or carbon with attached electrons.

In recent years research has significantly developed upon the dependence of breakdown current and stress on frequency (Sletten and Dakin [213] and Noisseir [216]. As an example fig. 19.23 shows the relationship $U = f(v)$ obtained by Sletten and Dakin [213].

Sletten and Dakin [213] performed a series of measurements with transformer oil using relatively large gap settings (about 1 cm) and alternating fields (60 c/s) with an applied voltage of about 450 kv. They determined the effect of impurities, the shape of electrodes and gas bubbles, and also observed luminescence on electrode surfaces.

Chapter 20

Theory of Conduction in Dielectric Liquids at High Fields and Electrical Breakdown

20.1. *Basic Theoretical Problems*. In Part 3, the most important basic theories of conductivity in dielectric liquids were considered. The theories of Jaffe, Plumley, Frenkel, Rice, the author's and that based on semi-conductor theory were among those discussed. In many cases these theories can be adopted to explain conductivity in very high-stress fields.

Below we shall consider a few theoretical aspects which are most characteristic of the problem of conductivity in very high-strength fields and those involved in electric breakdown. Such aspects include studies of the effects of:

(1) Electron emission from the cathode to the liquid due to the electric field applied.

(2) The electrode material, its shape and especially the radius of the electrode curvature.

(3) Multiplication of electrons due to collision and avalanche ionization.

(4) The field distribution between the electrodes.

(5) The mechanism of charge transfer carriers (electrons and ions) in the liquid.

(6) The possibility of the dissociation of liquid molecules due to collisions with electrons and ions accelerated by the electric field.

(7) The temperature.

(8) Cavitation and gas and vapour bubbles.

(9) Suspended matter in the liquid.

20.2. *Energy Levels of Electrons at the Metal–liquid Boundary*. The treatment of electron emission from metals has been developed in detail with the help of the modern theory of metals and especially by the quantum theory of metals and Fermi–Dirac statistics. The distribution of electron energy levels in a metal in the vicinity of its surface can, according to this theory, be represented in the form shown in fig. 20.1. Inside the metal the majority of the electrons are bound to individual atoms and are at separate discrete energy levels of these atoms. The outermost, or valence, electrons of the atom can move freely. Their energy levels are also quantized but they form a conduction band in the metal, and no electron can have an energy greater than the Fermi level at absolute zero. To force an electron completely out of the metal work must be done, known as the work-function $e\varphi$, which could be explained physically as the energy that must be added to an electron for it to be emitted while over-coming the electrostatic forces of the positive ions of the crystalline lattice at the surface. The work function can be defined as the difference between the energy of an electron at rest some distance outside the metal surface and the Fermi energy level. The work function can be reduced by temperature and high fields. In this way the emission of electrons can be increased. Denoting the energy of an electron at rest outside the metal surface by W_p at distance x from the surface, then:

$$W_p = \frac{e^2}{4\varepsilon x}, \tag{20.1}$$

where ε is the permittivity of the substance outside the metal.

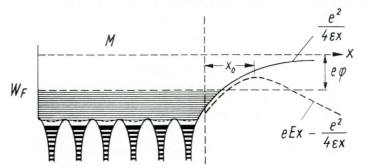

Fig. 20.1. Energy distribution at a metal–liquid interface in the presence of an electric field. W_F denotes Fermi energy level; $e\varphi$, electron work function from the metal; M, metal.

At a temperature T the electron current from the metal is given by Schottky's [70] or Richardson's [71] equation:

$$I = AT^2 \exp\left(-\frac{e\varphi}{kT}\right), \tag{20.2}$$

where A denotes a constant characteristic for a given metal and k is Boltzmann's constant.

When there is a substance of a dielectric constant ε outside the metal and an electric field of intensity E also acts, the energy distribution alters and is given by:

$$W_p' = -eEx + \frac{e^2}{4\varepsilon x}. \tag{20.3}$$

At a distance $x_0 = \frac{1}{2}\sqrt{(e/\varepsilon E)}$ from the metal surface there is a maximum potential energy which reduces the work function of an electron passing from the metal to the liquid by the value $e\Delta\varphi$, where:

$$\Delta\varphi = \sqrt{\left(\frac{eE}{\varepsilon}\right)}. \tag{20.4}$$

Baker and Boltz [5] modified Schottky's equation and obtained an equation for the emission current I from the metal surface into the liquid at a temperature T and field intensity E in the form:

$$I = AT^2 \exp\left\{ -\frac{e[\varphi - \sqrt{(eE/\varepsilon)]}}{kT} \right\}. \tag{20.5}$$

Fowler and Nordheim [72] derived an equation for emission current from the metal caused by an electric field E (cold emission) having the form:

$$I = AE^2 \exp\left(-\frac{b}{E} \right), \tag{20.6}$$

where A and B are functions of φ. This equation can be applied to very high fields.

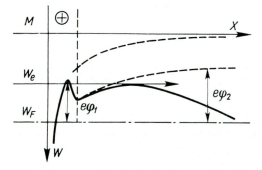

Fig. 20.2. Potential barrier near the cathode for positive ion layer interaction. M, denotes metal; W_F, Fermi energy level; W_e, electron energy; $e\phi_1, e\phi_2$, electron work functions. Ref.: Lewis [1].

Lewis [1] also suggested that irregularities of the cathode surface and the layer of positive ions next to the cathode may also affect the work function of the electrons passing from the cathode into the liquid. The relationships between the energy levels are shown in fig. 20.2. As can be seen, the main difference in the distribution of potential energies near the cathode is in the formation of two maxima, the second one being less than the first one. Due to a tunnelling effect the electrons can pass the first narrow maximum and enter the liquid more easily. Equation (20.6) is thus modified by a new factor D—an electron tunelling coefficient for the first barrier, giving in effect an increase in emission current.

House [11], Goodwin and Macfadyen [9], Dornte [7], Green [10], Morant [76, 131] and Lewis [1] have considered the possibility of explaining the electron emission from a metal into a liquid by applying Fowler and Nordheim's theory. Taking into account the effect of the layer of positive ions in front of the cathode, they obtained an equation:

$$I = AE^2 \varphi_0^{-1} \exp\left[-6{\cdot}8 \times 10^7 (\varphi_2^{3/2} E^{-1} + \Phi) \right], \tag{20.7}$$

where A is a constant characteristic for a given metal, φ_0 denotes the work function, φ_2—the reduced work function due to the electric field, and Φ is a function of φ_0 and φ_2. The distribution of electron potential energies is shown in fig. 20.3.

The intensity of an electric field necessary to produce cold emission is of the order of 10^6 v/cm; for lower electric fields (even 10^5 v/cm) the cold emission is very small because for emission to take place, the width of the barrier XY must be of the order of 10 Å.

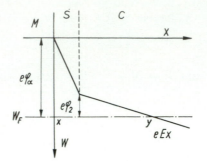

Fig. 20.3. Energy distribution near the cathode as in fig. 20.2. Barrier XY approximately 10 Å, S denotes positive ion layer; C, liquid; M, metal; W_F, Fermi energy level.

Le Page and Du Bridge [6] and House [11] also considered the possibility of the cold emission of a small number of electrons even in lower fields due to the tunnelling effect.†

20.3. *Distribution of Potential inside a Dielectric Material.*   Theoretical studies on the distribution of electrical potential inside a dielectric material have been carried out by Mott and Gurney [73], Lane [74], Fowler [72],

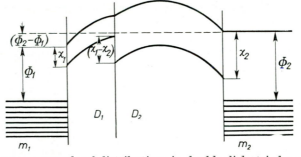

Fig. 20.4. Electron energy level distributions in double dielectric layer between the plates of a condenser. Φ_1, Φ_2 denote electron work functions from the metal; χ_1, χ_2, electron energies in lower part of the conductivity band. Ref.: Morant [76].

Skinner *et al.* [75] and Morant [76, 131].   Potential distribution in a dielectric filled with an atmosphere of electrons has been studied particularly carefully.   In fig. 20.4 the energy level distribution of electrons in two dielectrics placed between two metal electrodes is shown.

The expression for the potential distribution V in a liquid in terms of the distance x from one of the electrodes, according to Mott and Gurney,

† See also the new papers of the works with special tunnel cathode [275, 276, 307].

has the form:

$$V = -\frac{2kT}{e}\ln\left[\left(\frac{2\pi e\rho}{\varepsilon kT}\right)^{1/2}x+1\right],\tag{20.8}$$

where ρ denotes the electron density and equals:

$$\rho = 2e\left(\frac{2\pi mkT}{h^2}\right)^{3/2}\exp\left[\frac{-e(\Phi-\chi)}{kT}\right],\tag{20.9}$$

$e\Phi$ denotes the work function from metal to vacuum and χ—the depth of the potential well in the dielectric (see fig. 20.4).

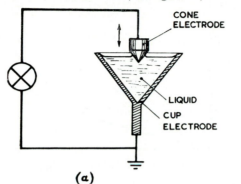

(a)

Fig. 20.5. Experimental arrangement for detecting space-charge potential at metal–hexane interface. Ref.: Morant [218].

The potential difference V_d between two electrodes separated at a distance d is:

$$V(d) = \frac{1}{e}(\Phi_2-\Phi_1).\tag{20.10}$$

In the case shown in fig. 20.4 for low dielectric materials:

$$\Phi_1 = \Phi_2+kT\ln\left(\frac{\varepsilon_1}{\varepsilon_2}\right).\tag{20.11}$$

Morant [218] in his experimental work (a diagram is shown in fig. 20.5) demonstrated the change of potential distribution with the space charge in a liquid.

In his experimental work he used a metal cone electrode immersed in hexane in a metal cup. By changing the position of the cone it was possible to measure the change of potential of this electrode. In fig. 20.5 a physical picture of the process is shown, and fig. 20.6 gives the relationship between the electrode potential and its position. It should be noted here that a change in the capacity of the system may also have an effect on the measured value of the potential.

Potential distribution has also been studied by Taris and Guizonnier [230] and Zein El-Dine *et al.* [225, 283, 293, 296, 297].† The results are discussed in another part of this work.

† See in particular the last paper of Hill and House [296] with using the Kerr electro-optic effect.

20.4. *Collision Ionization.*   Beside electron emission from the cathode surface into the liquid, the phenomena of collision ionization and avalanches in the liquid itself are also important. The theory of these phenomena is based on the theory of ions in gases, which has been thoroughly studied and confirmed. A review of these studies is given in the monograph by von Engel [17].

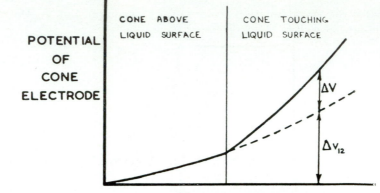

Fig. 20.6. The potential of the cone electrode in fig. 20.5 as a function of its vertical position. Ref.: Morant [218].

The theory of ions in gases was applied to dielectric liquids by Schumann [16] and Nikuradse [2] a long time ago [1932–34], who obtained very interesting results by comparing theoretical equations with experimental results. There is, however, a general tendency not to apply this theory to liquids because of the great differences in other properties of the gas and liquid phases.

The basis of the ionization theory is the assumption that a free electron in a gas or liquid can be accelerated by an electric field and along a 1 cm path produce α_i positive and α_i negative ions, where:

$$\alpha_i = z_i \frac{1}{v}, \tag{20.12}$$

where v is the velocity of the electron and z_i—the number of secondary electrons produced in 1 sec due to ionization. Taking $P(E)$ as the probability of a collision leading to ionization, $f(E)$ as the emission spectrum, and λ as the mean free path of electrons in the liquid, then:

$$z_i = \int_{eV_i}^{\infty} \frac{v}{\lambda} P(E) f(E) \, dE, \tag{20.13}$$

where v/λ is the number of collisions per second.

Another phenomenon which appears to occur when free electrons move in a gas or liquid is the formation of negative ions due to the possibility of electron attachment to electrically neutral atoms or molecules. If the number of electrons captured in this way along a 1 cm path of neutral atoms or molecules is denoted by n_e, it is possible to write down an expression for the number of electrons n_e and negative ions n_- formed along a path x cm long as follows:

hence:

$$dn_e = (\alpha_i - \eta)\, n_e\, dx,$$

$$\frac{n_e}{n_0} = \exp\left[(\alpha_i - \eta)\,\delta\right];$$

hence:

$$dn_- = \eta n_e\, dx = \eta n_0 \exp\left[(\alpha_i - \eta)\,\delta\right] dx,$$

$$\frac{n_-}{n_0} = \frac{\eta}{\alpha_i - \eta} \exp\left[(\alpha_i - \eta)\,\delta\right] - \frac{\eta}{\alpha_i - \eta}.$$

$$\frac{I}{I_0} = \frac{n_e + n_-}{n_0} = \frac{\alpha_i}{\alpha_i - \eta} \exp\left[(\alpha_i - \eta)\,\delta\right] - \frac{\eta}{\alpha_i - \eta}, \tag{20.15}$$

where δ is the distance between electrodes and n_0—the electron density on the cathode surface. These expressions are mainly used for determining the coefficients α_i and η.

When η is very small compared with α_i, for very high stresses the expression (20.15) may be reduced to:

$$\frac{I_0}{I} = \exp\left(\alpha_i \delta\right), \tag{20.16}$$

or to a first approximation (for $\alpha_i \delta \ll 1$):

$$\frac{I}{I_0} = 1 + \alpha_i \delta,$$

where I_0 depends on the electron emission from the cathode, and α_i is Townsend's coefficient.

These expressions help to explain why the $I = f(\delta)$ curves show a tendency towards saturation, which is greater for lower values of the field (see fig. 20.7). Nikuradse [2], by selecting the appropriate coefficients α_i and η, obtained good agreement between theoretical and experimental results. For toluene, for example, he found that α_i increases from 34 in a field of 150 kv/cm to 960 at 310 kv/cm, and η increases over the same range of fields from 67 to 915; the ratio η/α_i decreases from 2 to 0·95. According to Nikuradse, the coefficients α_i and η satisfy the following relationships:

$$\begin{aligned} \alpha_i &= a(E - E_1)^2, \\ \eta &= b(E - E_2)^2, \end{aligned} \tag{20.17}$$

where a and b are constants. In both cases there are certain critical values for E_1 and E_2, and as a rule the values for E_2 are lower than those for E_1.

In recent studies [132] concerned with the ionization coefficient α_1 and electron attachment η in gases, the theory of ions in gases has been extended leading to the following expression:

$$\frac{I}{I_0} = \frac{\dfrac{\alpha_i}{\alpha_i-\eta}\exp\left[(\alpha_i-\eta)\,\delta\right]-\dfrac{\eta}{\alpha_i-\eta}}{1-\dfrac{\gamma\alpha_i}{\alpha_i-\eta}\{\exp\left[(\alpha_i-\eta)\,\delta\right]-1\}}, \tag{20.18}$$

where γ denotes the secondary ionization coefficient (produced by heavy ions), other symbols having the same meaning as before.

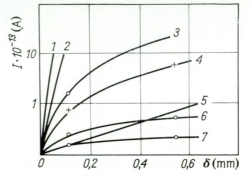

Fig. 20.7. Electrode separation (in mm) dependence of intensity of current I (in A × 10^{-13}). Ref.: Nikuradse [2].

Nr	1	2	3	4	5	6	7
E [kv/cm^{-1}]	310	260	310	260	150	180	150
α_{ef}	960	484	—	—	33·6	—	—

Ref.: Nikuradse [2].

The breakdown criterion is given by the expression:

$$\frac{\gamma\alpha_i}{\alpha_i-\eta}\{\exp\left[(\alpha_i-\eta)\,\delta\right]-1\} = 1. \tag{20.19}$$

In the case when γ equals zero, the expression (20.18) has a form similar to (20.15).

Prasad et al. [77–80], in their work on the measurements of the $I = f(\delta)$ function using a new method, found with the help of eqn. (20.18) the collision cross section and the cross section for electron attachment in various gases. They also explained the various processes involved, the formation of ions, dissociation of molecules, and the secondary ionization process, etc. This may be useful in further studies of the mechanism of ionic motion in liquids. The theory of electron attachment in liquids has been discussed by Crowe [33], Stacey [81], LeBlanc [82] and Terlecki [37] (see Chapter 16.4) and recently by Essex, Secker [247], Adamczewski [280], Gallagher [291] and others.

20.5. *Breakdown Mechanism. Hypotheses.* A number of hypotheses have been put forward to explain the experimental results obtained in breakdown studies, attributing them to changes in liquid properties and experimental conditions. Some of the hypotheses will be discussed below, both those which have been generally accepted and those which give agreement between theoretically calculated values and experimental results. It is accepted that electrons emitted from the cathode surface are accelerated by the field and retarded by collision with liquid molecules. The mechanism of slowing down electrons in liquids is different from that in gases (see Hippel [52], Lewis [1], etc.). In gases, an electron loses its energy mainly in ionization processes and in excitation collisions with atoms or molecules. In a liquid, since the density is several hundred times greater than that of gases, the electron is also acted upon by electrostatic forces (polarization) and therefore influenced by its surroundings. The mean free path of an electron is much smaller (about 10^{-7} cm), and it can acquire less energy than in gases later losing it in collisions with molecules. The energy lost by an electron is transformed into atomic vibrations along chemical bonds, rotation of individual groups of atoms and vibrations of the whole molecular chain, but most of it is changed into heat. The energies are much smaller than excitation energies of atoms or ionization energies, and therefore the acceleration of electrons to velocities corresponding to excitation energies of molecules or ionization energies of 10–20 ev is more difficult and requires much stronger fields.

Only in certain cases (e.g. in the presence of some kind of luminescent material) can it cause electroluminescence or electron multiplication.

The numerical values for these energies can be classified as follows:

(1) Ionization energy of atoms or molecules: 10–24 ev.

(2) Excitation energy for atoms or molecules: 2–7 ev.

(3) Energies needed to cause vibration of individual chemical bonds or whole 'crystalline' or 'pseudo-crystalline' structures of molecules: (0·111–0·366 ev).

The problem of electron retardation in solids has been studied by Hippel [52], Fröhlich [83], Seeger and Teller [84] and in liquids by Frenkel [85], Crowe *et al.* [86], Lewis [1, 13], Adamczewski [36] and others.

Lewis [13] investigated this problem in more detail for a group of saturated hydrocarbons in view of its connection with breakdown theory. In these liquids the vibration and bonding energies correspond to the frequencies given in table 20.1. An electron moving in these liquids therefore encounters several potential barriers which correspond to free molecular vibrations (fig. 20.8).

According to Lewis, when an electron has an energy which corresponds to natural frequencies of a molecule, an energy exchange may take place quite easily.

E_{bd} must satisfy the following relationship:

$$eE_{bd}\lambda = Kh\nu, \tag{20.20}$$

25

where e is the electronic charge, λ—the mean free path of an electron in the liquid, $h\nu$—one of the vibration energies of molecule and K—a constant.

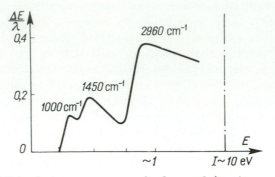

Fig. 20.8. Loss ΔE in electron energy over its free path λ as it crosses the vibration potential barriers in hydrocarbon molecules. Energy axis scale only approximate. I denotes ionization energy. Ref.: Lewis [13].

Table 20.1. Characteristic frequencies for hydrocarbon bonds

Group	Stretching	$h\nu$ (ev)	Group	Bending	$h\nu$ (ev)
CH_1, CH_2 CH_3 $C—C$	2960 890–1000	0·37 0·11–0·14	CH_3 CH_1, CH_2 Benzene ring (CH)	1000 1450 ~ 3000	0·12 0·185 $\sim 0·37$

After Lewis [13].

The mean free path of an electron can be evaluated from the equation:

$$\frac{1}{\lambda} = \frac{\rho}{M} N \sum_i n_i Q_i, \qquad (20.21)$$

where ρ is the density of the liquid, N—the number of molecules per cm³, M—the molecular weight and the expression $\sum_i n_i Q_i$ gives the sum of active cross sections for inelastic collisions with those parts of the molecule which play an important part in slowing down electrons, i.e. individual carbon atoms or hydrogen atoms, groups such as CH_2 or CH_3, various types of chemical bonds, and vibration or rotation of certain parts of the molecules; n_i is the number of CH_2 or CH_3 groups in the molecule. Lewis also gives [1, 13] a general formula for E_{bd}:

$$E_{bd} = KN\nu e^{-1} \sum_i n_i Q_i. \qquad (20.22)$$

Figure 20.9 shows the loss in energy of an electron on its mean free path passing through a potential barrier in a molecule. The straight line AB shows the energy the electron gains from the field. In section AB (inside the curve) the electron loses its energy to a molecule and is slowed down to the velocity corresponding to point A; for energy greater than B the electron is again accelerated by the field to the ionization energy level I. In this case electron multiplication takes place.

The condition $B = I$ was accepted by Fröhlich and Lewis as the breakdown criterion. On this basis Lewis [13] developed a theory of breakdown for a group of saturated hydrocarbons. He assumed for the critical barrier

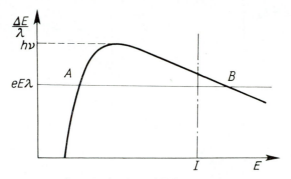

Fig. 20.9. Variation of $\Delta E/\lambda$ with E (where ΔE denotes the change in electron energy, λ the mean free path) at crossing the potential barrier portion. AB denotes energy gain in the electric field; I, the ionization energy. Ref.: Lewis [13].

for electrical breakdown the potential barrier which corresponds to a stretching vibration frequency of 2960 cm^{-1} for CH, CH$_2$ and CH$_3$ groups. He obtained the expression for breakdown stress E_{bd} for this group of liquids in the form:

$$E_{bd} = kN(n_1 Q_1 + n_2 Q_2 + n_3 Q_3), \qquad (20.23)$$

where n_i is the number of groups of type i in the molecule and Q_i—the active cross section for energy exchange for group i. The expression is based on the theory discussed in § 19.4.

Lewis [13] also derived some general relationships which are shown in fig. 20.10; he did not obtain agreement for all the experimental results however.

Adamczewski [36] has proposed a very simple theory connecting experimental results obtained for breakdown values with the structure of the liquid, and obtained extremely close agreement between theoretical values and the results obtained by various workers. Since such close agreement between theoretical experimental results has not been given by other theory so far and as the basic assumptions of the physical mechanism are very simple, the author considers that it would be useful to discuss the theory in more detail. It seems possible that it could form the basis for a more exact theory of breakdown.

The theory is based on the assumption that the active cross section for the retardation of electrons accelerated by the field is proportional to the longitudinal geometric cross sectional area of the molecule. For example, a saturated hydrocarbon molecule could be considered as a cylinder or rectangular prism with a fixed base and a length increasing proportionally to the number of CH_2 groups (fig. 20.11); its active cross section will therefore increase with the number of C—C bonds along the axis of the molecule. Thus the active cross section will increase in proportion to $(n-1)$ where n is the number of carbon atoms in the molecule.

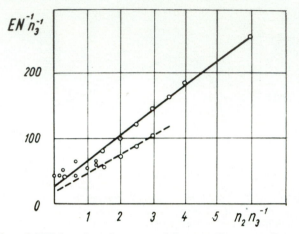

Fig. 20.10. Plot of $EN^{-1}n_3^{-1}$ against $n_2 n_3^{-1}$ for paraffins according to eqn. (20.23). Solid line is for pulse voltage, broken line for steady voltage. The open circles refer to a liquid neglecting the effective cross section Q. Ref.: Lewis [1, 13].

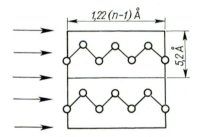

Fig. 20.11. Electron retardation by hydrocarbon molecules. Ref.: Adamczewski [36].

Taking the radius of the cylinder to be r and the axial projection of a C—C bond to be h_0 ($h_0 = 1.54 \sin 55° = 1.23$ Å) it is quite easy to calculate the active cross section of a single molecule, viz.:

$$Q = 2rh_0(n-1) = Q_0(n-1), \tag{20.24}$$

where $Q_0 = 2rh_0$ and is constant for the whole series of straight-chain saturated hydrocarbons. The active cross section per cm³ will be:

$$S = Qp = Q_0(n-1)\frac{\rho}{M}N = A(n-1)\frac{\rho}{M}, \tag{20.25}$$

since the number of molecules per cm³ is equal to $p = (\rho/M)N$, where ρ is the density, M—the molecular weight of the liquid, N—Avogadro's number and $A = 2rh_0 N$ which is constant for the whole series of the liquids. In each collision an electron loses an amount of energy equal to $E_0 = h\nu$.

The breakdown criterion is based on the assumption that the energy gained by the electron on its mean free path λ due to the work done L by

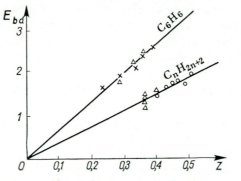

Fig. 20.12. The dependence of breakdown strength E_{bd} (in mv cm^{-1}) on the effective cross section Z of liquid molecules for benzene and aliphatic hydrocarbons. The straight lines are Adamczewski's [36] equation, and the experimental values from works by Sharbaugh *et al.* [32].

an electric field E $(L = eE\lambda)$ is not less than the energy which retards the electron on this path:

$$eE\lambda = E_0 = h\nu. \qquad (20.26)$$

Since:

$$A(n-1)\frac{\rho}{M} = \frac{1}{\lambda}, \qquad (20.27)$$

we obtain:

$$eE = h\nu A(n-1)\frac{\rho}{M}. \qquad (20.28)$$

The expression (20.26) in fact has the same form as those given by other authors (Crowe [33] and Lewis [13]), but the expression for the mean free path (20.27) is characteristic of the author's assumptions. The final equation for breakdown stress in terms of the parameters characterizing a given liquid (ρ, M, n) has the following form:

$$E = E_{bd} = \frac{h\nu}{e}A(n-1)\frac{\rho}{M} = B(n-1)\frac{\rho}{M} = BZ, \qquad (20.29)$$

where:

$$A = 2rh_0 N, \quad B = A(h\nu/e), \quad Z = (n-1)\rho/M.$$

From the expression (20.28) we can see that for straight-chain saturated hydrocarbons E should be a linear function of $(n-1)\rho/M$ which may be easily found from tables or by experiment.

Figure 20.12 (bottom curve) shows the graph illustrating the relationship (20.29) for ten saturated hydrocarbons. The experimental results are seen to be in very good agreement with the straight line passing through the origin. The slope of the line is determined by quantity B which is the product of two unknown values r and $h\nu$ connected with the radius of the molecule and the quantum energy exchanged between the electron and the molecule.

The author wishes to show that the value of B can be assessed in a reasonable way using only established experimental data for the properties of the liquids, and without using any data on breakdown values. Thus the active cross section for the whole series has been established using the number of molecules per cm³ and their molecular structure ([36], Moore *et al.* [128]). The diameter of the cross section varies between 4·9 and 5·2 Å. The present author has performed calculations for saturated straight-chain and branched hydrocarbons using literature values for r, h_0 and $h\nu$. The number of molecules per cm³ was calculated from the density and the molecular weight, giving the following quantitative expressions for electrical breakdown:

For straight-chain saturated hydrocarbons:

$$E_{bd} = 0\!\cdot\!114 \times 4\!\cdot\!9 \times 10^{-8} \times 1\!\cdot\!23 \times 10^{-8} \times 6\!\cdot\!024 \times 10^{23} \frac{\rho}{M}\,(n-1)$$

$$= 4\!\cdot\!15 \times 10^{7} \frac{\rho}{M}\,(n-1)$$

$$= 2\!\cdot\!60 \times 10^{6} \frac{n-1}{n+2\!\cdot\!27}. \tag{20.30}$$

For saturated branched hydrocarbons:

$$E_{bd} = 1\!\cdot\!06 \times 10^{6} \frac{(2\!\cdot\!45 + 0\!\cdot\!35\,m)\,(n-m-1)}{n+2\!\cdot\!27}. \tag{20.30a}$$

Expression (20.30a) is a more general one than (20.30). It includes both the number of carbon atoms in the molecules, n, and the number of branches in the molecule, m. For $m = 0$ the expression (20.30a) takes the form of (20.30):

For aromatic normal hydrocarbons:

$$E_{bd} = 4\!\cdot\!34 \times 10^{6} \frac{n-4}{n-0\!\cdot\!427}. \tag{20.31}$$

For branched aromatic hydrocarbons:

$$E_{bd} = 1\!\cdot\!06 \times 10^{6}(4\!\cdot\!24 + 0\!\cdot\!35\,m) \frac{n-m-4}{n-0\!\cdot\!427}. \tag{20.31a}$$

These formulae are similar in that they are derived from the physico-chemical properties of the liquids only, without introducing any factors based on breakdown measurements, and give breakdown values in agreement with those obtained by experiment as can be seen from table 20.2, in which the values calculated using the above equations are compared with the experimental results obtained by Sharbaugh *et al.* [32].

Table 20.2. Breakdown stresses for hydrocarbons liquids–

n	Liquid	M	ρ	E 10^6 v cm Experimental	Theoretical
	Saturated hydrocarbons C_nH_{n+2}				
5	Pentane, C_5H_{12}	72·14	0·627	1·44	1·44
6	Hexane, C_6H_{14}	86·17	0·659	1·56	1·57
7	Heptane, C_7H_{16}	100·19	0·6838	1·66	1·68
8	Octane, C_8H_{18}	114·22	0·703	1·79	1·77
9	Nonane, C_9H_{20}	128·25	0·717	1·84	1·85
10	Decane, $C_{10}H_{22}$	142·27	0·730	1·92	1·91
14	Tetradecane, $C_{14}H_{30}$	198·37	0·762	2·00	2·08
6	*iso*Hexane, C_6H_{14}	86·17	—	—	—
6	2-Methylpentane	—	0·654	1·49	1·44
6	2,2-Dimethylbutane	—	0·649	1·33	1·21
6	2,3-Dimethylbutane	—	0·662	1·38	1·21
7	*iso*Heptane, C_7H_{16}	100·19	0·673	1·44	1·44
7	2,4-Dimethylpentane	—	—	—	—
8	*iso*-Octane, C_8H_{18}	114·22	0·692	1·40	1·44
8	2,2,4-Trimethylpentane	—	—	—	—
	Aromatic hydrocarbons				
6	Benzene, C_6H_6	78	0·879	1·63	1·56
7	Toluene, C_7H_8	92	0·866	1·99	1·98
8	Ethylbenzene, C_8H_{10}	106	0·867	2·26	2·28
9	*n*-Propylbenzene, C_9H_{12}	120	0·862	2·50	2·53
9	*i*-Propylbenzene, C_9H_{12}	120	0·864	2·38	2·18
10	*n*-Butylbenzene, $C_{10}H_{14}$	134	0·862	2·75	2·73
10	*i*-Butylbenzene, $C_{10}H_{14}$	134	0·867	2·22	2·44

† Theoretical values calculated from the equations (20.30) and (20.31) derived by the author [36].

Agreement between theoretical and experimental results is so striking that the author's criterion for breakdown can be assumed to be correct, quite regardless of other phenomena taking place during breakdown.

From table 20.2 it can also be seen that there are deviations for dimethyl-butane (in the *iso*hexanes). This is to be expected because butane at room temperature is a gas, but both compounds given in the table are *iso*-hexanes and are liquids at room temperature. It was pointed out above that the mechanism of breakdown in gases is different from that in liquids and therefore the theory does not include this case. The results for the pentanes are very interesting. All compounds belonging to this group have the same length of molecule, equal to that of normal pentane (fig. 20.13), but differ in the number of C—C branches, which simply increase the transverse cross section of the molecule. This can be easily checked by calculating the molecular volume for these compounds from the ratio M/ρ. The volumes increase in a systematic manner from pentane up to octane and are nearly equal to the volumes of the normal compounds having the same n. The breakdown values for these compounds are nearly equal, which is in agreement with the assumption that breakdown depends on the length of the molecular chain.

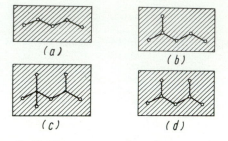

Fig. 20.13. Diagrams of effective cross sections for pentanes; (*a*) normal pentane, (*b*) methylpentane,   (*c*) 2,2,4-trimethylpentane,   (*d*) 2,4-dimethylpentane. Ref.: Adamczewski [36].

Some difficulties arise in the calculation of the length of aromatic molecules. It is assumed that the length of the molecule is determined by the number of C—C bonds along the molecule with only two bonds being counted for the benzene ring, the other bonds of the ring simply increasing the transverse cross section of the molecule. The length of branched benzene molecule is thus $(n-4)$ where n is the number of carbon atoms in the molecule. The expression for electrical breakdown stress for normal aromatic compounds can therefore be written in the form:

$$E_{bd} = B_2(n-4)\frac{\rho}{M},\qquad(20.32)$$

where B_2 is a new constant and the other symbols have the same meanings as before.   Figure 20.14 shows the general relationship between the electrical breakdown stress E_{bd} and P, which characterizes the active cross section of the molecules per cm³ of the liquid and is related to physicochemical properties of the liquid by:

$$P = r(n-m-a)\frac{\rho}{M},\qquad(20.33)$$

where $a = 1$ for saturated hydrocarbons and $a = 4$ for aromatic hydrocarbons; n as usual denotes the number of carbon atoms in the molecule, m is the number of branches and r—the radius of the molecule which is constant for all normal compounds (saturated or aromatic) and changes gradually according to the number and type of branches.

As can be seen from the graph almost all the experimental points lie on a straight line. The deviations which are observed in some cases have been discussed earlier. With the help of the expressions given above it is also possible to explain changes in breakdown values dependent upon temperature and hydrostatic pressure. In both cases the density is the most important factor or rather the number of C—C bonds that an electron encounters in the liquid.

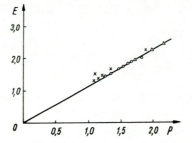

Fig. 20.14. The dependence of breakdown strength E_{bd} (in MV cm^{-1}) on the macroscopic effective cross section P for aliphatic and aromatic hydrocarbons. The solid line is from Adamczewski's theoretical equation and the experimental values were measured by Sharbaugh et al. [32].

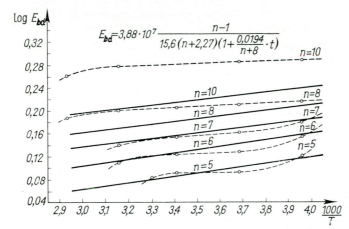

Fig. 20.15. General character of temperature dependence of breakdown strength E_{bd}. The numerical values have been multiplied by a constant coefficient in order to bring all the curves into the same area. The solid lines are from Adamczewski's [36] equation. Broken curves, from the experimental values obtained by Edwards [35], Goodwin [106] and Goodwin and Macfadyen [9].

The effect of temperature on breakdown stress for paraffin oils (including branched compounds) can be expressed in the following form:

$$E_{bd} = 3 \cdot 38 \times 10^7 \frac{n-1}{15 \cdot 6(n + 2 \cdot 27)\left(1 + \dfrac{0 \cdot 0194}{n+8} t\right)}. \tag{20.34}$$

The theoretical curves are given in fig. 20.15; the points show the experimental results taken from various sources. As can be seen, over certain temperature ranges the general character of temperature dependence is very accurately represented, and in some cases there is even numerical agreement between theoretical and experimental values.

The author [222, 280] now considers that the general form of relationships quoted above can be deduced from eqn. (16.67). A breakdown criterion based on the disappearance of activation energy, for example:

$$W = A - \gamma d E_{bd} = 0, \qquad (20.35)$$

hence:

$$E_{bd} = \frac{A}{\gamma d} = \frac{A Q (n-1) N_0 \rho}{M d}. \qquad (20.36)$$

There is a difference in the physical interpretation of this equation, namely that previously it was supposed that electrons pass through the molecules, but it is now considered that they pass along channels between the molecules; the basic dependence on molecular structure is not affected however.

Equation (20.30), after a small alteration, could be also applied to the results obtained by Lewis [39], Sakr and Gallagher [157]. Lewis in 1953 carried out a series of measurements using saturated hydrocarbons from pentane ($n = 5$) up to decane ($n = 10$), and recently Sakr and Gallagher completed this investigation using methane ($n = 1$), studying mainly the effect of the electrode material and electrode surface condition on the breakdown value.

For Lewis's results eqn. (20.30) gives good agreement when the constant value $2 \cdot 6 \times 10^6$ is changed to $1 \cdot 8 \times 10^6$ v. This equation cannot be used for methane because the numerator in the equation becomes zero, but it should be noted here that for $n = 2$ the values obtained from the equation are very close to those obtained experimentally by Sakr and Gallagher for methane. Physically it could be explained by the suggestion that although there are no C—C bonds in a single molecule, Van der Waals forces act between neighbouring carbon atoms of different molecules.

20.6. *Ward–Lewis Statistical Theory of Breakdown.* In one of their publications Ward and Lewis [38] proposed a statistical interpretation for the breakdown mechanism. Only the most important assumptions and conclusions based on this theory will be discussed below. In the initial paper the authors assume that the breakdown mechanism can be explained in the following manner: from the cathode a stream of electrons is emitted at a mean rate I per sec with a probability W that each electron can lead to breakdown. If a step-function voltage, creating a mean stress E, is applied to a liquid, the probability of breakdown occurring in time dt between the limits t and dt is:

$$P(E) dt = W I \exp [- W I (t - t_1)] dt, \qquad (20.37)$$

where t_1 is the formative time lag and $1/IW$—the mean statistical time lag. I and W will be functions of the electric stress and there will be a threshold stress E_0 below which $P(E) = 0$.

The probability of breakdown occurring when a pulse of duration T is applied to the liquid is:

$$p(E) = 1 - \exp\left[-WI(T - \tau_1)\right]. \qquad (20.38)$$

The probability $p(E) = 0$ if $E \leqslant E_0$ or $T < t_1$ (fig. 20.16).

In a study of breakdown strength it is usual to apply a succession of pulses of duration T and increase their magnitude by E until the value E_m is reached at which breakdown occurs. If at each level of the voltage N

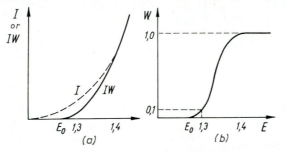

Fig. 20.16. The distribution of I and IW (a) and W (b) as functions of the electric field strength E (in MV cm^{-1}). E_0 denotes threshold value. Ref.: Ward and Lewis [38].

pulses are applied then the probability that breakdown will take place at stress E_m is:

$$P_m = 1 - (1 - p_m)^N = 1 - \exp\left[-I_m W_m T N\right]. \qquad (20.39)$$

It is more convenient to express the breakdown probability for this type of test in terms of a continuous variable $Q(E)$. If $Q(E) \, dE$ is the probability of breakdown occurring in the range E to $E + dE$ then:

$$Q(E) = \lambda p(E) \exp\left[-\lambda \int_{E_0}^{E} p(E) \, dE\right], \qquad (20.40)$$

where λ is the number of pulse applications in a unit interval of stress.

Experiments were carried out by the authors with hexane, using stainless-steel electrodes separated at a distance of 5×10^{-3} cm and step-function pulses lasting at maximum voltage for a time of 10^{-4} sec.

Some of their results are given in fig. 20.17 which shows the statistical nature of the time lag at different values of electric field. Figure 20.18 shows the effect of electrode preparation on the breakdown time lag. As can be seen, the time lag is significantly longer if a polished cathode is used rather than a roughened one.

Ward and Lewis also point out the different interpretation of τ_0 (fig. 20.19) resulting from their theory. Earlier investigators considered that τ_0 was a formative time lag but the statistical model shows that τ_0 is determined by the dependence of IW upon the electric stress. According to them, the quantity I is related to the processes involved in the electron emission from the cathode (material, preparation of the surface, tempera-ture) and determines the region AB in fig. 20.19. The quantity W is dependent upon the liquid and determines the region *BC* on the same graph. It is important to realize that experimental results of the type shown in fig. 20.19 cannot be related to ion transit times. The real

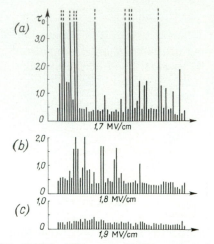

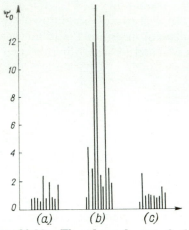

Fig. 20.17. Diagrams of the statistical time distribution against breakdown strength; (a) 1·7 MV cm⁻¹, (b) 1·8 MV cm⁻¹, (c) 1·9 MV cm⁻¹. Ref.: Ward and Lewis [38].

Fig. 20.18. The dependence of the statistical time distribution on breakdown strength in hexane; (a) normal cathode, (b) polished cathode. Electrode separation $\delta = 5 \times 10^{-3}$ cm. Electrodes were of stainless steel. Ref.: Ward and Lewis [38].

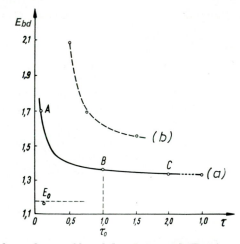

Fig. 20.19. Time dependence of breakdown strength E_{bd} (in MV cm⁻¹); *a*, Ward and Lewis [38] theory; *b*, experimental curve for hexane.

formative lag of breakdown can be determined only when the breakdown stresses are very high and the statistical time decreases to zero. In their opinion its value is about 10^{-7} sec, and if this is related to the transit time of a particle its mobility at 1·9 MV/cm is about 3×10^{-2} cm²/v sec. This is considerably greater than ion mobilities measured at low fields. As suggested by Ward and Lewis, the experimental determination of ion mobilities in higher fields appears to be an important problem.

A more comprehensive presentation of the statistical theory has been given in a later paper by the same authors. They prefer to present the statistical theory in terms of $f(E)$, the mean rate of breakdown events in a liquid under stress and subsequently consider $f(E)$ as the product IW in order to relate the statistical time lag to cathode and liquid functions respectively. They point out that in the light of the later work it is more reasonable to interpret I as the mean rate of localized bursts of emission from the cathode rather than the mean rate of electron emission. W is then the probability that a burst of emission will develop into a full discharge. Later work by other authors has also supported the statistical theory.†

Brignell [171] in 1966 demonstrated that electron multiplication in a liquid has a statistical character with particular reference to the difference for continuous and non-continuous changes.

Brignell based his mathematical calculations on new theories of the statistical distribution of impulses in proportional counters and in Geiger counters advanced by Scott [234], Burgess [235] and Campbell and Ledingham [236].

20.7. *Effect of Field Distribution on Breakdown.* In some theoretical work the effect of electric field distribution on electrical breakdown has been considered. Bragg *et al.* [31] developed a theory of breakdown based on the following assumptions.

An electron emission current j causes a change in potential distribution according to:

$$\frac{d^2 V}{dx^2} = -4\pi\rho\varepsilon^{-1}, \tag{20.41}$$

where $j = -\rho u E$, $0 \leqslant x \leqslant 1$, and ρ denotes the charge density, u—the mobility of the charge carriers and E—the electric field intensity. At the boundary conditions $V = 0$ and $E = E_c$ when $x = 0$ and $V = V_0$ when $x = 1$. Hence the following expression was obtained:

$$V_0 = \frac{\varepsilon u}{12\pi j}\left[\left(\frac{8\pi j l}{\varepsilon u} + E_c^2\right)^{3/2} - E_c^2\right], \tag{20.42}$$

which when differentiated gives the reduced anode field E_a. If the mean field strength $E_0 = V_0/l$ is small, the current density is small and there is no distortion in the field. The current obeys the cathode electron emission law. If E_0 increases, the field intensity near the cathode is reduced and increases in the vicinity of the anode when $j \sim E^2$. This theory has not been tested for very pure liquids but only for cathodes and solutions of electrolytes in benzene. Loeb [88], O'Dwyer [89] and Goodwin and

† See for example the papers [154, 245, 246, 247, 262, 272, 282, 294, 304].

Macfadyen [9] considered the effect of positive ion layer on field distribution when ionization occurs before breakdown, but the results obtained by them differed considerably from those obtained by experiment. If a is the ionization coefficient and u is the mobility of positive ions we can obtain:

$$E \frac{dE}{dx} = -\frac{4\pi j}{u\varepsilon} \left[\exp \left(\int_0^l \alpha \, dx \right) - \exp \left(\int_0^x \alpha \, dx \right) \right], \qquad (20.43)$$

by substituting $\alpha = \alpha_0 E^2$,

$$E^4 - E_c^4 = -\frac{16\pi j}{\alpha_0 u\varepsilon} [\mu \exp(\mu_l) - \exp(\mu + 1)], \qquad (20.44)$$

where:

$$\mu = \alpha_0 \int_0^x E^2 \, dx, \quad \mu_l = \alpha \int_0^l E^2 \, dx.$$

If we assume that the emission current from the cathode follows the Fowler–Nordheim law:

$$j = aE_c^2 \exp \left(-\frac{b}{E_c} \right),$$

we obtain:

$$E^4 - E_c^4 = -\frac{16\pi a E_c^2 \exp(-b/E_c)}{\alpha u\varepsilon} [\mu \exp(\mu_l) - \exp(\mu + 1)]. \qquad (20.45)$$

For the case when $E = E_c \exp(-px)$, for small values of x we obtain $\mu \ll \mu_l$ and:

$$p = \frac{4\pi a}{u\varepsilon} \exp \left[-\left(\frac{b}{E_c} + \mu_l \right) \right], \qquad (20.46)$$

where p expresses the field distortion and also some properties of the cathode and the liquid itself, and $\mu_l = \alpha l$.

A similar theory was advanced by Goodwin and Macfadyen who obtained a somewhat different expression for the field distribution:

$$E^2 - E_c^2 = -8\pi \frac{aE_c^2}{u\varepsilon} \exp \left(-\frac{b}{E_c} \right) [x \exp(\alpha l) - \alpha^{-1} \exp(\alpha x + 1)]. \qquad (20.47)$$

The expressions for the electric field and current intensities then have the following form:

$$j = \frac{euE_0}{2\pi} \exp(-\alpha l)(E_c - E_0) \qquad (20.48)$$

and:

$$E_c = E_0 + 2\pi j \exp \left(\frac{\alpha l}{\varepsilon u E_0} \right). \qquad (20.49)$$

For $E_0 = 10^6 \text{ v cm}^{-1}$ α is equal to about 10^4 which, however, is not in agreement with experiment.

Distortion in an electrical field has been demonstrated experimentally by Goodwin [106] by the Kerr effect in chlorobenzene. But in most non-polar liquids and particularly in hexane and benzene, this effect has not been detected, at least within the limits of experimental error.†

Substituting the values:

$$\gamma_c = \frac{E_c}{E_0}; \quad \eta = \frac{8\pi l j}{\varepsilon u E_0^2}, \tag{20.50}$$

in eqn. (20.38) where $E_0 = V_0/l$, the following equation is obtained:

$$i = \tfrac{2}{3}[(i + \gamma_c^2)^{3/2} - \gamma_c^3], \tag{20.51}$$

which expresses the relationship between the current density i and the intensity of the field at the cathode $i = (f\gamma_c, E_0)$.

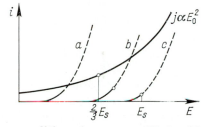

Fig. 20.20. Breakdown conditions in a space filled with negative ions: a, low-electron work function from the cathode, $E = \tfrac{2}{3}E_s$; b, medium-electron work function from the cathode, $\tfrac{2}{3}E_s < E_p < E_s$; c, high-electron work function from the cathode $E_p = E_s$. Ref.: Bragg $et\ al.$ [86].

By differentiating eqn. (20.42), the expression for the field intensity at the anode can be obtained:

$$E_l = -\left(\frac{\partial V}{\partial x}\right)_{x=l} = \gamma_l E_s. \tag{20.52}$$

For small E_0, j is small and there is no distortion of the field between the electrodes; $\gamma_c = \gamma_a = 1$, and the current obeys the emission law. With an increase in E_0, distortion of the field occurs (γ_c decreases, γ_a increases) and the current assumes values lower than those predicted by the emission law. In this case $\gamma_c \to 0$, and $\gamma_a \to 1\cdot5E_0$; the current density becomes independent of E_0, i.e. j is proportional to E_0^2.

The dependence of current density on the mean value of field intensity E_0 is shown in fig. 20.20. Curve a corresponds to a small work function for electrons to be emitted from the cathode surface; emission of electrons is thus rather high and breakdown may occur when $E_0 = \tfrac{2}{3}E_s$. Curve b corresponds to the case when work function is very large, the emission current is then very low and breakdown may occur only if $E_{bd} = E_s$. For values in between these two extremes, breakdown may occur for $\tfrac{2}{3}E_s < E_{bd} < E_s$. The continuous line in fig. 20.20 expresses the relationship between current and the mean value of the electric field intensity.

† See for example the paper of Hill and House [296].

20.8. *Swan's Theory.*   In 1961 Swan presented a theory [55, 130] based on previously published experimental results obtained by Watson and Sharbaugh [18], Swan and Lewis [21] and Sletten [24].

Swan attributes greater importance to field distortion at the cathode than have previous workers, and assumes a smaller effect of collision ionization which, according to him, may only appear at very high fields of the order of 1·2 mv/cm. He also assumes that the product αd varies between 0·1 and 1 and is not, as other workers assumed before (for example, Goodwin and Macfadyen), about 100. Electron emission from the cathode may take place in very narrow channels of a diameter of 10^{-4} cm, and at breakdowns is of the order of 10^{-9} Å, which corresponds to a current density of about 0·1 A/cm².

In Swan's final expression for field distortion E_0 at the cathode, there are terms connected with the mechanism of avalanche ionization $a = C \exp(-D/E_0^{1/2})$, and electron current intensity j_e:

$$E_c = E_0 + \frac{2\pi j_e C d^2}{\varepsilon u_+ E_0} \exp\left(-\frac{D}{E_0^{1/2}}\right), \qquad (20.53)$$

where E_0 is the field intensity between the electrodes and the other quantities have the same meaning as before. C and D are constants. Swan obtained an expression for the breakdown stress which depends on the values of the two terms, and, assuming numerical values for various quantities, showed agreement between published experimental results and his calculations principally for argon.

20.9. *Watson and Sharbaugh's Theory.*   According to Watson and Sharbaugh [132], the criterion of thermal breakdown is based on the assumption that the energy H needed to evaporate the liquid is:

$$\Delta H = m[c_p(T_b - T_0) + l_b], \qquad (20.54)$$

where m is the quantity of liquid evaporated, c_p—the mean specific heat of the liquid from room temperature T_0 to its boiling point T_b and l_b—the latent heat of evaporation. It can be said that

$$\Delta H = A E^n \tau_r, \qquad (20.55)$$

where τ_r denotes the time for which a liquid sample remains in the region of the highest stress. Liquid may also flow under the action of the field.

Breakdown is greatly dependent on hydrostatic pressure because T_b changes with pressure. The change of T_b for different liquids explains the dependence of breakdown on molecular structure. Since T_b depends on molecular structure. The effect of field is included in $n = \frac{3}{2}$.

Due to the high concentration of ions at the electrodes there is a possibility of local overheating of the liquid (the concentration is of the order of 10^7 cm³) which may lead to the formation of vapour bubbles in the liquid. In these bubbles electrons may be accelerated and result in ionization processes and avalanches.

20.10. *Theory of Breakdown due to Solid Particles and Other Impurities.*
Kok and Corbey [159] have advanced a quantitative theory of breakdown
due to solid particles and suspended colloidal matter in the liquid. They
assumed that these particles become polarized by the electric field, and
that electrostatic forces cause them to gather in the region of the highest
electrical field. In this manner bridges are formed which are often
initiated by irregularities on the surface of the electrodes.

The breakdown criterion is then obtained in the form:

$$t_b\, g^4\, r^7 (E_b{}^2 - E_0{}^2)^2\, N^2 = A \eta^2, \tag{20.56}$$

where N is the concentration of impurities, E_0—the breakdown stress
(extended breakdown), t_b—time required for breakdown to occur (impulse
duration) where $E_b{}^2 = 1/t_b$, g—the coefficient for irregularities on the
electrode surfaces, η—coefficient of viscosity and A—constant. Now:

$$(g^2 - 1)\, r^3\, E_0{}^2 = 2kT, \tag{20.57}$$

which for a long duration of the field corresponds to: $r^3 E_0{}^2 = \frac{1}{4} kT$. For
particles with a diameter of 20 Å, E_{bd} is 1·5 mv/cm and this gives agreement for *n*-alkanes. It also predicts an increase of E_{bd} with temperature,
but this is contrary to the experimental results.

Some experiments were carried out using aluminium particles ($r = 1\ \mu$m),
but theoretical results ($E_b = 10$ kv/cm) were different from those
obtained from experiment ($E_b = 4$ kv/cm). With hexane for $r = 1\ \mu$m,
the values of E_b obtained were $E_b = 100$–500 kv/cm greater than that
predicted by Kok and Corbey's theory.

20.11. *Kao's Theory.* Kao [160] derived a mathematical model of the
breakdown mechanism based on the formation of gas bubbles in liquids.
He assumed that bubbles may be formed in a liquid for the following
reasons:

(1) From the gas which accumulates in microscopic cavities and
hollows on the electrode surface.

(2) From the liquid itself by local evaporation on the surface of the
electrodes due to the action of electrical current, dissociation of molecules
through collisions with electrons or impurities, etc.

(3) Due to electrostatic forces overcoming the surface tension.†

The electrostatic forces in effect cause elongation of the bubble in the
electric field, and the energy accumulated in a cavity of volume v is:

$$W = \frac{1}{8\pi} \int_v (\varepsilon_1 - \varepsilon_2)\, E_2\, E_0\, dv, \tag{20.58}$$

where v is the volume of the bubble, E_0—the intensity of the applied field,
E_2—the field intensity inside the bubble and ε_1 and ε_2—the dielectric
constants of the liquid and bubble respectively.

† See also [295, 308].

26

With elongation of the bubble in an electric field its energy decreases and the bubble tends to elongate in the direction of the field (while keeping the same volume). Taking Paschen's law into account Kao obtained an expression for the breakdown stress in a bubble:

$$E_0 = \frac{1}{(\varepsilon_1 - \varepsilon_2)} \left\{ \frac{24\pi\sigma(2\varepsilon_1 - \varepsilon_0)}{r} \left[\frac{\pi}{4} \middle/ \left(\sqrt{\frac{v_b}{2rE_0}} - 1 \right) \right] \right\}^{1/2}, \qquad (20.59)$$

where σ denotes the surface tension of the liquid and r is the initial radius of the bubble.

This theory explains qualitatively the dependence of breakdown stress on temperature and hydrostatic pressure, but the quantitative results obtained from the theoretical considerations are far removed from those obtained experimentally.

20.12. *Krasucki's Theory.* A new theory of electrical breakdown of liquids has recently been developed by Krasucki [149, 302]. His theoretical considerations are based on his own experiments with a very viscous liquid (hexachlorodiphenyl). The viscosity of the liquid at 17·5°C is $6 \cdot 10^6$ P and it decreases rapidly with increasing temperature so that at, say, 75°C it is only 2 P. The liquid was carefully filtered and degassed. The electrodes were stainless-steel spheres of 5 mm diameter. Each measurement was made with newly polished electrodes and with a fresh quantity of purified liquid. The electrode gap was 87 μm. Negative 10/50 μs impulses were used for breakdown strength measurements and the times to breakdown were measured using direct voltages with the rise time of 10 μs.

Krasucki measured the dependence of breakdown strength (E_b) on temperature and showed that E_b decreased from 5 MV/cm at about 20°C to about 1·5 MV/cm at 75°C. At a constant stress of 1·3 MV/cm, the times to breakdown decreased from 10 sec at about 16°C to 10^{-5} sec at about 60°C.

Krasucki filmed through a microscope the development of the breakdown event with a cine camera at a speed of 16 frames per sec (figs. 20.21, 20.22, 20.23). His film records showed that, prior to breakdown, a bubble of vapour forms in the liquid and moves towards the anode. The vapour bubble forms between the electrodes at a point of maximum field concentration which, in liquids normally subjected to breakdown strength measurements, is at the surface of a particle impurity resting on the electrode surface. Krasucki investigated the time rate of growth of such a bubble and showed that in liquids like *n*-hexane, breakdown due to formation and growth of vapour bubble should occur in times of about 0·14 μs.

Krasucki analysed the conditions for vapour bubble formation in highly stressed liquids and derived from first principle an expression for the breakdown strength E_{bd} of a liquid. Krasucki showed that his equation predicts quantitatively the value of E_{bd} for *n*-hexane at room temperature, the decrease of E_{bd} for *n*-hexane with increasing temperature, the increase of E_{bd} for n-hexane with increasing hydrostatic pressure, and the increase of E_{bd} of aliphatic hydrocarbons with increasing molecular weight.

Fig. 20.21. Rotating test cell for measurement of elongation of gas bubbles in an electric field. Ref.: Garton and Krasucki [151].

20.13. *General Remarks*. As can be seen there are many different theories about electrical breakdown in dielectric liquids. In general it could be said that some of them explain the breakdown mechanism using a macroscopic interpretation such as heat production in certain places in the liquid especially at the cathode, the presence of impurities, colloidal suspensions, gas bubbles and vapour bubbles, non-uniform distribution of the electrical field, irregularities on the cathode surface, etc. Other theories consider the mechanism from a microscopic point of view and derive the breakdown criterion on the basis of molecular structure.

The first group of theories refer to experimental conditions in which liquids of a commercial grade are used, which are not properly cleaned and degassed, and when relatively large conduction currents and with long duration of the fields are applied. Such conditions are used in most of the industrial and commercial work, and for this reason these theories are acknowledged and find application.

Theories connecting the phenomenon of breakdown with molecular structure of the liquid tested refer to experimental conditions where it is possible to observe the physical mechanism of the phenomenon regardless

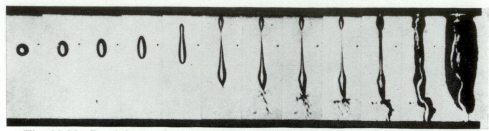

Fig. 20.22. Breakdown of a liquid dielectric (silicone fluid) due to instability of a water globule above the critical field (enlargements from 16 mm film). Magnification × 6. Ref.: Krasucki [119, 147].

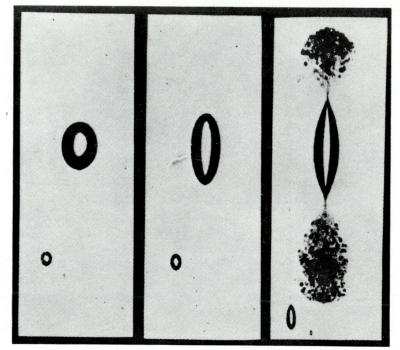

Fig. 20.23. Disruption of a water globule by formation of sharp points after instability. Ref.: Krasucki [119, 147].

of the purity of the liquid. Under such conditions it is possible to observe various measured values for different types of liquids, even for small variations in the properties. Some of the results published were obtained in this manner and furnish the basis for a physical theory, i.e. that of Lewis and the present author.

Initial physical processes in the macroscopic breakdown mechanism are difficult to observe under normal experimental conditions, but in the present author's opinion they form a basis for establishing a breakdown criterion, and their study could lead to very important future applications. The most significant aspects are sources of free electrons in the liquid, the mechanism of their solvation and transport, energy relationships with the surroundings, the effect of molecular structure, multiplication and avalanche of electrons and the amount of energy created during their transfer. An increase in the influence of these processes up to the level which can be detected and recorded by any measuring system establishes the point of transition from elementary physical processes to macroscopic phenomena and then to a routine measurement level. Studies on pre-breakdown phenomena are very important for this reason and should include studies of the dependence of breakdown on stress, time, temperature, frequency, etc.

Comparison of Experimental Results and Theoretical Studies

The liquids tested were of technical grade, very often transformer oils. In some work they were degassed and dried. Conductivity was from 10^{-8} to 10^{-12} Ω^{-1} cm^{-1}. In purified liquids the specific conductivity was 10^{-13}–10^{-16} Ω^{-1} cm^{-1}.

Experiment	Theory	
	Assumptions	Author
Exponential growth of current with increase in electrical field E type: $I = I_0 \exp [c(E - E_0)\,\delta]$, where δ is the electrode spacing	Breakdown mechanism based on: (a) Colloidal suspensions and additives	Frenkel
The current decreases during the action of the field	The main rôle is played by the amount of water in the liquid	Kao and Calderwood
The temperature variation of the variation of the current is expressed by the relationship: $I = I_0 \exp(-W/kT)$, W is of the order of 0·3–0·5 ev	(b) Variation of the activation energy in the electric field	Guizonnier *et al.*, Frenkel, Adamczewski
The change in breakdown stress is expressed by: $E_{bd} = E_0 t^{-\alpha}$, where α is a coefficient which decreases with increase in purity of the liquid		Nikuradse
Long duration of stress applied leads to thermal breakdown	(c) Bubbles at the electrode	Gemant, Watson, Sharbaugh
An increase in electrode spacing reduces the breakdown stress	(d) Bubbles in the liquid	Florenski, Wolkensztein

Experiment	Theory	
	Assumptions	Author
Increase in electrode curvature also reduces breakdown stress	(e) Chemical reactions	
In a non-uniform field (point–plate) the polarity effect is pronounced. Breakdown stress is greater when the point electrode is positive	(f) Thermal breakdown	Walter, Siemienow, Adler, Gunterschultze, Watson, Sharbaugh
An increase in hydrostatic pressure above the liquid surface increases the breakdown stress (from saturation). As the temperature is increased, a maximum of breakdown stress in observed (initial increase then a sharp drop, especially when near the boiling point). For high-frequency fields a maximum was observed for low frequencies	(g) Avalanche ionization	Peck, Schumann, Watson, Sharbaugh
Oscillograms and photographs of breakdown show small differences between breakdown in gases and in liquids	(h) Traces of water in liquids	Nikuradse, Gemant, Boning, Krasucki, Skowroński
The experimental results show that initial discharge and channels develop at the anode but main discharges develop from the cathode. Channels may be initiated at the anode and in the bulk of the liquid	(i) Effect of positive ions, effect of electrical wind. No theory as yet	Skowroński, Krasucki, Angerer, Chong, Sugimoto, Inuishi
Admixtures of various kind may lower the breakdown stress significantly	(j) Electron trapping	Angerer
Different electrode configuration:		
Spheres of diameter 12·5 mm at a separation of 2·5 mm		C.E.I.
Spheres of diameter 50 mm, hemispherical electrode of 40 mm		V.D.E.
Diameter in a hemispherical cup of 80 mm		Skowroński
Diameter with an electrode separation from 0·5–2·5 mm		
The distribution of potential in the liquid:		Guizonnier *et al.*
Linear for E from 0 to about 15–25 kv/cm		
Not uniform for E 25 kv/cm		Morant and Kahan

Short Comparison between Experimental Results and Physical Theories

Liquids studied (most frequently homologous liquids) were carefully purified, dried filtered, fractionally distilled and in most work degassed. The electrodes were polished, and in some work degassed. The conductivity of the liquids was between 10^{-18} and $10^{-19}\ \Omega^{-1}\,\mathrm{cm}^{-1}$.

Experimental results	Theory	
	Assumptions	Author
Exponential increase in current according to type of electric field	Cold emission mechanism	Shottky, Fowler, Nordheim, House
$I = A \exp(c_1 E)$ $\quad + B \exp[c_2(E - E_0)],$ where $c_2 > c_1$, or the type:	From cathode to liquid based on:	Baker, Boltz Le Page
$I = AE_k{}^2 \exp\left(-\dfrac{b}{E}\right)$ $\quad \times B \exp(a\delta),$		
where E_k is the field near the cathode and a represents the coefficients of ionization and electron attachment, A and B being constants depending on the cathode material and type of liquid tested	(a) Lowering of the work function (b) Strong fields	Du Bridge, Goodwin Macfadyen, Green
The $I = f(E)$ charateristics can be divided into three regions: (1) About 60–150 kv/cm (current $I = \sim 10^{-13}$ A)	Cathode emission	
(2) About 150–250 kv/cm (exponential increase of the current, fluctuations)		Kahan
(3) Above 400 kv/cm (current $I = 10^{-7}$–10^{-6} A)	Rearranging of molecules	
Current increases slowly with an increase in the field with a very rapid increase just before breakdown	(c) A layer of positive ions (d) Field distortion	Lewis, Ward Bragg, Sharbaugh, Crowe, Morant, Watson
Electroluminescence of the liquid, pure or with admixtures	(e) Excitation of admixture molecules by the electrons accelerated in the field	Tropper and Darveniza, Kalinowski and Dera Gosling, Angerer, Jones
Linear dependence of the intensity of light and electric current	(f) Avalanche ionization	Nikuradse, Schumann
Ionization currents increase linearly with intensity of the field and electrode separation for fields of 1–300 kv/cm $i = i_0 + cE$		

Experimental results	Theory	
	Assumptions	Author
or $\quad i = i_0(1 + \gamma E \delta)$. In a uniform field $(E = U/d)$: $i = i_0(1 + \gamma U)$.	(g) Dissociation of molecules of colloidal suspensions in the liquid	Onsager, Plumley
The whole $I = f(E)$ characteristic may be expressed by the formula: $$I = \frac{nqSd}{kT} \exp\left[\frac{A - \gamma dE}{kT}\right],$$ $$n = \frac{n_0}{1 + \alpha n_0 t},$$ or $\quad n = \dfrac{n_0}{1 + \alpha n_0 (d/uE)}$. (1) For E within the limits from 0 to the breakdown (inclusive) (2) Within a wide temperature region (3) The criterion of breakdown $\quad A - \gamma dE_{bd} = 0$.	(h) Variation of the activation energy in electric field	Frenkel, Adamczewski
Electric breakdown. With careful purification of liquids and clean electrodes it is possible to obtain breakdown values as high as 1·4–2·6 [MV cm^{-1}]	(1) The theory postulating the essential rôle of the electrode (cathode) in electric breakdown	House, Kao, Green,
The effect of the electrode surface is very pronounced as regards polishing and degassing, but different workers report different effects of electrode material, especially in impulse tests and uniform field distribution.	(a) Strong electron emission from the cathode causes local heating of the liquid and makes a vapour bubble to arise which elongates due to the action of the electric field	Lewis, Calderwood, Chong, Inuishi,
In non-uniform fields the effect of the work function was found with a plate cathode and a point anode Effect of the curvature of electrodes on breakdown value was found	(b) Desorption of gases from the metal surface with subsequent formation of the bubble	
Impulse tests showed that for impulse times less than the minimum value τ_0 the	(2) The theory postulating the essential rôle of the liquid	

Experimental results	Theory	
	Assumptions	Author
following relation holds: $$E_{bd} = \frac{(t-\tau_0)^c}{c_1},$$ where c_1, c and τ_0 are constants depending on the type of liquid	(a) The breakdown develops like in gases (collision ionization-production of avalanches)	Nikuradse, Schumann, Ushakov, Stielnikov,
The values of E_{bd} for t greater than τ_0 are constant and depend on the structure of the liquid:	(b) Breakdown develops when in applied field the electrons gain energy higher than they lose for excitation processes	Komelkov, Chong, Inuishi, Lewis, Adamczewski,
(a) They increase with increase in density and viscosity of the liquid	(c) Formation of bubbles due to high electric field (thermal breakdown)	Watson, Sharbaugh, Kao, Krasucki
(b) They depend on the structure of the liquid molecule and their ordering	(d) Formation of bridge from mechanical suspension (particles)	Kok, Corbey, Krasucki, Angerer
(c) They decrease with increase in temperature, but not uniformly (over first region of temperature the increase is proportional to changes in liquid density)	(e) The assumption of quasi-crystalline structure of the liquid. The electrons are slowing down:	Gzowski, Smith, Calderwood
(d) They increase slowly with increase in hydrostatic pressure and approach a constant value	On C—H bonds On C—C bonds The rôle of the length of molecules	Lewis, Adamczewski
For small gap settings, E_{bd} increases when the gap is reduced	(f) The rôle of activation energy in electron transfer	Adamczewski
Frequency dependence was found using high-frequency alternating fields	(3) The statistical interpretation of breakdown	Ward, Lewis, House,
The presence of impurities reduces the breakdown values but in some cases, for example if oxygen or SF_6 acceptor is added, an increase in breakdown voltage results		
Mobility of ions is of the order of $10^{-4}\,cm^2\,v^{-1}\,sec^{-1}$ and is independent of the field for a range of the latter from 0·4–300 kv/cm	(4) Effect of field distribution	Brignell, Sharbaugh, Crowe, Goodwin, Macfadyen, Swan, Lewis
Some additives may increase the breakdown stress	(5) Effect of impurities and additives	Zein, El-dine, Tropper, Angerer

References to Part IV

[1] Lewis, T. J., 1959, *Progress in Dielectrics*, Vol. 1 (London: Heywood), p. 97.
[2] Nikuradse, A., 1934, *Das Flüssige Dielektrikum* (Berlin: Springer).
[3] Skanavi, G. U., 1958, *Physics of Dielectrics* (Moscow: Goc. Izd. Phys.-Mat.), pp. 264–399 (in Russian).

[4] KOK, J. A., 1961, *Electrical Breakdown of Insulating Liquids* (London: Phillips Technical Library, Cleaver-Hume).

[5] BAKER, E. H., and BOLTZ, H. A., 1940, *Phys. Rev.*, **58**, 61; 1937, *Ibid.*, **51**, 275.

[6] LE PAGE, W. R., and DU BRIDGE, L. A., 1940, *Phys. Rev.*, **58**, 61.

[7] DORNTE, R. W., 1939, *J. appl. Phys.*, **10**, 514.

[8] PLUMLEY, H. J., 1941, *Phys. Rev.*, **59**, 200.

[9] GOODWIN, D. W., and MACFADYEN, K. A., 1953, *Proc. Phys. Soc. Lond.*, **B66**, 85, 815.

[10] GREEN, W. B., 1956, *J. appl. Phys.*, **27**, 921.

[11] HOUSE, H., 1957, *Proc. Phys. Soc. Lond.*, **B70**, 913; 1955, *Nature, Lond.*, **176**, 610.

[12] LEWIS, T. J., 1955, *Proc. Phys. Soc. Lond.*, **B68**, 504; 1953, *Ibid.*, **B66**, 425.

[13] LEWIS, T. J., 1955, *J. appl. Phys.*, **26**, 1405; 1956, *Ibid.*, **27**, 645.

[14] LEWIS, T. J., 1960, *J. Electrochem. Soc.*, **107**, 185.

[15] PAO, C. S., 1943, *Phys. Rev.*, **64**, 60.

[16] SCHUMANN, W., 1932, *Z. tech. Phys.*, **75**, 380.

[17] VON ENGEL, A., 1956, *Handbuch Phys.*, Vol. XXI, p. 504.

[18] WATSON, P. K., and SHARBAUGH, A. H., 1960, *J. Electrochem. Soc.*, **107**, 516.

[19] DARVENIZA, M., and TROPPER, H., 1961, *Proc. Phys. Soc.*, **78**, 854.

[20] DARVENIZA, M., 1960, *J. Inst. Petrol*, **46**, 84.

[21] SWAN, D. W., and LEWIS, T. J., 1961, *Proc. Phys. Soc.*, **78**, 448; 1960, *J. Electrochem. Soc.*, **107**, 180.

[22] YAMANAKA, C., and SUITA, T., 1953, *J. phys. Soc. Japan*, **8**, 277.

[23] INGE, L., and WALTER, A., 1931, *Arch. J. Elektr. D.D.Z.T.F.*, **7**, 756; 1932, *Ibid.*, **2**, 7; 1932, *Ibid.*, **26**, 6, 409; 1934, *Ibid.*, **4**, 9, 1669.

[24] SLETTEN, A. M., 1959, *Nature, Lond.*, **183**, 311.

[25] GEMANT, A., 1927, *Wiss. Veröf. Siem.-Kons.*, **5**, 87; 1929, *Z. tech. Phys.*, **9**, 398; 1925, *Z. Phys.*, **33**, 789.

[26] SCHWAIGER, A., 1925, *Elektrische Festigkeitslehre* (Berlin).

[27] EDLER, H., 1930, *Arch. Elektrotech.*, **24**, 37; 1931, *Ibid.*, **25**, 447.

[28] VOLKENSTEIN, F. F., 1934, *Breakdown in Dielectric Liquids* (Moscow: Goc. Izd. Tekn. Teor. Lit.) (in Russian).

[29] SALVAGE, B., 1951, *Proc. I.E.E.*, **98**, Pt. IV, 15.

[30] MAKSIEJEWSKI, J. L., and TROPPER, H., 1954, *Proc. I.E.E.*, **101**, Pt. II, 183.

[31] CROWE, R. W., BRAGG, J. K., and SHARBAUGH, A. H., 1954, *J. appl. Phys.*, **25**, 392.

[32] SHARBAUGH, A. H., CROWE, R. W., and COX, E. B., 1956, *J. appl. Phys.*, **27**, 806.

[33] CROWE, R. W., 1956, *J. appl. Phys.*, **27**, 156.

[34] SHARBAUGH, A. H., BRAGG, J. K., and CROWE, R. W., 1953, *J. appl. Phys.*, **24**, 814; 1955, *Ibid.*, **26**, 434.

[35] EDWARDS, W. D., 1954, *Electr. Ing.*, **26**, 36.

[36] ADAMCZEWSKI, I., 1957, *Zesz. Nauk. P. G.*, Ch. I, 3 (in Polish).

[37] TERLECKI, J., 1962, *Nature, Lond.*, **194**, 172.

[38] WARD, B. W., and LEWIS, T. J., 1960, *J. Electrochem. Soc.*, **107**, 191.

[39] LEWIS, T. J., 1953, *Proc. I.E.E.*, **100**, Pt. IIa, 141.

[40] ZEIN EL-DINE, M. E., and TROPPER, H., 1956, *Proc. I.E.E.*, **103**, Pt. C, 35.

[41] WEBER, K. H., and ENDICOTT, H. S., 1957, *Proc. I.E.E.*, **104**, Pt. C, 543.

[42] WILLIAMS, R. L., 1957, *Can. J. Phys.*, **35**, 134.

[43] MALKIN, M. S., and SCHULTZ, H. L., 1951, *Phys. Rev.*, **83**, 1051.

[44] HOOVER, W. G., and HIXSON, W. A., 1949, *Trans. A.I.E.E.*, **68**, 1047.

[45] WATSON, P. K., and HIGHAM, J. P., 1953, *Proc. I.E.E.*, **100**, Pt. IIa, 168.

[46] ADAMCZEWSKI, I., 1962, *Postępy Fiz.*, **13**, 253 (in Polish).

[47] WEBER, K. H., and ENDICOTT, H. S., 1956, *Trans. Am. I.E.E.*, **75**, Pt. III, 371.

[48] FRIESE, Y. 1921, *Siem. Konz.*, **1**, 41.

[49] SUZUKI, M., and FUJIOKA, R., 1940, *Electrotech. J.*, **4**, 274.

[50] RUHLE, F., 1941, *Arch. Electrotech.*, **35**, 490; 1943, *Phys. Z.*, **44**, 89.

[51] PEEK, F. W., (1929–30), *Dielectric Phenomena in High Voltage Engineering* (New York).

[52] HIPPEL, A., 1937, *J. appl. Phys.*, **8**, 815.

[53] MUSSET, E., NIKURADSE, A., and ULBRICH, R., 1956, *Z. angewandte Phys.*, **8**, 8.

[54] BOOTH, D. H., and JOHNSON, O. S., 1954, *Conference Int. Grands Res. Electr.*, Vol. 2, p. 122.

[55] SWAN, D. W., 1961, *Proc. Phys. Soc.*, **78**, 423.

[56] BIONDI, M. A., 1960, *Proc. 4th International Conference on Ionization Phenomena in Gases, Amsterdam*, Vol. I, p. 72.

[57] HURST, G. S., and BORTNER, T. E., 1959, *Radiation Research*, **1**, 547.

[58] KOMELKOV, B. S., 1950, *Izv. Akad. Nauk.*, **6**, 851 (in Russian).

[59] SKOWROŃSKI, J. I., 1962, Paper in C.I.G.R.E. session.

[60] LIAO, T. W., and ANDERSON, J. G., 1953, *Trans. Am. I.E.E.*, **72**, Pt. I, 641.

[61] SOMMERMAN, G. M. L., BUTE, C. J., and LARSON, E. L. C., 1954, *Trans. Am. I.E.E.*, **73**, Pt. I, 147.

[62] LEWIS, T. J., 1953, *Proc. I.E.E.*, **100**, Pat. IIa, 141.

[63] BASSECHES, H., and McLEAN, D. A., 1955, *Ind. Engng Chem.*, **47**, 1782.

[64] BUCHHOLZ, H. H., 1954, *E.T.Z.*, **A75**, 763.

[65] WATSON, P. K., and HIGHAM, J. B., 1953, *Proc. I.E.E.*, **100**, Pt. IIa, 188.

[66] NAGAO, S., 1956, *J. phys. Soc. Japan*, **11**, 1205.

[67] BATES, D. R., and KINGSTON, A. E., 1961, *Nature, Lond.*, **189**, 652.

[68] KOPPELMANN, F., 1935, *Z. Tech. Phys.*, **16**, 126.

[69] WALTER, A. F., and INGE, L., 1929, *D.Z.P.F.*, **6**, 49; 1931, *D.Z.T.F.*, **7**, 756.

[70] SCHOTTKY, W., 1939, *Z. Phys.*, **113**, 367; 1942, *Ibid.*, **118**, 539.

[71] RICHARDSON, E. W. T., 1953, *Proc. Phys. Soc.*, **A66**, 403.

[72] FOWLER, R. H., and NORDHEIM, L., 1928, *Proc. R. Soc.*, **A119**, 173.

[73] MOTT, N. F., and GURNEY, R. W., 1948, *Electronic Process in Ionic Crystals* (Oxford: Clarendon Press).

[74] LANE, M., 1925, *Handb. Radiol.*, **6**, 460.

[75] SKINNER, S. M., SAVAGE, R. L., and RUTHER, J. E., 1953, *J. appl. Phys.*, **24**, 438.

[76] MORANT, M. J., 1960, *Nature, Lond.*, **187**, 48.

[77] PRASAD, A. N., and CRAGGS, J. D., 1960, *Proc. Phys. Soc.*, **76**, 223.

[78] CRAGGS, J. D., and TOZER, B. A., 1960, *Proc. R. Soc.*, **254**, 229.

[79] PRASAD, A. N., and CRAGGS, J. D., 1961, *Proc. Phys. Soc.*, **77**, 385.

[80] BHALLA, M. S., and CRAGGS, J. D., 1960, *Proc. Phys. Soc.*, **76**, 369; 1961, *Ibid.*, **78**, 438.

[81] STACEY, F. D., 1959, *Aust. J. Phys.*, **32**, 105.

[82] LeBLANC, O. H., 1959, *J. chem. Phys.*, **30**, 1443.

[83] FRÖHLICH, H., 1958, *Theory of Dielectrics* (Oxford: Clarendon Press).

[84] SEEGER, R. J., and TELLER, E., 1954, *J. appl. Phys.*, **25**, 382.

[85] FRENKEL, J. I., 1947, *Kinetic Theory of Liquids* (Oxford: Clarendon Press).

[86] BRAGG, J. K., SHARBAUGH, A. H., and CROWE, R. W., 1954, *J. appl. Phys.*, **25**, 382.

[87] LAZAREV, A. M., and RESHEKTAJEV, I., 1936, *Tech. Fiz.*, *SSSR*, **3**, 36 (in Russian).

[88] LOEB, L. B., 1939, *Fundamental Processes of Electrical Discharge in Gases* (New York: Wiley).

[89] O'DWYER, J. J., 1954, *Aust. J. Phys.*, **7**, 400.

[90] WILLIAMS, R. L., and STACEY, E. D., 1957, *Can. J. Phys.*, **35**, 928.

[91] BALYGIN, I. E., 1956, *Zh. eksp. teor. Fiz.*, **3**, 355 (in Russian).

[92] BALYGIN, I. E., 1954, *Dokl. Acad. Nauk. SSSR*, **95**, 745 (in Russian).

[93] LEWIS, T. J., 1957, *Proc. I.E.E.*, **104**, Pt. B17, 493.

[94] MORANT, M. J., 1954, *J. appl. Phys.*, **25**, 8, 1053.

[95] MORANT, M. J., 1955, *Proc. Phys. Soc.*, **B68**, 513.

[96] EDWARDS, W. D., 1951, *Can. J. Phys.*, **29**, 310.

[97] HOUSE, H., 1954, *J. Scient. Instrum.*, **3**, 261; 1955, *Nature, Lond.*, **176**, 610.

[98] HOUSE, H., and MORANT, M. J., 1954, *J. scient. Instrum.*, **31**, 342.

[99] HEYLEN, A. E. D., and LEWIS, T. J., 1956, *Br. J. appl. Phys.*, **7**, 411.

[100] LEWIS, T. J., 1958, *Br. J. appl. Phys.*, **9**, 30.

[101] HANCOX, R., 1956, *Nature, Lond.*, **178**, 1305; 1957, *Br. J. appl. Phys.*, **8**, 476.

[102] GEMANT, A., 1940, *Phys. Rev.*, **58**, 904.

[103] WHITEHEAD, S., 1928, *Electrical Discharges in Liquids: Dielectric Phenomena* (London: Benn).

[104] GINDIN, L. G., MOROZ, L. M., PUTILOVA, I. C., and FRENKEL, J. I., 1950, *Dokl. Acad. Nauk. SSSR*, **74**, 1, 49 (in Russian).

[105] GELLER, M., 1956, *Phys. Rev.*, **101**, 1685.

[106] GOODWIN, D. W., 1956, *Proc. Phys. Soc.*, **B69**, 61.

[107] SHARBAUGH, A. H., and WATSON, P. K., 1959, *Nature, Lond.*, **184**, 2006.

[108] KELLNER, L., 1951, *Proc. R. Soc.*, **A64**, 521.

[109] SHIMANOUCKI, T., KAKIUTI, Y., and GANSE, I., 1956, *J. chem. Phys.*, **25**, 1245.

[110] SECKER, P. E., JESSUP, M., and HUGHES, J. F., 1966, *J. scient. Instrum.*, **43** 515.

[111] FRANZ, W., 1956, *Handbuch Phys.*, Vol. XVII, p. 155.

[112] STARK, K. H., and GARTON, C. G., 1955, *Nature, Lond.*, **176**, 1225.

[113] FLOREK, K., and SKOWROŃSKI, J. I., 1962, *Arch. Elektrotech.*, **11**, 565.

[114] CHONG, P., SUGIMOTO, T., and INUISHI, Y., 1960, *J. phys. Soc. Japan*, **15**, 1137.

[115] KALINOWSKI, J., and DERA, J., 1964, *Acta phys. pol.*, **25**, 205.

[116] GOSLING, C. H., 1963, Conference on Electronic Processes in Dielectric Liquids, Durham.

[117] MUROOKA, Y., NAGAO S., and TORYAMA, Y., 1963, Conference on Electronic Processes in Dielectric Liquids, Durham.

[118] ANGERER, L., 1963, *Nature, Lond.*, **199**, 62.

[119] KRASUCKI, Z., 1963, Conference on Electronic Processes in Dielectric Liquids, Durham.

[120] SHARBAUGH, A. H., and WATSON, P. K., 1963, Conference on Electronic Processes in Dielectric Liquids, Durham.

[121] BRIGNELL, J. E., and HOUSE, H., 1963, Conference on Electronic Processes in Dielectric Liquids, Durham.

[122] WATSON, P. K., and SHARBAUGH, A. H., 1963, Conference on Electronic Processes in Dielectric Liquids, Durham.

[123] DARVENIZA, M., 1963, Conference on Electronic Processes in Dielectric Liquids, Durham.

[124] BRIÈRE, G., 1963, Conference of Electronic Processes in Dielectric Liquids, Durham.

[125] MORANT, M. J., 1963, *Br. J. appl. Phys.*, **14**, 469.

[126] WATSON, P. K., 1963, *Nature, Lond.*, **199**, 646.

[127] BALYGIN, I. E., 1964, *Electric Strength of Dielectric Liquids*, M. L. (Moscow: Energia) (in Russian).

[128] MOORE, R., GIBBS, P. and EYRING, H., 1953, *J. phys. Chem.*, **57**, 172.

[129] LEWIS, T. J., 1964, *Advmt. Sci.*, March, p. 501.

[130] SWAN, D. W., 1962, *Br. J. appl. Phys.*, **13**, 208.

[131] MORANT, M. J., and HOUSE, H., 1956, *Proc. Phys. Soc.*, **49**, 14.

[132] SHARBAUGH, A. H., and WATSON, P. K., 1962, *Progress in Dielectrics*, Vol. 4 (London: Heywood), p. 199.

[133] LOEB, L. B., 1955, *Basic Processes of Gaseous Electronics* (Berkeley: University of California Press).

[134] LLEWELLYN JONES, F., 1957, *Ionization and Breakdown in Gases* (London: Methuen).

[135] DAKIN, T. W., and BERG, D., 1962, *Progress in Dielectrics*, Vol. 4 (London: Heywood), p. 151.

[136] MEEK, J. M., and CRAGGS, J. D., 1953, *Electrical Breakdown of Gases* (London: Oxford University Press).

[137] PAPOULAR, R., 1965, *Phénomenes électriques dans les Gaz* (Paris: Dunod) (in French).

[138] JONES, E., and ANGERER, L., 1966, *Nature, Lond.*, **210**, 1219.

[139] HEYLEN, A. E. D., 1958, *J. chem. Phys.*, **29**, 813.

[140] HEYLEN, A. E. D., 1957, *Nature, Lond.*, **180**, 703.

[141] HEYLEN, A. E. D., 1957, Doctorate Thesis, University of London.

[142] HEYLEN, A. E. D., and LEWIS, T. J., *Br. J. appl. Phys.*, **7**, 411.

[143] HEYLEN, A. E. D., and LEWIS, T. J., 1958, *Can. J. Phys.*, **36**, 721.

[144] CROWE, R. W., 1966, *J. appl. Phys.*, **37**, 1515.

[145] FARRALL, G. A., and CRAIG MILLER, H., 1965, *J. appl. Phys.*, **36**, 2966.

[146] CRAIG MILLER, H., 1966, *J. appl. Phys.*, **37**, 784.

[147] KRASUCKI, Z., 1966, *E.R.A. Reports L/T*, pp. 376, 393, 400.

[148] NOSSEIR, A., HAWLEY, R., and CLOTHIER, N., 1965, *Nature, Lond.*, **206**, 389, 494.

[149] KRASUCKI, Z., 1966, *Proc. R. Soc.*, **A294**, 393.

[150] KRASUCKI, Z., CHURCH, H. F., and GARTON, C. G., 1960, *Brit. Electr. Res. Ass. (E.R.A.)*, **107**, 598.

[151] GARTON, C. G., and KRASUCKI, Z., 1964, *Proc. R. Soc.*, **A280**, 211.

[152] CHADBAND, W. G., and WRIGHT, G. T., 1965, *Br. J. appl. Phys.*, **16**, 305.

[153] SMITH, C. W., KAO, K. C., CALDERWOOD, J. H., and McGEE, J. D., 1966, *Adv. Electronics Electron Phys.*, **22**, 1003.

[154] GZOWSKI, O., WŁODARSKI, R., HESKETH, T. R., and LEWIS, T. J., 1966, *Br. J. appl. Phys.*, **17**, 1483.

[155] GZOWSKI, O., 1967, *Zesz. Nauk. P. G. Fiz.*, Vol. 1 (in Polish).

[156] USHAKOV, B. J., 1965, *J. Tech. Phys. Sov.*, **35**, 1844 (in Russian).

[157] SAKR, M. M., and GALLAGHER, T. J., 1964, *Br. J appl. Phys.*, **15**, 647.

[158] SAMUEL, A. H., HALLIDAY, F. O., KEAST, A. K., and TAJMTY, S. I., 1964, *Science*, **144**, 839.

[159] KOK, J. A., and CORBEY, M. M. G., 1957, *Appl. Sci. Res. Hague*, **B6**, 285; 1958, *Ibid.*, **B7**, 257.

[160] KAO, K. C., 1965, *Nature, Lond.*, **208**, 279.

[161] WATSON, P. K., and SHARBAUGH, A. H., unpublished work (see also [132]).

[162] METCALF, W. S., 1965, *J. scient. Instrum.*, **42**, 742.

[163] MINORU, U., and MASAKASU, I., 1964, *Mem. Fac. Engng Nagoya Univ.*, **16**, 1.

[164] SCHOTT, H., and KAGHAN, W. S., 1965, *J. appl. Phys.*, **36**, 3399.

[165] KIŻEKIN, M. P., 1966, *Z. tech. Phys.*, **36**, 338.

[166] GEBALLE, R., and HARRISON, M. A., 1952, *Phys. Rev.*, **85**, 372.

[167] GEBALLE, R., and REEVES, M. L., 1953, *Phys. Rev.*, **92**, 867.

[168] STIEKOLNIKOV, J. S., and USHAKOV, B. J., 1965, *J. Tech. Phys. Sov.*, **35**, 1692 (in Russian).

[169] LEWIS, T. J., and WARD, B. W., 1962, *Proc. R. Soc.*, **269**, 233.

[170] GALLAGHER, T. J., and LEWIS, T. J., 1964, *Br. J. appl. Phys.*, **15**, 491; 1964, *Ibid.*, **15**, 929.

[171] BRIGNELL, J. E., 1966, *Proc. I.E.E.*, **113**, 1683.

[172] BEDDOW, A. J., and BRIGNELL, J. E., 1965, *Electronics Letters*, **1**, 253; 1966, *Ibid.*, **2**, 142.

[173] WARD, B. W., and LEWIS, T. J., 1963, *Br. J. appl. Phys.*, **14**, 368.

[174] TORIYAMA, Y., SATO, T., and MITSUI, H., 1964, *Br. J. appl. Phys.*, **15**, 203.

[175] BRIGNELL, J. E., and HOUSE, H., 1965, *Nature, Lond.*, **206**, 1142.

[176] MEGAHED, I. A. Y., 1965, *Nature, Lond.*, **205**, 686.

[177] PICKARD, W. F., 1965, *Progress in Dielectrics*, Vol. 6 (London: Heywood), p.1.

[178] OKANO, K., 1965, *J. appl. Phys. Japan*, **4**, 292.

[179] KAHAN, E., 1964, Doctorate Thesis, University of Durham.

[180] RUDENKO, N. S., and TSVETKOV, V. I., 1966, *Sov. Phys., Tech. Phys.*, **10**, 1417 (in Russian).

[181] GRAY, E., and LEWIS, J. T., 1965, *Br. J. appl. Phys.*, **16**, 1049.

[182] HALPERN, B., and GOMER, R., 1965, *J. chem. Phys.*, **43**, 1069.

[183] HART, J., and MUNGALL, A. G., 1958, *Trans. Am. I.E.E.*, Pt. 3, 1295.

[184] ZAKY, A. A., ZEIN, M. E., EL-DINE, M. E., and HAWLEY, R., 1964, *Nature, Lond.*, **202**, 687.

[185] BIERE, G. B., 1964, *Br. J. appl. Phys.*, **15**, 413.

[186] KOK, J. A., 1964, *Rigidité diélectrique des Liquides isolants* (Paris: Dunod), p. 152 (in French).

[187] DEKKER, A. J., 1962, *Solid State Physics* (London: Macmillan), p. 540.

[188] STRATTON, R., 1961, *Progress in Dielectrics*, Vol. 3 (London: Heywood), p. 233.

[189] MASON, J. H., 1959, *Progress in Dielectrics*, Vol. 3 (London: Heywood), p. 3.

[190] CROITORU, Z., 1965, *Progress in Dielectrics*, Vol. 6 (London: Heywood), p. 103.

[191] CROITORU, Z., 1960, *Bull. Soc. Fr. Electr.*, **1**, 362.

[192] O'DWYER, J. J., 1964, *The Theory of Dielectric Breakdown of Solids* (London: Oxford University Press).

[193] KALINOWSKI, J., 1967, *Acta phys. pol.*, **31**, 3.

[194] DRUYVESTEYN, M. Y., 1936, *Physica*, **3**, 65.

[195] TERLECKI, J., 1966, *Acta phys. pol.*, **29**, 743.

[196] HEYLEN, A. E. D., and LEWIS, T. J., 1963, *Proc. R. Soc.*, **271**, 531.

[197] CROWE, R. W., 1966, *J. appl. Phys.*, **37**, 1515.

[198] SLETTEN, A. M., and LEWIS, T. J., 1963, *Br. J. appl. Phys.*, **17**, 883.

[199] SMITH, C. W., KAO, K. C., CALDERWOOD, J. H., and McGEE, J. D., 1966, *Nature, Lond.*, **210**, 192.

[200] SMITH, C. W., KAO, K. C., and CALDERWOOD, J. H., 1966, Conference on Electrical Insulation and Dielectric Phenomena, Pennsylvania, Publ. 1484 N.A.S., N.R.C. Washington (1967), p. 45.

[201] BENNETT, R. G., and CALDERWOOD, J. H., 1962, *Adv. Molec. Spectr.* (Oxford: Pergamon Press), p. 1187.

[202] ANGERER, L., 1965, *Proc. I.E.E.*, **112**, 1025.

[203] YAHAGI, K., KAO, K. C., and CALDERWOOD, J. H., 1966, *J. appl. Phys.*, **37**, 4289.

[204] ANGERER, L., and TROPPER, H., 1964, Conference on Electrical Insulation, Cleveland, Ohio.

[205] KAO, K. C., and CALDERWOOD, J. H., 1965, *Proc. I.E.E.*, **112**, 597.

[206] HUQ, A. M. Z., and TROPPER, H., 1965, *Br. J. appl. Phys.*, **15**, 481.

[207] KUCHIŃSKII, G. G., 1966, *Zurn. Techn. Fiz.*, **36**, 1297; 1967, *Sov. Phys. Tech. Phys.*, **11**, 964.

[208] KUFFEL, E., and RADWAN, R. O., 1966, *Proc. I.E.E.*, **113**, 1863.

[209] KUFFEL, E., 1959, *Proc. Phys. Soc.*, **74**, 297.

[210] KUFFEL, E., 1961, *Proc. I.E.E.*, **108**, 295, 308.

[211] ABDULLAH, M., and KUFFEL, E., 1965, *Proc. I.E.E.*, **112**, 1018.

[212] SLETTEN, A. M., and LEWIS, T. J., 1963, *Br. J. appl. Phys.*, **14**, 883.

[213] SLETTEN, A. M., and DAKIN, T. W., 1963, *Scientific Paper*, March **21**, 63, 131 (Westinghouse Res. Lab.), p. 3.

[214] BROWN, D. R., 1964, *Nature, Lond.*, **202**, 868.

[215] GZOWSKI, O., 1966, *Nature, Lond.*, **212**, 185.

[216] NOISEIR, A., 1963, *Nature, Lond.*, **198**, 1295.

[217] BRIGNELL, J. E., 1963, *J. scient. Instrum.*, **40**, 576.

[218] MORANT, M. J., 1960, *J. electrochem. Soc.*, **107**, 671.

[219] KOPYLOW, G. N., 1964, *Sov. Phys. Tech. Phys.*, **8**, 962 (in Russian).

[220] TINKO, C. A., PENNEY, G. W., and OSTERLE, J. F., 1965, *Proc. I.E.E.*, **53**, 141.

[221] STUETZER, O. M., 1960, *J. appl. Phys.*, **31**, 136.

[222] ADAMCZEWSKI, I., 1968, Colloque International—journées d'electronique de Toulose.

[223] SCHONE, G., 1963, *Electronics*, **10**, 335.

[224] CHATTERTON, P. A., 1966, *Proc. Phys. Soc.*, **88**, 231.

[225] ZEIN, M. E. EL-DINE, ZAKY, A. A., HAWLEY, R., and CULLINGFORD, M. C., 1964, *Nature, Lond.*, **201**, 1309.

[226] POHL, H. A., and SCHWAR, J. P., 1959, *J. electrochem. Soc.*, **30**, 69; 1960, *Ibid.*, **5**, 383.

[227] POHL, H. A., 1951, *J. appl. Phys.*, **22**, 869; 1958, *Ibid.*, **29**, 1182.

[228] ANGERER, L., 1963, *Nature, Lond.*, **199**, 62.

[229] SCHOTT, H., and KAGHAN, W. S., 1965, *J. appl. Phys.*, **36**, 3399.

[230] TARIS, F., and GUIZONNIER, R., 1966, *Rev. Gen. Electr.*, **75**, 1295.

[231] HUGHES, J. F., and SECKER, P. E., 1965, *Electronic Letters*, **2**, No. 5.

[232] KAHAN, E., and MORANT, M. J., 1965, *Br. J. appl. Phys.*, **16**, 943.

[233] BROWN, D. R., 1964, *Nature, Lond.*, **202**, 4935.

[234] SCOTT, L., 1966, *Electronics Letters*, **2**, 110.

[235] BURGESS, R. E., 1966, *Electronic Letters*, **2**, 166.

[236] CAMPBELL, J. L., and LEDINGHAM, K. W. D., 1966, *Br. J. appl. Phys.*, **17**, 769.

[237] KUFFEL, E., 1962, Doctorate Thesis, University of Manchester.

[238] HAKIM, S. S., and HIGHAM, J. B., 1961, *Nature, Lond.*, **189**, 996.

[239] SKOWROŃSKI, J., 1966, *Proc. I.E.E.*, **113**, 1106.

[240] FARBER, H., and KIHM, R. T., 1966, *Conference on Electrical Insulation, Buck Hill Falls* (New York: National Academy of Science), p. 57.

[241] KAO, K. C., 1966, *Conference on Electrical Insulation, Buck Hill Falls* (New York: National Academy of Science), p. 44.

[242] HAYASHI, M., 1966, *Mem. Fac. Engng Kyoto Univ.*, **28**, 318 (in Japanese).

[243] COE, G., HUGHES, J. F., and SECKER, P. E., 1966, *Br. J. appl. Phys.*, **17**, 885.

[244] TERLECKI, J., 1968, Science Society de Gdańsk (GTN), in press (in Polish).

[245] BEDDOW, A. J., 1968, Doctorate Thesis, University of London.

[246] BRIGNELL, J. E., 1964, Doctorate Thesis, University of London.

[247] ESSEX, V., and SECKER, P. E., 1968, *Br. J. appl. Phys. (J. Phys. D.)*, **1**, 63.

[248] OSTROUMOV, G. A., 1954, *Zh. Tekh. Fiz.*, **24**, 1915.

[249] OSTROUMOV, G. A., 1962, *Soviet Physics J.E.T.P.*, **14**, 317.

[250] KAHAN, E., and MORANT, M. J., 1965, *Br. J. appl. Phys.*, **16**, 943.

[251] CONSTANTINIDES, A. G., 1967, *Electronic Letters*, **3**, 3, 125.

[252] FELICI, N. J., 1959, *Direct Current*, **4**, 3.

[253] SIMON, A. W., 1961, *Rev. scient. Instrum.*, **32**, 6, 758.

[254] TOBAZEON, R., 1966, Doctorate Thesis, University of Grenoble.

[255] SECKER, P., 1968, Instr. Phys. Exhibition, Alexandra Palace, England, March.

[256] COELHO, R., 1966, *C.R.A.S. (Sciences Phys.)*, **8**, (April).

[257] FELICI, N., 1967, *Rev. Gen. Electr.*, **76**, 786.

[258] FELICI, N., AYANT, Y., and TOBAZEON, R., 1968, *C. r. Acad. Sci., Paris*, **266**, 51.

[259] JEANMAIRE, C., 1964, *Rev. Gen. Electr.*, **74**, 499.

[260] JEANMAIRE, C., 1964, Doctorate Thesis, University of Paris.

[261] PRUET, H. D., 1965, Doctorate Thesis, University of California.

[262] BRIGNELL, J. E., 1966, *Proc. I.E.E.*, **113**, 1683.

[263] HILL, C. E., and HOUSE, H., 1967, *Electronic Letters*, **3**, 78.

[264] GEMANT, A., 1964, *Ions in Hydrocarbons* (New York: Academic Press).

[265] HUMMEL, A., and ALLEN, A. O., 1967, *J. Chem. Phys.*, **46**, 5, 1602.

[266] LEVINE, J. L., and SANDERS, T. M., 1967, *Phys. Rev.*, **154**, 137.

[267] SCHMIDT, W. F., 1966, Inaugural Dissertation, Berlin.

[268] LENKEIT, S., and EBERT, H. G., 1964, *Naturwissenschaften*, **51**, 237.
[269] DUTREIX, J., DUTREIX, A., and BERNARD, M., 1963, *Méd. Biol. Ther.*, **4**, 13.
[270] DUTREIX, J., DUTREIX, A., and BERNARD, M., 1964, *Phys. Méd. Biol.*, **7**, 69.
[271] BERNARD, M., 1964, Doctorate Thesis, University of Paris.
[272] BRIGNELL, J. E., and HOUSE, H., 1965, *Nature, Lond.*, **206**, 1142.
[273] CROITORU, M. Z., 1960, *Bull. Soc. franç. Elect.*, **8**, 342.
[274] BARKER, R. E., and SHARBAUGH, A. H., 1965, *J. Polymer Sci.*, C **10**, 139.
[275] SAVOY, E. D., and ANDERSON, D. E., 1967, *J. appl. Phys.*, **38**, 3245.
[276] SMITH, C. W., KAO, K. C., and CALDERWOOD, H., 1961, *Revue générale de l'Electr.*, **76**, 810.
[277] BRIÈRE, G., FELICI, N., and FILLIPINI, J. C., 1965, *C. r. Acad. Sci., Paris*, **261**, 5097.
[278] GALLAGHER, T. J., and LEWIS, T. J., 1966, *Br. J. appl. Phys.*, **17**, 34.

Appendix

ON September 17–20, 1968, in Grenoble (France) there was held an International Colloquium upon the "Phenomena of Conduction in Dielectric Liquids" organized by the French National Centre of Scientific Research (Professor N. J. Felici).

The papers held on the Colloquium concern mostly with problems discussed in this monograph. Therefore they are listed below and are included in notes and text.

[279] ADAMCZEWSKI, I., 'Liquid-filled Ionization Chambers as Dosimeters' (p. 21).
[280] ADAMCZEWSKI, I., 'The Mechanism of Electric Charge Transfer in Natural and Ionized Dielectric Liquids' (p. 13).
[281] ATTEN, P., and GOSSE, J. P., 'Régime transitoire de Conduction lors d'une Injection unipolaire dans les Liquides isolants' (p. 19).
[282] BEDDOW, A. J., BRIGNELL, J. E., and HOUSE, H., 'Breakdown Time-lags in n-Hexane' (p. 8).
[283] BLOOR, A. S., and MORANT, M. J., 'Probe and Charge Transit Studies of the Potential Distribution in Dielectric Liquids' (p. 16).
[284] BRIÈRE, G., CAUQUIS, G., and ROSE, B., 'Comportement du Radical anion du Nitrobenzène dans le Nitrobenzène' (p. 11).
[285] BRIGHT, A. W., and MAKIN, B., 'Preliminary Experiments on Electrostatic Generators operating in a Polar Liquid Medium' (p. 19).
[286] BUTTLE, A., and BRIGNELL, J. E., 'A Method of measuring the Mobility of Photo-Injected Charges' (p. 10).
[287] CHYBICKI, M., 'Ionization Currents induced by Corpuscular Radiations in Liquid Dielectrics' (p. 4).
[288] COELHO, R., 'Phénomènes prédisruptifs en Tension continue dans les Liquides cryogéniques (p. 2).
[289] FELICI, N. J., 'L'Utilisation des Liquides fortement polaires comme Isolants' (p. 13).
[290] FORSTER, E. O., and LANGER, A. W., 'A.C. Conductivity of some Organo Lithium Complexes in Organic Solvents' (p. 13).
[291] GALLAGHER, T. J., 'Breakdown and Conduction in Liquid Argon' (p. 7).
[292] GUIZONNIER, R., 'Injection de Porteurs par Surfaces d'electrodes' (p. 6).
[293] GUIZONNIER, R., 'Nature des Porteurs de Charges' (p. 5).
[294] GZOWSKI, O., 'Statistical Nature of Electrical Breakdown in n-Hexane' (p. 9).
[295] GZOWSKI, O., LIWO, J., and PIĄTKOWSKA, J., 'Triggering of Electrical Breakdown in n-Hexane' (p. 4).
[296] HILL, C. E., and HOUSE, H., 'The Problems in using the Kerr Electro-optic Effect to Measure the Field Distribution in Non-polar Liquids' (p. 16).

[297] HUCK, J., 'Variation du Champ et des Charges dans un Milieu diélectrique: Variation de la Permittivité différentielle avec le Tension de Polarisation' (p. 17).

[298] HUYSKENS, P., 'Le Rôle de la Liaison Hydrogène dans la Formation d'Ions dans les Liquides isolants' (p. 15).

[299] JACHYM, B., and JACHYM, A., 'The Influence of Temperature on the Natural and Induced Conduction of Non-polar Liquids' (p. 10).

[300] JANUSZAJTIS, A., 'On the Applicability of the Theory of Initial Recombination to the Liquid-filled Ionization Chambers' (p. 6).

[301] KALINOWSKI, J., 'Electron Drift in Non-polar Organic Dielectric Liquids' (p. 16).

[302] KRASUCKI, Z., 'High-field Conduction in Liquid Dielectrics' (p. 12).

[303] MATHIEU, J., BLANC, D., and PATAU, J. P., 'Mesure de l'Énergie moyenne nécessaire pour créer une Paire d'Ions dans des Diélectriques Liquides irradiés par des Rayons γ' (p. 27).

[304] METZMACHER, K. D., and BRIGNELL, J. E., 'Application of the Statistical Model for Liquid Breakdown' (p. 15).

[305] POLLARD, A. F., and HOUSE, H., 'A Fast Ellipsometer for Electric Field Measurement' (p. 6).

[306] SHARBAUGH, A. H., and BARKER, R. E., JR., 'Ionic Impurity Conduction in Organic Liquids' (p. 13).

[307] SILVER, M., CHOI, S. I., and SMEJTEK, P., 'Transient Space Charge Limited Currents by Injection into Organic Liquids from a Tunnel Cathode' (p. 8).

[308] SMITH, C. W., and CALDERWOOD, J. H., 'Studies of Pre-breakdown Phenomena in Liquid Dielectrics using an Image Intensifier' (p. 25).

[309] STEPHENSON, M. I., and TAYLOR, G. C., 'The Measurement of the Viscosity of Insulating Liquids subjected to Strong Electric Fields' (p. 15).

[310] TERLECKI, J., 'Ionization Chamber filled with a Mixture of *n*-Hexane and Carbon Tetrachloride for Phantom Measurements of Photon Radiation (X, γ)' (p. 5).

[311] TOBAZÉON, R., 'Comportement du Nitrobenzène pur sous Champs intenses continus et alternatifs' (p. 16).

[312] SĂVEANU L., and MONDESCU D., 'Aspects de l'influence des champs magnétiques sur la conductibilité électrique des diéléctriques liquides'.

Table

Comparison of the physical units of various systems with those of the SI system

Lengths	$1\,\mu = 1\,\mu m = 10^{-6}\,m$; $1\,m\mu = 1\,nm = 10^{-9}\,m$
Density	$1\,g\,cm^{-3} = 10^3\,kg\,m^{-3}$
Pressure	$1\,atm = 101\,325\,N\,m^{-2}$
Energy	$1\,ev = 1{\cdot}602 \times 10^{-19}\,J$
Charge	$1\,e = 1{\cdot}602 \times 10^{-19}\,C = 1{\cdot}602 \times 10^{-19}\,As$
Electric strength	$1\,v\,cm^{-1} = 10^2\,v\,m^{-1} = 10^2\,N\,C^{-1}$
Boltzmann's constant	$k = 8{\cdot}6167 \times 10^{-5}\,ev\,deg^{-1} = 1{\cdot}3805 \times 10^{-23}\,J\,deg^{-1}$
Viscosity	$1\,P = 0{\cdot}1\,N\,s\,m^{-2}$
Activation energy	$1\,ev/particle = 23{\cdot}03\,kcal/mol$
Specific resistance	$1\,\Omega\,cm = 10^{-2}\,\Omega\,m$
Specific conductance	$1\,\Omega^{-1}\,cm^{-1} = 100\,s\,m^{-1} = 100\,\Omega^{-1}\,m^{-1}$
Mobility coefficient	$1\,cm^2\,v^{-1}\,s^{-1} = 10^{-4}\,m^2\,v^{-1}\,s^{-1}$
Recombination coefficient	$cm^3\,s^{-1} = 10^{-6}\,m^3\,s^{-1}$; $cm^3\,(ues)^{-1}\,s^{-1} = 3 \times 10^3\,m^3\,C^{-1}\,s^{-1}$
Diffusion coefficient	$1\,cm^2\,s^{-1} = 10^{-4}\,m^2\,s^{-1}$
Activity	$1\,Ci = 3{\cdot}7 \times 10^{10}\,s^{-1}$
Dose	$1\,rad = 10^{-2}\,J\,kg^{-1}$
Dose rate	$1\,rad\,s^{-1} = 10^{-2}\,w\,kg^{-1}$
Exposure	$1\,roentgen\,(r) = 1\,R = 2{\cdot}58 \times 10^{-4}\,c\,kg^{-1}$
Exposure rate	$1\,r\,s^{-1} = 1\,R\,s^{-1} = 2{\cdot}58 \times 10^{-4}\,c\,kg^{-1}\,s^{-1}$
Ionization constant	$1\,R\,m^2\,Ci^{-1}\,h^{-1} = 1{\cdot}93446 \times 10^{-12}\,c\,m^2\,kg^{-1}\,s^{-1}$
Stopping power	$1\,ev\,cm^{-1} = 1{\cdot}602 \times 10^{-17}\,J\,m^{-1}$
Mass stopping power	$1\,ev\,cm^2\,g^{-1} = 1{\cdot}602 \times 10^{-20}\,J\,m^2\,kg^{-1}$
LET	$1\,ev\,cm^{-1} = 10\,kev\,\mu m^{-1} = 1{\cdot}602 \times 10^{-17}\,J\,m^{-1}$
Kerma	$1\,erg\,g^{-1} = 10^{-4}\,J\,kg^{-1}$
Cross section	$1\,barn = 10^{-24}\,cm^2 = 10^{-20}\,m^2$

Author Index

Abdullah M. 414

Abe R. 327

Adamczewski I. 5, 23, 24, 50, 51, 57, 58, 59, 70, 122, 129, 143–145, 149–153, 155, 156, 159, 161, 172, 176, 183, 186, 188–190, 193, 194, 207, 213, 214, 216, 219–224, 235–239, 253–259, 263, 269–271, 279, 280, 282–284, 289–293, 308–311, 315–318, 321, 322, 324–333, 342, 355, 384, 385, 387–389, 392, 393, 405, 408–410, 415, 416

Adams G. E. 103, 123

Aglintsev K. K. 122, 135, 269, 325

Aguirre P. 329

Aizu K. 327

Alexander P. 79, 122

Allen A. O. 96, 98, 123, 124, 329, 414

Allis W. P. 330

Allnat A. R. 32, 36, 58, 230, 276, 328

Altenburg K. 58

Anderson D. E. 369, 416

Anderson J. G. 411

Andrade E. N. 26, 32, 43, 57

Angerer L. 156, 347, 367, 368, 406, 407, 409, 412, 413, 414, 415

Atkins K. R. 327

Atten P. 333, 416

Attix F. H. 304, 305, 328, 332

Aukerman L. W. 166, 331

Ayant Y. 415

Bach N. A. 110, 123

Bacq Z. M. 79, 122

Baker E. H. 336, 337, 378, 407, 410

Bałygin I. E. 335, 411, 412

Bär F. 57

Barabanov E. N. 124

Barker R. E. 32, 36, 37, 58, 416, 417

Barlow A. J. 58

Barret S. 330

Barth H. 134

Basseches H. 372, 410

Basset K. H. 250, 328

Bässler H. 169, 176, 181, 182, 183, 184, 260, 261, 330

Bates D. R. 327, 372, 410

Baxendale J. W. 123

Beackey H. D. 124

Bearli I. 314, 326

Beddow A. J. 413, 415, 416

Bennet R. D. 325, 414

Berg D. 353, 415

Bernard M. 329, 416

Bernasik S. vii

Bethe H. A. 72, 122

Beverley W. B. 331

Beynon J. H. 124

Bhalla M. S. 330, 332, 410

Białobrzeski Cz. 143, 144, 145, 149, 316, 321, 325

Bianchi E. 145, 250

Bierlejew T. J. 135, 325

Bijl H. 186, 187, 188, 194, 224, 234, 235, 326

Biondi M. A. 369, 410

Birks J. B. 20, 57, 332

Blanc D. 140, 141, 149, 150, 156, 158–161, 183, 302, 309–311, 316–319, 322, 326, 328–333, 417

Bloor A. S. 416

Boag, J. W. 89, 96, 99, 100–104, 123, 158, 269, 328

Bogoroditskii N. P. 5, 59

Boltz, H. A. 336, 337, 378, 407

Boltzman L. 257, 275, 418

Bonch-Bruyevich A. N. 135, 325

Booth D. H. 367, 411

Boning 406

Born M. 125

Bornstein R. 21, 24, 57, 327

Bortner T. E. 410

Boyer J. 158, 326, 329

Bragg J. K. 69, 70, 335, 397, 399, 407, 410, 411

Briere G. 331, 333, 412, 416

Bright A. W. 416

Brignell J. E. 333, 356, 357, 409, 412–417

Brinkelhoff J. 331

Bronson G. E. 145

Brown D. R. 414, 415

Brown S. C. 135, 145, 185, 244, 271, 325–327, 372

Bruschi L. 231, 331

Bueche F. 49, 58
Buchholz, H. H. 372, 410
Burges R. E. 397, 415
Burhop E. H. S. 91, 95, 123
Burton M. 124
Bute C. J. 411
Buttle A. 333, 416

Calderwood J. H. 156, 331, 332, 336, 339–341, 358, 405, 408, 409, 413, 414, 416, 417
Caldwell S. H. 134
Callaghan L. 124
Camenade P. 332
Campbell J. 122, 133–137, 325, 397, 415
Cardaillac D. 330
Careri G. 228, 231, 232, 327, 330
Carmichael J. B. 37, 58
Carnevalle E. H. 47, 58
Carr H. V. 328
Cauquis G. 333, 416
Chadband H. G. 338, 413
Chapiro A. 91, 109, 122
Charalambus S. 332
Charlesby A. 5, 58, 65, 68, 69, 75, 79, 91, 96, 109, 117–120, 122–124
Chatterton P. A. 353, 415
Cheney G. T. 331
Cherenkov P. A. 70
Choi S. 142, 328, 417
Chong P. 142, 162, 199, 204, 205, 211, 216, 228, 327, 328, 360, 408, 409, 412
Church H. F. 413
Chybicki M. 172–175, 281, 328, 331, 333 416
Ciborowski S. 96, 122, 123
Clancy T. M. 175, 329
Clausius R. 31, 277
Clothier N. 357, 413
Coe G. 156, 330, 415
Coelho R. 291, 329, 415, 416
Cohen M. H. 266
Collison E. 124
Compton A. H. 62, 63, 64, 65, 78, 128, 133, 325
Compton D. M. J. 144, 148, 301, 331
Condon E. U. 93
Constantines A. G. 415
Corbey, M. M. G. 122, 401, 409, 413
Coulson C. A. 6, 14, 15, 57, 59
Cox E. B. 410
Craggs J. D. 145, 329–332, 410, 411, 413
Craig Miller H. 413
Cremer E. 264, 328
Croitoru M. Z. 327, 331, 414, 416

Crowe R. W. 145, 150, 156, 274, 335, 354–356, 360, 384, 389, 407, 409, 410, 411, 413, 414
Cullingford M. C. 415
Cunsolo S. 330
Curie P. 149, 325
Curtiss G. F. 59

Dainton F. S. 99, 108, 123
Dakin T. W. 353, 375
Darveniza M. 344, 345, 346, 407, 410, 412
Davidson N. 204, 326
Davies H. T. 28, 58, 228, 230, 252, 276, 277, 327, 328, 331
Davison W. H. T. 110, 124
Day F. H. 313, 327
Debye P. 31, 58, 300
Dekker A. J. 354, 414
Delard R. 332
Dera J. 245, 246, 248, 327, 407, 412
Devonshire A. F. 35, 37, 58
Dewhurst H. A. 210, 326
Di Castro, C. 330
Dienes G. J. 59
Donaldson E. E. 329
Donnely R. J. 327
Doolitle A. K. 26, 43, 48, 57, 58
Doolitle D. B. 26, 43, 48, 57, 58
Dornte R. W. 161, 327, 336, 379, 410
Douglass D. C. 248, 249, 327
Dries B. 330
Druyveysteyn M. J. 331, 349, 414
Du Bridge L. 134, 135, 325, 326, 336, 339, 380, 407
Dupre F. 330
Dutreix A. 416
Dutreix J. 416

Ebert H. G. 332
Edeles H. 330
Eder M. 148, 325
Edler H. 353, 410
Edwards W. D. 355, 360, 361, 365, 366, 393, 410, 412
Egloff G. 5, 21, 57
Einstein A. 243, 249, 251, 252
El-Sayed M. A. 331
Elert M. 99, 123
Endicott H. S. 363, 367, 410
Engel A. 145, 269, 327, 330, 339, 381, 410
Essex H. 91, 123, 384, 415
Eyring H. 5, 26, 27, 32, 43, 44, 45, 46, 47, 57, 58, 262, 263, 276, 327

Farmer F. T. 296, 328
Farrall G. A. 353, 413
Fasoli U. 326
Feather N. 73
Felici N. J. 156, 331, 415, 416
Fellows C. H. 111, 124
Fermi E. 377–380
Filipini J. C. 416
Finch H. W. 136
Fishman E. 326
Flammersfeld A. 74
Florek K. 412
Flory P. J. 37, 49, 58
Forner S. N. 105, 123
Forster E. O. 142, 176, 178, 179, 183, 261, 265, 328, 333, 416
Fowler J. F. 141, 238, 296–300, 327, 328, 344, 348, 379, 380, 398, 407, 410
Fox T. G. 49, 58
François H. 331, 332
Frankiewicz E. L. 124
Franz W. 412
Freeman G. R. 176, 261, 265, 329
Frenkel J. 5, 26, 32, 43, 57, 101, 150, 284, 286–289, 293, 327, 377, 385, 405, 408, 410, 412
Fricke E. F. 306, 307
Friese H. 410
Fröhlich H. 204, 296, 326, 327, 385, 387, 410
Frost L. 326, 327
Fujioka R. 367, 368, 411
Funabashi H. 332
Futrell J. H. 111, 124

Gaeta F. S. 175, 291, 327, 328
Gallagher T. J. 384, 394, 413, 416
Ganse I. 412
Garton C. G. 120, 124, 357, 367, 403, 412, 413
Gazda E. 149, 156, 238–241, 250, 251, 322, 326, 328, 329, 332
Geballe R. 413
Geiger H. 69, 145
Geller M. 412
Gemant A. 353, 405, 406, 410, 412, 414
Generałova W. W. 331
Gibaud R. 150, 156, 176, 326, 329
Gibbs, P. 368
Gindin L. G. 412
Glasser D. 71, 122
Glasstone S. 58
Goodwin D. W. 336, 338, 339, 355, 359–361, 364, 366, 379, 393, 397
Gomer R. 228, 414

Gordon S. 407
Gosling C. H. 345, 372, 407, 412
Gosse J. P. 333
Gray L. H. 69, 70, 105, 107, 123, 218, 329, 414
Green W. B. 291, 336–338, 350, 355, 379, 407, 408, 410
Grindlay J. 59
Groenweghe L. C. D. 58
Grunberg L. 37, 39, 58
Guizonnier R. 156, 176, 179, 183, 329, 333, 340, 381, 405, 406, 415, 416
Gumiński K. 6, 57
Gunterschultze 406
Gurney I. D. 17, 18
Gurney R. N. 57, 294, 327, 380, 411
Guzman J. 26, 32, 43, 57
Gzowski O. 50, 58, 135, 141, 150, 156, 162–164, 172, 173, 175, 176, 183, 192, 196–199, 204–210, 212–214, 216, 222, 228, 263, 325, 326, 328, 329, 359, 409, 413, 414, 416

Hadden S. T. 41, 42
Hahn K. L. 247
Haissinsky M. 6, 59, 91
Hakim S. S. 374, 415
Hall L. 43, 58
Hallyday F. O. 413
Halpern B. 228, 414
Hancox R. 412
Handley R. 59
Hannay N. B. 300, 328
Harrison M. A. 413
Hart E. J. 101, 102, 123, 414
Hasse H. R. 272, 326
Hawley R. 357, 413–415
Hayashi M. 415
Heftel J. 58
Henshaw D. G. 36, 37, 58
Hermans J. J. 58
Herrington E. G. 55, 56, 58, 59
Herzberg G. 93, 122
Herzfeld W. F. 43, 58
Hesketh T. R. 413
Heylen A. E. D. 145, 328, 330, 332, 354, 412–414
Higham J. B. 365, 366, 372, 374, 410, 411, 415
Hill C. E. 381, 415, 416
Hippel J. A. 112, 124, 210, 326, 385, 411
Hirai N. 43, 44, 45, 46, 47, 58
Hirschfelder J. O. 5, 57
Hixson W. A. 364, 366, 410
Hoffman G. 132, 144, 151, 187, 321, 325

Hoover W. G. 364, 366, 410
Hornbeck J. A. 273, 327
House H. 291, 329, 336–339, 341, 342, 344, 350, 369, 379, 380, 381, 407–410, 412, 413, 415–417
Howe S. 228
Huck J. 417
Hudson R. L. 105, 123
Hughes J. F. 156, 330, 350–352, 412, 414, 415
Hummel A. 212–216, 329, 414
Hurst G. G. 410
Huq A. M. Z. 357, 414
Hustrilid K. 112, 124
Huyskens P. 333, 417

ICRU 85
Inge L. 353, 373, 410, 411
Inuishi Y. 162, 199, 204, 205, 211, 216, 228, 327, 328, 408, 409, 412
Isida S. I. 21, 32, 37, 38, 39, 43, 57
Isihara A. 58, 59
Ivanov W. J. 150, 158, 326, 332

Jachym A. 180, 183, 223, 311, 312, 330, 331, 333, 417
Jachym B. 149, 156, 161, 176, 179, 180, 183, 222–224, 228, 260, 265, 291, 293, 297, 310, 325, 326, 329, 331, 333, 417
Jacobi W. 322–324, 330
Jaeger R. G. 79, 123, 314, 327
Jaffé G. 58, 143, 144, 149, 150, 157, 158, 161, 162, 176, 177, 186, 193, 194, 234, 235, 278–282, 293, 322, 325, 377
Jahns A. 322–324, 330
Janniuk G. 58
Januszajtis A. 156, 161, 281, 309, 310, 314–316, 318, 320, 321, 326, 328, 329, 331, 333, 417
Januszkiewicz E. 309, 320, 326
Jarnagin R. C. 168, 170, 171, 330
Jeanmaire C. 415
Jessup M. 156, 330, 412
Jeżewski M. vii
Joffe A. F. 294
Jones E. 29, 35, 276, 347, 407, 413
Jonson O. S. 367, 411
Jortner J. 107, 123, 231, 266, 270, 330

Kaghan W. S. 58, 349, 413, 415
Kahan E. 141, 328, 333, 336, 341, 343, 356, 406, 407, 414, 415
Kakiuti I. 412
Kalinowski J. 156, 161, 163, 171, 183, 330, 331, 333, 346–348, 407, 412, 414, 417

Kallmann H. P. 168–170, 172, 330
Kamerlingh-Onnes H. 39
Kao K. C. 156, 181, 311, 331, 332, 336, 339–341, 401, 405, 408, 409, 413–416
Kaplov G. N. 58
Karagounis G. 6, 11, 57
Karrer P. 5, 9, 10, 11, 12, 57
Kasztelan G. 83, 84, 122, 329
Kavada A. 168, 170, 171, 330
Kawarabayashi T. 327
Keast A. K. 413
Keen J. P. 101, 123
Kekule F. A. 30
Kellner L. 57, 411
Kihm R. T. 415
Kingston A. E. 239, 327, 372, 411
Kisinger J. B. 58
Kittel C. 6, 59, 294, 327
Kiżekin M. P. 413
Kneser H. O. 43, 58
Kohlrausch L. 5, 57
Koepp R. 332
Kok J. A. 20, 57, 335, 353, 401, 410, 413, 414
Komelkov B. C. 369, 409, 411
Kondratiev V. N. 6, 59
Kono R. 58
Koppelmann F. 373, 411
Koppl F. 5, 57
Kopylov G. N. 415
Korpalska I. 252, 328
Kozłowski K. 255, 258, 330
Kramers H. A. 150, 281, 328
Krasnansky Y. J. 330
Krasucki Z. 156, 346, 358, 402–404, 406, 409, 412, 413, 417
Kręglewski J. 59
Kroh J. 96, 98, 123
Kuchiński I. I. 414
Kuffel E. 145, 330, 332, 354, 414, 415
Kuhn H. 15, 16, 17, 57
Kuper C. G. 124
Kurata M. 21, 32, 37, 38, 39, 43, 57

Ladu M. 309, 332
Laidler K. J. 58, 91, 93, 95, 123
Lancaster G. F. viii
Landolt H. 21, 24, 57, 327
Lane, M. 380, 411
Langer A. W. 333, 416
Langevin P. 185, 187, 188, 190, 219, 271, 272, 326
Lang-il Choi 328
Larsch A. E. 204, 326
Larson E. L. 411

Lauer E. J. 330
Lauritsen E. 133
Lazarev A. 353, 411
Lea D. E. 74, 105, 107, 108, 122, 150, 158, 281, 282, 327
Leach S. 329
LeBlanc O. H. 183, 194, 199, 200, 201, 204, 224, 228, 274, 326, 384, 411
Ledingham K. W. D. 397, 415
Lekner J. 228, 266
Lenkeit S. 332, 416
Lennard-Jones J. E. 29, 35, 36, 37, 58
Le Page W. R. 326, 336, 407, 410, 413
Levine J. L. 415
Levy M. 37, 58
Lewis T. J. 5, 59, 141, 156, 217, 218, 328, 329, 330, 332, 335, 336, 339, 353–355, 362–366, 369, 379, 384–389, 394–397, 400, 404, 407–414, 416
Liao T. W. 369, 411
Lindemann F. A. 133, 151
Litovitz T. A. 28, 47, 58
Liwo J. 416
Llewellyn Jones F. 326, 332, 353
Loeb L. B. 145, 234, 269, 328, 353, 397, 411, 412
Lorquet J. C. 17, 57
Lower S. K. 331

McCall D. W. 248, 249, 326
McCubbin W. L. 17, 18, 57
McCullogh, R. L. 58
Macedo P. B. 28, 58
Macfadyen K. A. 335, 336, 338, 339, 355, 359–361, 364, 366, 379, 393, 398, 400, 407, 409, 410
McGee J. D. 332
Mache H. 269, 326
Macy R. 59
MacLean D. A. 372, 411
Magge J. L. 96, 105–107, 123, 124, 158, 332
Makin B. B. 416
Makoto N. 124
Maksiejewski J. L. 355, 363–366, 410
Małecki J. 284, 329
Malkin M. G. 363, 410
Maraviglia B. 331
Marcus R. A. 105, 123
Marion J. B. 314
Masakasu I. 413
Mason J. H. 354, 416
Massey H. S. W. 91, 95, 123
Matheson M. 28, 58, 230, 329, 331
Mathieu J. 156, 158, 254, 309, 326, 328–333, 417

Mayer P. 330
Mazzoldi P. 330, 331
Meek J. M. 353, 413
Megahed I. A. Y. 357, 414
Metcalf W. S. 413
Metzmacher K. D. 417
Meyer L. 194, 195, 229, 252, 326–328, 330, 331
Mickiewicz M. K. 331
Miller A. A. 115, 117, 124
Miller Craig H., 353, 413
Miller L. S. 228
Miller N. 72, 122, 353
Millikan R. A. 144
Minoru U. 413
Mitsui H. 413
Moeller C. 72, 122
Moffitt, W. E. 59
Mohler H. 80, 94, 122, 123
Molnar J. P. 330
Molony O. W. 57, 59
Mondescu D. 417
Moore R. 390
Morant M. J. 5, 59, 141, 161, 164, 165, 204, 326, 328, 329, 333, 335, 336, 356, 379–382, 406, 407, 411, 412, 414–416
Moroz L. M. 412
Mott, N. F. 92, 123, 294, 326, 380, 411
Moulinier P. M. 328
Mozumber A. 124, 332
Muller F. 329
Mungall A. G. 414
Murooka Y. 345, 372, 412
Musset E. 367, 411

Nagao S. 326, 372, 411, 412
Naghizadeh J. 252, 329
Nakanishi K. 59
Nernst W. 243, 251, 328
Nicholas J. F. 59, 264, 265, 326, 328
Nikuradse A. 5, 50, 53, 57, 142, 144, 291, 325, 335, 336, 339, 341, 342, 382–384, 405–407, 409, 411
Nissan A. H. 37, 58
Nordheim L. 344, 348, 379, 398, 407, 411
Nosseir A. 357, 376, 413, 414
Nowak W. 142, 149, 156, 161, 176, 184, 211, 212, 224, 242, 328

O'Connor D. 133–137, 325
O'Dwyer J. J. 301, 331, 354, 397, 411, 416
O'Kano, Koi 414
Onsager L. 143, 284, 322, 326, 408
Oppenheimer J. R. 123
Ostroumov, G. A. 217, 415

Pao C. S. 50, 53, 58, 130–132, 141, 149, 150, 156, 157, 176–178, 285, 291, 325, 336, 410, 413
Papoular R. 330, 353
Patau J. P. 329, 331–333, 417
Pauling L. 6, 59
Pawłowski C. 112, 115, 116, 124
Peek F. W. 353, 406, 411
Pekar S. J. 99, 123
Pellicioni M. 309, 332
Penney G. W. 415
Pfann W. G. 53, 54, 55, 59
Phahle A. M. 331
Piątkowska J. 416
Piciotti E. 309, 332
Pickard W. F. 331, 416
Pierre G. 331
Pigoń K. 59, 328
Piotrowski K. 49, 58
Pitts E. 166, 171, 330
Platzmann R. L. 105, 107, 123
Plumley H. J. 143, 156, 176, 177, 284– 286, 293, 325, 336, 338, 350, 377, 408, 410
Pohl H. A. 415
Polak L. C. 109
Poll R. A. 331
Pollard A. F. 417
Ponchiev A. W. 109, 123
Popov W. 331
Powell C. 194, 326
Prasad A. N. 331, 354, 384, 411
Price W. J. 89, 122, 135, 269, 325
Prigogine I. 37, 58
Protopopov O. O. 331
Pruet H. D. 415
Przybylak I. 309, 315, 317, 327
Pszona S. 332
Purcell J. R. 328
Putiłova I. H. 412

Radwan R. O. 330, 332, 354, 414
Ramacciotti 309, 332
Raschektaev I. 353, 411
Read J. 123
Reeves M. L. 413
Reid R. C. 5, 57
Reiff E. 194, 195, 326, 330, 331
Reiss K. H. 143, 150, 161, 166, 192–194, 285, 325
Rice S. A. 32, 36, 43, 58, 142, 230, 231, 252, 266, 270, 276, 327–330, 377
Richardson E. W. T. 281, 284, 327, 378, 411
Riehl N. 182, 183, 260, 261, 330

Roberts D. 57, 59
Roesch W. C. 329, 332
Rogoziński A. 143, 144, 325
Rose D. J. 327
Rose B. 244, 333, 416
Rosiński K. 59, 328
Rossi B. 250, 325, 328
Rossini F. D. 59
Rotblat J. 88, 122
Rowlison J. S. 59
Rudenko N. S. 416
Ruetschi P. 28, 264, 328
Ruhle F. 367, 368, 411
Ruppel H. 166, 167, 331
Ruther J. E. 411

Sacr M. M. 394, 413
Salvage B. 353, 364, 410
Samuel A. H. 358, 413
Sanders T. M. 415
Santini M. 331
Sato T. 175, 195, 196, 326, 413
Savage R. L. 411
Saveanu L. 417
Savoye E. D. 416
Scaramuzzi F. 327
Schaffer E. 222
Schintlemeister J. 325
Schlecht 183
Schmidt W. F. 162, 309, 322, 329, 331, 332, 415
Schnyders H. 330, 331
Schockley W. 294
Schoepfle C. S. 111, 124
Schott H. 58, 349, 413, 415
Schott L. 58, 397, 415
Schottky W. 339, 344, 378, 407, 411
Schrödinger E. J. 92
Schultz H. L. 410
Schültz W. 309, 322, 329
Schultze P. 124
Schumann W. 327, 339, 382, 406, 407, 409, 410
Schwaiger A. 353, 410
Schwar J. P. 415
Secker P. E. 156, 217, 218, 330, 350–352, 384, 412, 415
Seeger R. J. 385, 411
Seyer W. F. 143, 329
Sharbaugh A. H. 5, 58, 59, 156, 335, 336, 344, 345, 353, 363, 365, 372, 389, 390, 393, 400, 405–407, 409–413, 416, 417
Sherwood T. K. 5, 57
Shimanoucki T. 412
Shockley W. 327

Shone G. 415
Shonhorn H. 58
Smejtek P. 417
Silver M. 176, 289, 330, 417
Simha R. 41, 42, 58
Simon A. W. 415
Simpson J. H. 296, 327
Skanavi G. J. 5, 58, 335, 353, 409
Skiner S. M. 380, 411
Skowroński J. J. 353, 369–371, 406, 415
Sletten A. M. 328, 369, 375, 376, 400, 410–412, 414
Smith C. W. 123, 372, 373, 407, 409, 413, 416, 417
Smith D. E. 124
Smoluchowski R. 327
Sneddon I. N. 123
Sommermann G. M. L. 369, 411
Sosnowski L. 240, 300, 301, 328
Soudain G. 331, 332
Spear W. E. 228
Spiegel V. 331
Stacey F. D. 150, 210, 274–276, 326, 384, 411
Standhammer P. 143, 329
Stark K. H. 120, 124, 367, 412
Starodubcev C. B. 331
Staub H. H. 325
Steginsky B. 59
Stein G. 107
Steinke E. G. 144, 147, 326
Stephenson M. I. 417
Stewart G. W. 57
Stiekolnikov J. S. 409, 413
Stokes G. 169, 172, 192, 211, 223, 250, 276
Stratton R. 354, 414
Strong S. L. 58
Stuart S. A. 328
Stuetzer O. M. 415
Sugimoto T. 204, 328, 412
Sugita K. 329
Suita T. 327, 350, 410
Sullivan A. M. 314, 326
Sułocki J. 183
Summerfield G. C. 58
Sutherland W. 234, 326
Suzuki M. 367, 368, 411
Swan D. W. 162, 164, 165, 210, 231, 311, 326, 328, 331, 363, 400, 409, 411, 412
Szapiro M. M. 70
Szczeniowski S. 122
Szivessy G. 222
Szulze R. 113, 124, 149, 296, 328
Ścisłowski W. 296, 297, 328

Świętosławska-Ścisłowska J. 219, 220, 222, 326
Świętosławski W. 149

Tajnucty S. I. 413
Talroze W. L. 124
Tamura M. 59
Taris F. 381, 415
Tatevski B. M. 5, 32, 34, 35, 57
Taylor G. C. 417
Taylor W. J. 58, 61
Telang M. S. 58
Telegdi V. L. 148, 325
Teller E. 385, 411
Terlecki J. 50, 57, 58, 59, 135, 141, 149, 150, 156, 161–164, 183, 194, 196–204, 210, 212, 214, 222, 224, 274, 281, 309, 325, 326, 329, 331–333, 343, 344, 360, 361, 380, 410, 415–417
Terry G. C. 330
Theobald J. K. 331
Thomas A. 331
Thomas D. G. A. 325
Thomas J. H. 329
Thomas L. H. 58, 135
Thompson J. O. 327
Thomson J. J. 57, 289, 327
Tilikher M. D. 5, 59
Tobazeon R. 415, 417
Toriyama Y. 326, 412, 413
Torres L. 330
Townsend J. S. 383
Tozer B. A. 329, 332, 411
Tropper H. 156, 335, 344–346, 355, 357, 362–367, 407, 409, 410, 411
Tsutoma W. 124
Tsvetkov V. I. 416
Turriekii M. 331
Tyndall A. 194, 326

Ulbrich R. 411
Ullmaier H. A. 311, 329, 332
Umiński T. 245, 246, 248, 326, 327
Ushakov B. J. 375, 409, 413

Vermande P. 149, 150, 155, 156, 158, 309, 328–330
Vermeil C. 162, 165, 329
Victoreen J. A. 81, 82, 122, 137
Vogel H. 330
Volkenstein F. F. 410

Wagner G. 124
Walden P. 169, 192, 207, 211, 212, 216
Walter A. 353, 373, 406, 410, 411

Wannier G. H. 273, 327
Ward B. W. 156, 363, 394–397, 407, 409, 410, 413
Watson P. K. 5, 59, 156, 175, 216, 217, 329, 335, 336, 344, 345, 353, 363, 365, 366, 372, 400, 405–407, 409–413
Wazer J. R. 37, 58
Weber H. H. 363, 367, 410
Weiss J. 99, 123
Whitfield G. D. 124
Whitehead S. 412
Wierl R. 57
Wigner R. 93
Wilcox W. R. 53, 54, 59
Wiliams N. T. 48, 58, 123
Williams R. L. 274, 275, 326, 363, 410, 411
Willets F. W. 330
Wilson T. 269, 278, 328
Winger R. 58
Witt, H. T. 166, 167, 331

Wizevich J. 204, 326
Włodarski R. 413
Wolkensztein F. F. 405
Wollan E. O. 325
Wright G. T. 358, 413

Yahagi K. 331, 416
Yamanaka C. 327, 350, 410

Zając W. 142, 156, 161, 183, 212
Zaky A. A. 329, 415, 416
Zanstra H. 280, 327
Zdanowicz W. 55, 59
Zeessoulees N. 331
Zein El-Dine M. E. 362, 367, 381, 409, 410, 415, 416
Zielczyński M. 332
Zinovieva K. N. 229, 328
Zündi V. L. 148, 325
Żarnowiecki K. 332

Subject Index

Absorbed dose 85, 305, 313
— —, unit 418
Absorption coefficient, of hexane 81, 84
— —, of saturated hydrocarbons 82
— —, of water 81, 84
— effective cross-section 76, 77
— law 76, 78–80, 310
— linear coefficient 76
— macroscopic coefficient 77
— mass coefficient 84
— microscopic coefficient 77
— real true coefficient 80
— spectra, of α-radiation 67, 68
— —, of β- or electron-radiation 68–75
— —, of hydrated electrons 100, 102–104
— —, of neutrons 62, 117
— —, of protons 68, 71
— —, of radiation 61, 67, 76
— —, of X- and γ-radiation 76–84
Accelerator, as source of radiation 101, 103
Acetylene hydrocarbons 6
Activation energy 27, 120, 181, 182, 184, 207, 209, 214, 249, 257–261, 289, 290, 297, 339, 378, 380, 405, 408, 418
— —, for diffusion 249, 252, 253, 257
— —, of chemical processes 261
— —, of competing mechanism 263–265, 297
— —, of complex processes 263–265
— —, of ion mobility 258, 259
— —, of ion recombination 258
— —, theory 258, 261–265
Admixtures, influence on breakdown 367–369
—, — on conductivity 143, 157, 171, 179, 209, 340, 401, 405, 406, 409
Aliphatic hydrocarbons 122
Alpha particles, absorption in air 67, 68
— —, — energy 68
— —, — in plate emulsions 67
— —, — range 75
— —, — in water 67
— radiation sources 62
Aluminium, absorber curve 79
—, absorption coefficient 79, 81
Ambipolar diffusion 243, 244

Anthracen 6, 56, 166, 171, 183
Apparatus for distillation of liquid 50, 53
— — investigating, current in liquids 125, 150, 176, 340, 343, 381
— — —, diffusion of ions 245, 246
— — —, electrical breakdown 355, 357, 371, 372
— — —, electro-optical 416
— — —, Kerr effect 416
— — —, liquid conductivity 125, 150, 159, 176, 343, 381
— — —, mobility of ions 176, 193, 202
— — —, photoelectric current 167, 168
— — —, pulse radiolysis 101, 103
— — —, recombination of ions 150, 238
— — —, Schlieren method 372, 374
— — —, statistical phenomena of breakdown in liquid 294
— —, zonal melting 53
Application, of hydrocarbon radiochemistry 120
—, of liquid ionization chambers 302, 303, 308–325
—, of radiochemistry 120
—, of similarity principle 269
Argon, liquid 228, 230, 273, 276
Aromatic substances 14, 18, 122, 142, 391–393, 361, 390
Atomic orbitals radiation 10–11
Attachment, electrons 276, 354, 383, 384
Autodiffusion (self-diffusion) 243–249
—, measuring method 244, 247
Automatic measuring 56, 188, 357
Avalanche ionization 359, 377, 382–384, 400, 406, 407

Background in ionization chambers (residual current) 166
Barn definition 418
Basic theoretical problems 266, 270
Benzene 6, 14, 15
—, electrical conductivity 169, 178, 183
—, fragments 114
—, mobility of ions 211, 212
—, molecule structure 13–15
BEPO-reactor (as source of radiation) 117

Beta particles, absorption 68, 69, 71–75
— —, energy 62
— —, range, in the air, 74, 75
— —, — in aluminium 73
— —, — in the tissue 74, 75
— —, sources 62, 307
— —, spectrum 62
— —, standards 307
— —, stopping power 72
— radiation, influence on the conductivity 161, 172, 173, 175
— —, sources 62
Blackening of films by radiation 313
Boiling temperature of hydrocarbons 21, 22, 40
Bond energy 13, 14, 30
Bragg–Gray curve 70
Branches of the chain 4, 32, 35
Branching 3, 4, 391, 392
Breakdown, channels 358, 370
—, criterion 384, 394
—, effect of admixtures 357
—, — of electrodes 353, 361, 370
—, mechanism in gases 353, 354, 382–384
—, — in liquids 353, 358–369, 372–385, 390–393, 400
—, phenomena 353–369, 377, 385, 400
—, pulse strength, time dependence 357, 360, 361, 396
—, theory 374, 375, 384, 385–409
—, voltage 356, 363, 370, 375
Bremsstrahlung 71
Bronson resistor 134
Bubble model 350, 401, 402, 405
— theory 401, 402, 405
Bulk viscosity 43–46
Butadiene 12
Butane, 113

Calculation methods for hydrocarbons 21–25, 32–35
Calorimetric method of dosimetry (dose ranges) 89
Capture of electrons 274–276
Carbon atom, model 7, 10
— tetrachloride (CCl_4) 179, 183
Cathode influence, on the breakdown 394, 398–400, 403, 407, 408
— —, on the conduction 337, 339, 348–350, 361–369, 371, 377–380, 398–400, 403, 407, 408
Cavity models 377
Cell theory of liquids 35
Cerrium dosimeter 307

Chamber for breakdown investigation 355, 371, 403, 404
—, ionization, filled with liquid 127–132, 168, 172, 176, 196, 200, 218, 309–312, 315–321
— — — — —, as dosimeters 303, 308–325
— —, standard 38, 304, 306
Change of density with temperature 23, 24
—, of melting temperature 22
—, of specific volume temperature 34, 35
—, of viscosity with temperature 24–27
Characteristics of dielectric liquids 1
Charge sensibility of electrometers 133
— transfer 217
Chemical bonds 13, 14, 18
— —, energy 13, 14, 18
— dosimeters (dose ranges) 89, 306–307
— molecule 4–6, 11–20
— reactions 97
— —, after irradiation 91, 97
— —, effect of radiation 89, 91, 97
— —, interaction 120
Chromatographic measurements 168, 341
Circuits input, for breakdown measurements 357, 372, 403
— —, for conductivity measurements 126, 128, 150, 167, 176
— —, for ion diffusion measurements 245
— —, for ion mobility measurements 150, 198, 200, 201
— —, for ion recombination measurements 150, 238
— —, for pre-breakdown measurements 357, 372
— —, on breakdown 357, 371, 372, 403
Coefficient of absorption, calculating 78–82
— — —, dependence wavelengths of radiation 81
— — —, for different liquid hydrocarbons 82
— — —, linear 76
— — —, mass 77, 79, 81
— — —, of ionizing radiation 62, 76–82
— — —, — — —, α-radiation 68
— — —, — — —, β-radiation 68, 69, 75
— — —, — — —, for hexane 81, 84
— — —, — — —, for water 81, 84
— — —, of X- and γ-radiation 62, 64, 76–82

Coefficient, true 78, 80
— of viscosity, temperature variation 24–28, 46, 48, 208, 211, 254
— — —, variation with molecule structure 24, 25, 48, 254
Coincidence measurements of ionization showers and bursts 145
Collecting electrode 126, 128
Collection time of ions 127, 128
Collision ionization 382–385, 400
Columnar ionization 278, 279, 282, 322
— recombination 278, 322
Complex conductivity 263–265
Compton effect 62–63, 78
— scattering 62–63, 78
— — coefficient 78
Conduction mechanism 143–147, 158, 171, 172, 181, 254, 266, 377, 405–409, 416, 417
Conductivity, comparison of experimental and theoretical results 405–409
—, in liquefied gases 228
— of dielectric liquids, ionized 149, 267
— — — —, natural 125, 139, 144
— — — —, specific 2, 3
— — — —, in high fields 335–352
— — — —, in low fields 125, 267, 288
— — semiconductors 294–302
— — —, theory 294–301
— rotational effects 231, 232
Continuum dielectric models 380
Cosmic rays, effect on conduction 143–145
— —, — on ionization bursts 144, 145
Coulomb forces of ion repulsion 251
Critical temperature of hydrocarbons 21, 40
Crosslinking 91
— and degradation 91, 120
Cross section for absorption, macroscopic 77
— — — —, microscopic 77
— — for collision 106, 155, 382–384
— — for dissociation 217, 392
— — for ionization 155, 382–384
Crystallization method 53
Curie, unit 418
Current, field characteristics 146, 149, 151, 160, 161, 177–179, 292, 293, 336, 337
— fluctuation 356
—, intensity 131, 293
—, time dependence in gases 146, 147
—, — — in liquids 140, 141

Current, voltage characteristics 146, 268, 292, 293, 296
Cyclohexane, conductivity 111, 179–181, 183
Cylindrical chambers 130

Dark current (residual) 166, 299
Decan 163
Degassing of liquids 368, 369, 408, 409
Degradation 91
Delayed fluorescence 170–172
Delta (δ) electrons 106, 174, 281
Density of liquids 20, 23, 24
Dependence of breakdown strength on, density 359, 362
— — — — —, different factors 359
— — — — —, frequency 372–375
— — — — —, kind of liquids 351, 369
— — — — —, liquids 359
— — — — —, number of breakdowns, 363
— — — — —, pressure 359, 366
— — — — —, temperature 359, 364–366
— — — — —, time 357, 360
— — — — —, viscosity 359, 365
— — — — —, molecule structure 254, 257
— — — — —, temperature 254, 257
— — — — —, viscosity 258
— — ion mobility on, field strength 203, 205
— — — — —, molecule structure 241, 254, 256
— — — — —, temperature 207, 208, 210–213
— — — — —, viscosity 208, 209–211, 214, 258
— — — recombination on, field strength 278–282, 284
— — — — —, molecule structure 241, 254, 256
— — — — —, temperature 240, 254, 256
— — — — —, viscosity 237, 240, 258
Detectors of radiation 89, 306–307
Dielectric liquids, characteristics 2, 5, 17, 23–25, 34, 40
— —, current-field characteristics 149–153, 161, 163, 169, 173, 175, 177–179, 292, 293, 309–312, 335
Diffraction radiation 119
Diffusion 243
Diffusion, coefficient 243, 245, 257

Diffusion, dependence on temperature 248

—, —, on viscosity 248, 249

—, in liquefied gases 252

—, measuring methods 244–247

—, of hydrocarbons 249

—, of ions 243, 250, 254

—, of neutral molecules 243, 249

—, types 243

Dissociation 91, 94, 286, 349, 354, 377

—, of ions 94, 284

—, of molecules 91, 94, 284, 349, 354, 377

Distances between atoms 4, 14

Distillation methods 50–53

Distribution, of breakdown voltage 375, 376

—, of electron cloud 9–12

— factor, of atoms and radicals 108, 109

—, of electron cloud in atom 9, 10

—, of energy levels in dielectric 18, 378–382

—, of energy losses in hexane 215, 348

—, of field in liquid dielectric 129, 269, 340, 381, 397, 399, 400

—, of ions in column 279, 281

—, of potential 92, 181, 269, 286, 287

—, of radiation 90

Division of the ionizing radiation 62

Dose calculation 86

— range for different methods 89

— rate 87, 418

— —, calculation 87, 88, 308, 323

— —, measurement 87, 88, 308, 323

— units 85, 418

Doses from reactors 117

Dosimeter, liquid scheme 319–321

Dosimetric quantities 89, 307

Dosimetrical methods 88, 89, 306, 308

Dosimetry of radiation 85, 303, 306

Double ionization chamber 172

— layer, electric 269, 380

Durham conference 165

Dye motion in cell 218

Dynamic condenser 136

— (vibrating) capacitor 136

Effect, of admixtures 209, 367, 374, 406, 407, 409, 417

—, of ionizing radiation 61

—, of oxygen 354, 368

—, of pair electron creation 64

—, of temperature, on breakdown 359, 364–366, 393, 400–402, 406

Effect, of temperature, on conductivity 139–142, 177–181

— — —, on density 23, 24

— — —, on diffusion of ions 252

— — —, — — —, of molecules 245–249

— — —, on mobility of ion 195, 207–215, 221, 223, 227–232

— — —, on recombination of ions 240, 241

— — —, on viscosity 24–28, 43–45

Effective absorption coefficient 82

— atomic number of medium 311, 312

— mass of electron 277

— wave length 83, 84

Effects of radiation interaction 61, 349

Efficiency of liquid-filled ionization chambers 310, 311, 318, 416, 417,

Electric double layer 145, 149

— strength (unit) 418

Electrical breakdown 353

— —, dependence on density 362

— —, — — gap 363

— —, — — irradiation 120

— —, — — pressure 357

— —, — — structure of the molecule 359, 362, 385–393

— —, — — temperature 357, 359, 364–366

— —, in gases 353, 354

— —, in high frequency field 372–376

— —, in hydrocarbons 353, 372, 391

— —, in liquids 353, 354, 358, 391

— —, in oils 369–371, 376

— capacity, of ionization chamber 127, 128

— —, of system 127, 128

— circuit with ionization chamber 126

— conductivity, at high fields 335

— —, in dielectric liquids 125, 139, 335

— —, in gases 145

— —, of aromatic hydrocarbons 142

— — , of hexane 142, 149–156, 160–165, 172–174, 183, 336–338, 341–345, 351

— — , of ionized gases 145

— —, of saturated hydrocarbons 142, 149–156, 160–165, 172–174, 183, 335, 336

— ionization conductivity 145, 149

— natural (dark) conductivity 125, 139

Electrode collecting 126, 128

— guard 126

Electrodes, effect on breakdown 394, 398–400, 403, 407, 408

Electroluminescence 346–349, 407

Electrometer, charge sensitivity 133

Electrometer, electrostatic 133
—, dynamic 135–137
—, Lindemann 133, 134
—, logarithmic 137, 138
—, sensitivity 133
—, string 133
—, types 133
—, valve 133, 134, 135
—, vibrating, capacitance-type 133, 136, 137
—, voltage sensitivity 133
Electron, attachment 276, 354, 382–384
—, charge 418
—, cold emission from the cathode 161, 336, 339, 350, 359, 361, 362, 377–379, 400, 407, 408
—, emission from cathode 161, 176, 336–339, 350, 352, 359, 361, 362, 377–379, 400, 407, 408
—, energy loss of, 214, 215
—, — — —, by radiation 214, 215
—, hot emission from the cathode 176, 352, 377–379
—, hydrated 99, 101–105
—, in liquids 96, 99, 101–105, 214, 215, 385–394
—, injection 175, 176, 199
—, mobility in liquids 277
—, model, free 101, 274
—, multiplication 354
—, pairs, creation 64
—, π 11–15, 181, 182
—, range 74, 215, 216
—, solvated 96, 101, 165
—, solvation energy 100
—, shell's structure 6–15
—, specific ionization 68
—, spin resonance 90
—, traps 274, 298
Electrons, δ 174, 281
—, hydrated 99, 101–105
—, ion recombination 234
—, solvated 101
Electrostriction 349
Elipsometer 417
Emission spectrum 347
Emulsion photographic in dosimetry 89, 312–317
Energetic division of radiation 62
Energy bands 18, 378–380
— distribution, in carbon atom 10
— —, in dielectric and the metal 303, 378–380, 386
— —, in molecule 14, 17, 18, 91
—, free 263

Energy, levels, of carbon atom 10
—, —, of electrons at the metal 377–380
—, —, of molecule 16–18, 91
—, loss, per ion pair 303, 310, 311, 385, 417
—, losses 69, 386
—, needed for ion pair formation 86, 303, 310, 311, 385, 417
—, of α-particles 62
—, of β-particles 62, 75
—, of Compton electron 63–65
—, of deuterons 70
—, of electrons 62, 69, 75
—, of γ-radiation 62, 314
—, of ionization 385
—, of neutrons 62
—, of photoelectrons 63–69
—, of polarons 100
—, of X-radiation 62, 305
—, standards of radiation 314
—, transfer 305, 385, 405
—, — mechanism 385–387
—, unit 418
Enthalpy 42, 262, 263
Entropy 44, 262, 263
Ethyl acetate, coefficient of self-diffusion 248
Ethylene 11
Excitation energy 18, 93, 385
— —, transfer 385
Excited, atom 96, 385
—, molecules 93, 96, 97, 354, 385, 407
—, —, influence on the conductivity 171
Exposure 85, 313, 418
—, rate 87, 418

Feather's formula 73
Fermi energy level in metals 378, 380
Ferrous sulphate dosimeter 89, 306–307
Field, distribution 381, 382, 398–400
—, —, in the liquid 340, 381, 382
—, electric 138, 334, 381, 382
—, high 334, 354, 369
—, low 139
—, stress pulse 355–357
Flammersfeld's formula 74
Fluorescence 94, 346–349, 407
—, delayed 170–172
—, quenching 93
—, spectra 170, 346–348
Fission of atomic bond (linkage) 91, 94, 120, 284, 385
Fractional crystallization 57
Fragments, of benzene molecule 114
—, of n-butane molecules 113

Franck–Condon principle 92–93
Free electron model 101, 274–276
— radical reaction 109–110
— volume 37, 48, 49
Frequency, of current fluctuations 159, 160
—, of ionization bursts 144, 145
—, of molecule bonds 386
—, of molecule collisions 26, 263, 271
Friction, molecular 277

G- (radiation yield) 98, 215
G-value, hydrocarbons 111–113, 120
—, water 98
Galvanometer 130
Gamma-rays, absorption 76
— —, detection by ionization chambers 154, 157, 177, 178, 305, 309, 311, 315–319, 324
— —, half-thickness (half-value layer) 78
— —, photoelectric effect 62
— radiation sources 62, 314
Gas discharge 147
Gaseous ionization chambers 145–148, 303–306
Gases, effect on, the breakdown 368, 377, 401, 408
—, — —, the conductivity 139, 141
—, electric conduction 139, 145–148
—, ionization 145–148
—, radiation yield 111
—, yield in radiolysis 111
Generator, high voltage, with liquid 350, 351
Glass, as dosimeter 89
Grenoble conference 416, 417
Guard ring for ionization chamber 128

Half-value layer, for β-radiation 75
— — —, for X- and γ-radiation 78
— — —, γ-rays 78
— — —, X-rays 78
Health physics 1
Helium, liquid 175, 228, 229, 232, 273, 277, 278
Heptane, diffusion 249, 257
—, electrical breakdown 360, 362, 364, 365, 366, 388, 389, 391
—, ion mobility 191, 200, 205–210
—, isomers 391
—, purification 50
—, recombination 237, 240, 241, 256
—, temperature, boiling 40
—, —, melting 40

Hexane, density 23, 24
—, electric natural conductivity 139–144
—, electrical breakdown 391
—, ion diffusion 250–252
—, — mobility 191, 203, 205, 210, 213, 255–258
—, — recombination 236, 237, 241, 256
—, ionization conductivity 151–155, 160–162, 173, 183, 336–338, 341–344, 345, 351
—, isomers 4, 391
—, molecular structure 3, 4, 5
—, physiochemical properties 40
—, purification 50
—, temperature, boiling 40
—, —, critical 40
—, —, melting 40
—, viscosity 24
High voltage electrode 355
— — liquid generator 350, 351
— — pulse 356, 357
Hole conductivity 172, 295
Hybrid 10–12
—, distribution 10–12
—, electron density distribution 11–12
Hydrated electron 96, 99–105
— —, hydration energy 100
— —, thermodynamic properties 102
Hydrocarbon, breakdown 353, 391
—, conductivity 154, 155, 183, 191, 208, 209, 211
—, cyclic 6
—, density 23, 24
—, liquids, diffusion of ion 250
—, —, mobility of ions 191, 208, 211
—, oils 17
—, radiation effects in 91
—, recombination of ions 237, 241
—, saturated 3, 4, 5, 21, 23, 34, 40, 50, 91, 111–113, 191, 227, 254, 257, 391
—, unsaturated 5
—, viscosity 24, 25, 27
Hydrocarbons yield in the radiolysis 91
Hydrogen atom 7
— —, model 7
—, bonds 417
— hyperoxide (H_2O_2) formation 97, 98
— — —, radiation yield 97, 98
— — —, reactions 97, 98

Induced current 145
Influence of admixture 139–143, 171, 179, 340, 367, 368
Injection of electrons into dielectric liquids 161, 172

Insulators 2, 126

Interaction, of α-radiation 61, 62

—, of β-radiation 61, 62

—, of cosmic rays 143

—, of ionizing radiation, chemical effects 94, 96, 105, 109, 112, 120

—, — — —, physical effects 61, 62

—, of neutrons 117, 309, 321, 322

—, of ultra-violet radiation 161–172

—, of X- and γ-radiation 62

Interatomic distance 4, 14

Intrinsic conductivity 3, 140

— volume variation 118

Ion, creation 86, 94, 96, 105, 106, 266, 385, 417

—, density 266

—, diffusion 250–253

—, mobility 185

—, pair 62, 67

—, recombination 233, 266

—, theoretical mobility 254, 266, 270

Ionization bursts 144, 145

— by collision 282–284

— chambers, double 172

— —, filled with liquid 126–132, 168

— —, — — —, as dosimeters 308–325

— —, gaseous 303–306

— —, guard ring 127–129

— —, insulators 126

— —, integrating type 127, 128

— —, parallel-plate 127, 129

— —, pulse type 127, 159

— —, schematic diagrams 127, 129–132

— —, sensitivity 310, 311, 318

— —, standard 304, 306

— —, tissue-equivalent 323, 324, 417

— —, various kinds 126, 127, 129–132

— current, at different temperature 175, 177–184, 289–293

— —, at high fields 344, 346

— —, at low fields 139

— — in, dielectric liquids induced, by α-particles 158, 160

— — —, — — —, by β-particles 172, 173, 174, 175

— — —, — — —, by cosmic radiation 143–145

— — —, — — —, by δ-electrons 174

— — —, — — —, by injected electrons 161, 176, 351, 352

— — —, — — —, by neutrons (indirectly) 158

Ionization current, in, dielectric liquids induced, by reactor radiation 117

Ionization current, in, dielectric liquids introduced, by X- and γ-radiation 149–158, 175–181, 229

— — —, — — —, in argon 230

— — —, — — —, in different hydrocarbons 155, 180

— — —, — — —, in helium 175, 228–232

— —, in gases 145, 148

— —, in hexane 149–158, 180, 181

— —, in krypton 148

— —, in liquids 149

— —, in neon 148

— —, in other dielectric liquids 155, 175, 181–184, 191, 208–212, 222, 335

— —, in solids 294–302

— —, in xenon 230

—, density 68

—, effects, in hexane 111, 112, 154, 155

—, —, in other dielectric liquids 155, 175, 181–184, 191, 208, 209, 211, 212, 335

—, —, in solids 115, 119, 120, 294–302

—, —, in water 96–98, 104–109

—, energy 385

—, in time and space 105–109

—, potential 86, 385

—, pulses 282

—, space distribution 106–109

—, specific 68

—, time distribution 106–109

—, work of 86, 385

Ionized, atoms 62, 86, 88, 90

—, gases 145–148

—, liquids 90, 149

—, molecules 90, 91, 96

Ionizing radiations 62

— —, dosimetry 85

Iso-heptane 391

Iso-hexane 391

Iso-octane 177, 178

Isotopes, radioactive, table (energy standards) 314

Kerma (Kinetic Energy Released in Material) 85, 305, 418

Kerr effect 381, 416

Kinetic spectrophotometry 103

Krypton, liquid 228, 230, 276

Lauritsen electroscope 133

Law of, diffusion of ions 250–253

— —, — — molecules 243, 245, 249, 150

— —, Lennard–Jones 29, 35–37, 58

Law of, mobility of ions 190, 209, 270, 272–277, 288
— —, Navier–Stockes 169, 172, 192, 211, 233, 250, 276
— —, radiation absorption, for β-particles 73–75
— —, — —, for neutrons 117, 321, 322
— —, — —, for X- and γ-radiation 76–80
— —, recombination of ions 233, 234, 237, 241, 297–299
— —, Walden 169, 192, 207
LET (Linear Energy Transfer) 68, 418
Life time of polaron 100
Lindemann electrometer 133, 134, 151
Linear attenuation and absorption coefficient 76
— coefficient of absorption 76–78
— energy transfer (LET) 68
Liquefied gases 228
Liquid dosimeter 303, 306, 308
— — yield 89, 307, 323
— filled ionization chambers 308
— generator 350, 351
— models 28, 32, 35–40, 43–49, 286–289, 386–392
— movement in electric field 217
— triode 350, 352
Liquids, bulk viscosity 43
—, molecular properties 2, 21, 32
—, purification 50–57, 139–142
—, shear viscosity 24, 43, 48
Logarithmic dosimeter 137, 321
Low electric field, conductivity at 125
Lucite (dose ranges) 89
Luminescence 344–348, *see also* Fluorescence, Phosphorescence
—, decay and concentration 170–172
—, delayed 170–172
—, ultra-violet induced 346–348
Luminous-voltage characteristics 346–348

Macroscopic absorption coefficient 77
— mass coefficient 78–84
— properties of liquids 21
— radiochemical changes 91, 96, 108, 112, 120
Mass absorption coefficient 77–78, 81–84
— attenuation coefficient, for hexane 81, 84
— — —, for other hydrocarbons 82
— — —, for water 79, 81, 84
— stopping power, for α-particles 67
— — —, for β-particles 67, 70, 75

Mass stopping power, for electrons 67, 70, 75
— — —, for neutrons 117, 321, 322
— — —, for protons 67, 71
— — —, for X- and γ-rays 78–84
Mean free path of electrons in liquids 101, 106, 107, 109
— range 72, 74, 216
Measuring of, breakdown 353–376
— —, diffusion coefficient 244–249
— —, $I = f(E)$ characteristics 146, 149–152, 161, 163, 173, 292, 309, 310, 336, 344
— —, ion mobility 185–232
— —, — recombination 235–242
Mechanism, of breakdown in liquids 385
—, of conductivity 266, 278, 281, 286, 290, 293, 294
—, of diffusion, of ions 250–253
—, — —, of molecules 243–250
—, of electron attachment 382–384
—, of energy transfer 305, 385, 405
—, of interaction of ionizing radiation 90
—, of ionization 61, 62, 94, 146, 385
—, of mobility of ions 270–277, 288
—, of radiolysis 96–99, 109, 306–307
—, of recombination of ions 233–235, 241
Melting temperature for hydrocarbons 21, 22, 40, 117
Methylacetate, coefficient of self-diffusion 248
Microscopic absorption cross-section 77
Mobility as function, of electric field strength 203, 205, 208, 273
— — —, of molecular weight for gases 273
— — —, of molecule structure 209, 254
— — —, of temperature, 207, 208, 210, 211, 213, 254, 255
— — —, of viscosity 205, 208, 209–211, 214, 258
—, of drift 232
—, of electrons 170, 277
—, of ion in liquefied gases 228–232
—, of ion theory 270, 273
—, of negative ions 170, 185, 191, 199, 202–205, 219, 222, 254, 255, 258
—, of positive ions 185, 191, 199, 206, 219, 222, 254, 255, 258
Models, of bubble 400, 404
—, of cavity 377
—, of continuum dielectric 295, 298, 380
—, of liquid 28, 32, 35–40, 43–49, 286–289, 386–392

Models, of polaron 99, 100
—, of quasi-free electron 99–100
Molar refraction 31
— —, for hydrocarbons 31
Molecular, chains 3, 5, 19
—, mass (weight) 115–117
—, moment of inertia 247
—, orbitals 11–18
—, potential 16
—, structure 3–6, 11–16, 19, 21
Molecule, excitation 97
—, length 4, 16
—, moment of inertia 93
—, potential 11–18
—, structure 3, 4, 11–16
Mylar (dose range) 89

n-decane 23, 25, 40
n-heptane 23, 24, 40
n-hexadecane 23, 40
n-hexane 3, 5, 23, 24, 40
n-nonane 23, 24, 25, 40
n-octane 23, 25, 40
n-pentane 3, 23, 24, 40, 111, 155, 390
Naphthalene 6, 18, 57, 169, 183, 212
Navier–Stokes equation 169, 172, 192,
    211, 233, 250, 276
Negative ion formation 94, 96, 97, 104
Neon, liquid 273
Neutron, detection by proton recoil in
    ionization chamber 309, 310, 321
—, dosimetry with liquid-filled ioniza-
    tion chamber 309, 310, 321
—, fast 117, 321, 322
—, flux from reactors 62
—, thermal 117
Neutrons, dose 117, 309, 321–324
—, dosimetry 321, 324
—, flux 117, 309, 324
—, induced conduction 309, 321, 324
—, sources 117
Nitrobenzene 417
Nomogram, for ion mobility 224, 227,
    255, 258
—, for ion recombination 258
—, for molecule diffusion 257
—, for viscosity coefficient of liquids
    256
Nonane 155, 209
Nuclear emulsion 324
— magnetic resonance 90, 247
— radiation 62, 117, 314
— reactions as radiation sources 62, 117,
    314
Numerical method for hydrocarbons 32

Oak Ridge, reactor 117
Octacosane 115
Octadecane 40
Octane 23, 25, 40, 155, 191
Olefins 18
Orbitals, of atoms 8–11
—, of molecules 11–17
Organic liquids 3, 168
Oscillogram for motion of ions 199, 201,
    202, 205, 206
— of electric current pulse 201
Oxidation, of hydrocarbons 121
—, reaction 121, 122
Oxygen effects 121, 122, 354, 368
— influence on breakdown of liquids 368
— — — conduction of liquids 368

Pair production 62, 64
Parachor 28, 30, 31
—, of the hydrocarbons 30, 31
Paraffin, electrical conductivity 3, 115,
    116, 183, 296
—, hydrocarbons 3–5
—, molecule structure 16, 116
—, oils 17
Parallel-plate ionization chamber 127,
    129, 131, 132, 168, 172, 196, 197, 200,
    201, 218, 267, 317, 319, 343, 349
Pentane 3, 23, 24, 111, 155, 390
Permeability (dielectric constant) 31,
    284, 397, 400, 401, 402
Phosphate glass (dose ranges) 89
Photocathode 90, 161
Photodisintegration of nuclei 324
Photoelectric coefficient 78
— effect 62, 63, 161
Photoelectron 63, 64, 65
—, maximum range 65
—, mean range 65
Photoemulsions 67, 324
Photographic film (dose ranges) 89, 313
Photoionization 163, 165, 166
Photoluminescence 161, 344–348
Photon 62–66, 76
—, energy 62–66
—, flux 76–78
Physical effects of radiation 61, 67, 112
Physicochemical properties of hydro-
    carbons 3–6, 11–17, 21–23, 34, 40
Pictures of electric breakdown 358, 370,
    373, 374
Pi-electrons (π) 11, 12, 14, 15, 181, 182
— current 161–171
Pile (reactor) 117
Polarization electric 53

Polaron, energy levels 100
—, model 99, 100, 277
Polyethylene, ionization effect, 75, 119, 120, 225, 298–300, 349
Polymerization 91, 165
Polymers irradiated 111–122
Polystyrene, as insulator 126
POPOP 345–348
Potential distribution, in condenser 377, 382
— —, in dielectric 380–382
— —, in molecule 16
—, ionization 86
Pre-breakdown 353, 356, 373, 374
—, measurements 356–359, 372–374
—, mechanism 356–359, 373–374
Prigogine theory 37, 58
Products, of benzene radiolysis 114
—, of hexane radiolysis 111, 112
—, of hydrocarbon radiolysis 91, 111, 114
—, of water radiolysis 96
Protons, recoil ranges 115
Pulse height analysis 102, 103
— of ionization 159, 160, 282
— radiolysis, detection of free radicals 101–104
— —, method 101–104
— —, solvated electrons 101–104
— voltage 356, 357, 395
Purification, methods of liquids 50
—, of aromatic hydrocarbons 50, 57
—, of ionization chambers 53, 129, 343, 344
—, liquid, electrical methods 139
—, —, physical methods 50, 139–142
—, of oils 18, 19
—, of saturated hydrocarbons 50, 139–142

Quality factor 89
Quantum fluids 230, 231, 330
— theory, of atom 7–13
— —, of liquid 231, 330
— —, of molecule 11–18, 91–93
Quasi-free electron model 99, 100, 301
Quenching of discharge 93

Rad-dose unity 85, 418
Radiation, absorption 76
—, chemistry 90, 94–98, 108–122, 306, 307
—, detectors (dose ranges) 89, 303
—, dose 85
—, — rate 85, 88
—, dosimetry 85

Radiation, of hydrocarbons 91, 110–114
—, of polymers 112–122
—, of water 96–98
—, sources 62, 117, 314
—, yield 98, 110–114, 120, 121, 307
Radicals, creation 97, 109
—, of hydrocarbons 109
—, of polymers 109–122
—, of water 97
—, reactions 97
Radioactive iostopes 172, 173, 244, 246, 314
— — (γ-standards) 314
Radiolysis, chemistry 94, 122
—, of hexane 111, 112
—, of hydrocarbons 91, 110–114
—, of water 96
—, pulse 101–104
Radium element 153, 154
—, needle 153, 154
Range, of α-particles 75
—, of β-particles 73–75
—, of deuterons 70
—, of electrons 72–75, 216
—, — —, in various substances 216
—, of protons 75
Rate, dose 87, 418
—, Oak Ridge, as gamma and neutrons flux source 117
—, unity 117
Reactor, as radiation source 117
—, BEPO, as gamma and neutrons flux 117
Recoil electrons (δ-electrons) 174
— protons 115
Recombination, columnar 278
—, dependence on, the temperature 240, 241, 256
—, — —, the viscosity 235, 237, 241, 256, 258
—, effect 233
—, losses 233
—, kinds 233
—, of electrons and ions 233
—, of ions, coefficient 233, 418
—, — —, influence on conductivity 233, 266, 267, 278, 281, 293
—, — —, methods of measurements 233, 235, 238
—, — —, phenomena 233, 256
—, — —, theory 233, 256
—, — —, various kinds 234
References to part I 57
— — — II 122
— — — III 325

References to part IV 409
Resistors for electrometers 126, 134
Rice–Alnatt theory 276
Roentgen, dose unity (exposure) 85, 418
—, equivalent physical (rep) 85
—, per second (exposure rate unity) 85, 418
— rays, as ionizing agent 62, 64, 76–84, 85
Rotation of liquid 230, 231
Rotational effects 13, 91, 230, 231
Rotons 231, 232

Saturated hydrocarbons, boiling temperature 21, 34, 40, 191
— —, breakdown 385, 391
— —, — mechanism 385
— —, density 23, 34, 155, 191
— —, diffusion, of ions 250
— —, —, of molecules 245, 257
— —, melting temperature 22, 40
— —, mobility of ions 227, 254, 255
— —, molecule structure 4, 5, 19
— —, physico-chemical properties 21, 34, 40, 155, 191
— —, purification 50
— —, radiation yield 91, 111–113
— —, recombination of ions 191, 235–237, 241
— —, viscosity 24, 25, 27, 191
Saturation current, in gases 146, 147
— —, in liquids 149, 155
Scattering, of electrons 65, 71, 72
—, of lights 62
—, of neutrons 322
—, of photons 62–66
—, of protons 71, 115
Scavengers of the radicals 99
Scheme of apparatus, for breakdown measurements 355, 357, 371, 372
— — —, for measuring diffusion coefficient 245, 246
— — —, — — field distribution 381
— — —, — — ion mobility 150, 159, 167, 168, 176, 193, 196–198, 200, 201
— — —, — — recombination of ions 150, 238
— — —, for pulse radiolysis 101, 103
—, of the carbon atoms distribution in molecule 4, 5, 11, 12, 14–16, 19
—, of distillation apparatus 50–51
—, of dosimeter 316–321
—, of electrometer 133–137
—, of logarithmic dosimeter 321
Schlieren method 358, 372–374

Schrödinger equation, for atoms 7
— —, for molecules 92
Secondary electrons 65, 383, 384
Self (natural) conductivity 139, 144
Semiconductors, conductivity 294
—, theory 297–299
Sensitive volume, defined by guard rings 127, 128
Sensitivity, of dosimetrical films 89, 313
—, — — methods 89, 313
—, of ionization chamber 89
Silicone oils 3, 17, 19, 20
Silicones 3, 17, 19, 20
Similarity principle 269
Solvated electrons 101, 105
Sources of γ-radiation energy 314
Space charge 269, 381, 382
— —, experimental method of measuring 381, 416
— —, in liquids 269, 381, 382, 397–400
— —, theory 267, 269, 378–381, 397–400
— (volume) distribution of electrons, in atoms 7, 9–11
— — — — —, in molecules 11–18
Specific energy loss 68–70
— ionization 68–70
— resistivity, of gases 145–147
— —, of insulators 126
— —, of liquids 3, 125, 144
— —, of solids 126
— volume, variation with irradiation 118
Spin, concentration 247
—, echo experiment 247
Standards of radiation energy 314
Statistical methods 363, 375, 394–397
— spread of results 363, 394–397
Stopping power, for α-particles 67, 68
— —, for β-particles 67–75
— —, for deuterons 70
— —, for electrons 67–75
— —, for ionizing particles 67–75
— —, for protons 68, 71, 75
— —, in air 68, 72
— —, in liquids 69, 72
— —, in nuclear emulsions 67
— —, unit 418
Sulphur hexofluoride, breakdown 354
Superfluidity 228–230, 232
Surface tension 28
Table, of absorption coefficients 81, 84
—, of activation energy 209, 249, 252, 259, 261, 262
—, of autodiffusion 248, 249
—, of binding energy 14, 34, 42

Table, of breakdown, in liquids 391
—, — —, strength of hydrocarbons 391
—, of comparison of experimental and theoretical results 405–409
—, of density of hydrocarbons 23, 24, 34, 35, 155, 191
—, of electrical conductivity of liquids 171
—, of electrometers 133, 137
—, of electron stopping power for various substances 72
—, of energy, of radiation 62
—, — —, standards 314
—, of enthalpy 42
—, of fragments of ions and molecules 109, 111–114
—, of insulators 126
—, of ionization work energy 86
—, of melting, boiling and critical temperature of hydrocarbons 40
—, of mobility of ions 191, 202, 208, 211, 221, 222, 229, 230, 273
—, of molecule structure 182
—, of parachors 30
—, of radiation yields 98, 111, 112, 114, 120, 121, 307
—, of ranges for electrons 74
—, of recombination coefficient of ions 191, 241
—, of SI system 418
—, of viscosity of hydrocarbons 24, 25, 191, 208, 211, 222, 249
Teflon 126
Temperature influence, on autodiffusion 245, 248, 249, 252
— —, on breakdown of liquids 120, 364–366
— —, on conductivity of liquids 175–184, 337–342
— —, on density of liquids 23, 24
— —, on diffusion 245, 248, 249, 252
— —, on ion, mobility 170, 195
— —, — —, recombination 240, 241, 254, 256
— —, on viscosity of liquids 24, 25, 26, 27
Tetralin, effect of radiation 121
Theory, of Adamczewski 24, 269, 282, 289–293, 387–394, 405
—, of Angerer 406
—, of Barker 36
—, of Bragg 397–398
—, of Crowe 274
—, of Doolitle 48
—, of Eyring 26

Theory, of Feymann 277
—, of Forster 142
—, of Fowler 297–299
—, of Fox and Flory 48
—, of Frenkel 26, 286–288, 405
—, of Froehlich 296
—, of Goodwin and Mcfadyen 398–399
—, of Gray 105, 107, 108
—, of Gzowski 359
—, of Hiraia and Eyring 43
—, of Jaffe 278–280, 322
—, of Kao 401, 402, 405
—, of Kok and Corbey 401
—, of Kramers 281
—, of Krasucki 402
—, of Kuchiński
—, of Kurata and Isida 37
—, of Lea 105–108, 281, 282
—, of Lewis 379, 385–387
—, of Magee 105–107
—, of Małecki 284
—, of Morant 379
—, of Nikuaradse 382–384, 405
—, of Onsager 284, 322
—, of Platzman 105–108
—, of Plumley 284–286
—, of Reiss 143
—, of Rice and coworkers 43, 276
—, of Schumann 382–384, 406–408
—, of Sharbaugh and Watson 344, 345, 405–407, 409
—, of Simha and Hadden 41
—, of Skowroński 370, 371, 406
—, of Sletten and Lewis 369
—, of Sosnowski 300, 301
—, of Stacey 274–276
—, of Swan 400
—, of Tatevsky and coworkers 32
—, of Ushakov 375
—, of Ward-Lewis 394–397
—, of Williams 48
—, of Zanstra 280
—, of activation energy 263
—, of ambipolar diffusion 243, 244
—, of conduction, in gases 266–268
—, — —, in liquid dielectrics 266, 293, 377
—, gas bubbles formation 350, 401–403
—, of ion, diffusion 243, 266
—, — —, mobility 270–277
—, — —, recombination 233, 278
—, of ionization conductivity 266, 293
—, of liquids (cell-theory) 35
—, — — (tunnel-theory) 35, 36
—, of similarity 269

Theory, of viscosity of liquids 24, 42–49
Thermodynamics of liquids 32, 261–265
— processes of ion transfer in liquids 261, 266, 293
Thermoluminescence (dose range) 89
Time and space distribution of ions 108, 109
— dependence of current 140, 141
— distribution of ionizing effects 108, 109
Toluene 183
Townsend avalanche in liquid 382–384
Trapped electrons 274
Traps, electron 274
Triode liquid 350, 352
Tunnel theory of liquids 35
Type of ionizing radiations 62

Ultra-violet excitation 161–165, 170, 171, 347
— induced luminescence 170, 171, 344–349
— irradiation 162–165, 193
—, light 162, 163, 347
—, spectrum 162, 166–168, 347
Unity, of dose 85, 418
—, of dose rate 87, 88, 418
—, of exposure (dose) 85, 418
—, of first collision dose (Kerma) 85, 418
Unsaturated hydrocarbons 5, 6, 13–15, 114, 178, 183

Vaseline oil 3, 222
Velocity of chemical reaction 120
Vibrating–capacitance electrometer 135–137
Vibrations of atoms in molecules 385
Victoreen's formulae 82

Viscosity, as a function, of molecule structure 24–25, 48–49
—, — — —, of temperature 24–28, 43–49
—, formulae 24–28, 43–49
—, of hydrocarbons 24–25
—, influence on mobility 186, 190, 192
—, of liquids 24–28, 43–49
—, variation with irradiation 118

Water, absorption coefficient 81, 84
—, influence on breakdown 367, 406
—, — — conductivity 50
—, irradiation of 96–98
—, pulse radiolysis 101–104
—, radiation yield 98
—, removing 50–53, 139
—, stopping power 69, 72
Wave function 7, 9
— —, of carbon atom 10
— length of X-radiation 62, 81, 84, 152
Wigner function 93

X-ray, absorption 62–66, 76–84, 90
—, apparatus 152
—, bremsstrahlung 71
—, creation 62, 71
—, diffraction 116, 119
—, energy 62
—, filters 83
—, mechanism of absorption 62–66, 76–84, 90
—, sources 62
—, tubes 152
Xenon, liquid 228, 230, 276

Yield of radiation 98, 110–114, 120, 121, 215, 307

Zanstra's curves 280
Zonal melting 53